# Fruit Science
## Culture and Technology

## Other Volumes

**VOLUME 1**

**Section 1: Fruit Area and Production Scenario**

**1.** Fruit Area and Production Status/*J.S. Bal*

**Section 2: Fruit Plant Fundamentals Culture and Technology**

**2.** Planning and Planting of an Orchard /*Jaswinder Singh Brar and Krishan Kumar,* **3.** Suitability of Climate and Soil/*Jaswinder Singh Brar,* **4.** Soil Cultural Practices Management/*S.K. Upadhyay,* **5.** Fruit Plant Canopy Management/ *N. Sharma, Chaitanya R. Belsare and D.P. Sharma,* **6.** Nutritional Needs of Fruit Plants/*P.P.S. Gill,* **7.** Water Need of Fruit Plants/*J.S. Chandel and Chaitanya R. Belsare,* **8.** Physiology of Fruit Plants/*J.S. Bal and K.S. Gill* **9.** Growth and Development of Fruits/*N. Sharma and Preet Pratima*

**Section 3: Fruit Plant Propagation Technology**

**10.** Propagation Methods/*V.K. Wali, Amit Jasrotia and Parshant Bakshi* **11.** Propagation Techniques and Rootstocks/*V.K. Wali, Amit Jasrotia and Deep Ji Bhat,* **12.** Tissue Culture and Biotechnology/*P. Mohanpuria,* **13.** Seed Germination and Dormancy/*D.P. Sharma and Rimpika,* **14.** Propagation Media Mixtures and Containers/*S.S. Gill and J.S. Bal,* **15.** Care of Propagated Plants/ *J.S. Bal,* **16.** Protected Cultivation/*K.S. Gill and Rachna Arora*

**Section 4: Post-harvest Fruit Technology**

**17.** Fruit Maturity and Ripening/*K.S. Thakur and Satish Sharma,* **18.** Fruit Grading and Packaging/*B.V.C. Mahajan and J.S.Bal,* **19.** Treatments for Enhancing Shelf Life/*B.V.C. Mahajan,* **20.** Storage Techniques for Fruits/*B.V.C. Mahajan,* **21.** Post-harvest Transportation of Fruits/*Jagmohan Singh* Colour Plates

**VOLUME 2**

**Section 1: Culture and Technology of Major Tropical Fruits**

**1.** The Mango/*V.K. Singh,* **2.** Banana/*K. S. Gill and N. K. Arora* **3.** Guava/ *J.S. Bal,* **4.** Papaya/*S.K. Bhatia and R.B. Kumatkar,* **5.** Pineapple,.. *P.P. Joy T.A. Rashida Rajuva and R. Anjana*

**Section 2: Culture and Technology of Common Tropical Fruits**

**6.** Sapota/*Rajbir Singh Boora,* **7.** Cashewnut/*Adavi Rao Desai*

**Section 3: Culture and Technology of Minor Tropical Fruits**

**8.** Custard Apples/*A.P. Gaikwad and S.R. Lohate,* **9.** Jackfruit/*P.M. Haldankar, Y.R. Parulekar, R.A. Patil and N.V. Dalvi,* Colour Plates

## VOLUME 4

# Fruit Science

## Culture and Technology

### Volume 3: Sub-Tropical Fruits

**J. S. Bal** Ph.D. (Hort.)
Former Professor of Horticulture
Department of Fruit Science
Punjab Agricultural University
Ludhiana-141004

NEW INDIA PUBLISHING AGENCY
New Delhi – 110 034

**NEW INDIA PUBLISHING AGENCY**
101, Vikas Surya Plaza, CU Block, LSC Market
Pitam Pura, New Delhi 110 034, India
Phone: + 91 (11) 27 34 17 17 Fax: + 91 (11) 27 34 16 16
Email: info@nipabooks.com
Web: www.nipabooks.com

Feedback at feedbacks@nipabooks.com



ISBN: 978-93-86546-22-7
978-93-86546-20-3 (Vol. 1)
978-93-86546-21-0 (Vol. 2)
978-93-86546-23-4 (Vol. 4)
978-93-86546-24-1 (Set)



Composed and Designed by NIPA

***Dedicated to***

My son Dr Naveet Singh Bal and daughter Dr Harit Kaur Bal who brought great laurels to parents with consistent hard work and self supporting by earning scholarships throughout their higher education

# Foreword

Cultivation of fruits is a labor intensive occupation engaging the workers throughout the year. Thus, it is ideally suited to the over populated countries like India where 87 per cent operational holdings are less than one hectare. The economic returns from well manage fruit crops is better than any of the agricultural crops. Many fruit crops have quite high returns such as kinnow, banana, pineapple, papaya and grapes. Depending upon the management, production of 25 – 45 tonnes of fruits can be obtained per hectare and this would compete with any other agricultural crop cultivation. Traditional rice-wheat rotation has high water requirement. Underground water table is receding continuously from the last over three decades. There is therefore, an urgent need to switch over to crops that require lesser water to arrest further depletion of underground aquifer. Among the available options, the cultivation of fruit crops is one of the best alternatives to get better returns ensuring nutritional security of the people also.

The total area under fruits in the world is 57.5 million hectares with annual production of 654 million metric tonnes. India is producing 13.6 per cent of fruits in the world. The average productivity of fruits in India at present is 12.3 mt/ha only. There appears to be good scope of increasing fruit production with appropriate scientific management. The requirement for fruits is increasing continuously due to fast growth of population in urban areas. The fruits in the market being priced very high at present these have going out of reach of common people. Increase in the area and productivity of fruits is therefore essential to meet even the minimum required per capita consumption of fruits in the country.

Profitable fruit growing involves many aspects which need special attention and care on a regular basis. The basic problems faced by the fruit growers in on these aspects are discussed in this book, "*Fruit Science Culture and Technology Vol 2.Tropical Fruits; Vol 3.Sub-Tropical Fruits and Vol 4.Temperate Fruits*". The chapters are designed in three broad sections namely *culture and technology of major fruits, common fruits and minor fruits.* All the aspects of different fruits are elaborated with the help of tables, figures and photographs. Eminent scientists working in different Agricultural Universities, ICAR Institutions and Research Stations have contributed chapter related to their areas of specialization.

Dr. J.S. Bal has took good initiative, compiled and edited very helpful information for the overall benefits of horticultural students doing graduation and post-graduation, fruit growers, technical experts and others associated with fruits science. Dr Bal has my appreciation for bringing out such a useful publication on different aspects of fruits growing in India.

**S.S. Johl**
Chancellor
Central University of Punjab

# Preface

India is the second largest country in the world in area and production of fruits. The annual production of fruits is 91.4 million tonnes from an area of 6.4 million hectares. Climatically India is favorable for the production of a variety of tropical, sub-tropical and temperate fruits. Most of the fruits are grown on wide range of agro- climatic conditions with varied production.

A large variety of fruits are grown in our country. Mango, citrus, banana, guava and apple are the main fruits. Large number of other fruits are grown in good acreage. The fruits are grown throughout the country from temperate zone to coastal tropical region. The major fruit growing states in the country are Maharashtra, Andhra Pradesh, Karnataka, Gujarat and Uttar Pradesh.

In previous book "Fruit Science Culture and Technology Vol 1" basic information on various horticultural aspects and in "Fruit Science Culture and Technology Vol 2" tropical fruits were discussed in detail. In present book "Fruit Science Culture and Technology Vol 3" different sub-tropical fruits grown in the country are discussed in three sections denoted to (i) Culture and Technology of Major Sub-Tropical Fruits (ii) Culture and Technology of Common Sub- Tropical Fruits and (iii) Culture and Technology of Minor Sub-Tropical Fruits. The various aspects of fruit cultivation mainly covered are nutritive and cultural significance; origin, history, and distribution ; taxonomical and botanical description ; climatic and soil adaptability; propagation technology and rootstocks; plant and fruit physiology; planning and planting; recommended and popular cultivars; soil cultural practices technology - water need, nutritional need, weed control, inter culture; plant cultural practices technology- training and pruning, fruit thinning ,fruit quality improvement , use of plant growth regulators; special problems; harvesting and production of fruits; post-harvest fruit technology; insect-pests and diseases management ; marketing and export potential. Section-1 covers 2 leading sub- tropical fruits of the country. Similarly, section- 2 covers 4 and section-3 covers 6 sub- tropical fruits in order of their importance.

I am grateful to living legend Dr. J.S. Jawanda, Ph.D (Hort.) Ohio (USA), Prof. & Head (Rtd.), Department of Horticulture, PAU Ludhiana for his great encouragement throughout my professional career in doing best for students and fruit growers. I can never forget the politeness of my wife Dr. Kanwaljit Kaur Bal for bearing period of loneliness during this busy task.

I am highly thankful to all the scientists working in different Universities/ Institutions and Research Stations for their contributions in this book which will be highly beneficial to the graduate and post-graduate students in Fruit Science, fruit growers, scientists and extension workers.

10, June 2017

**J.S. Bal**

# Contents

# List of Contributors

**A.D. Huchche**
ICAR-Central Citrus Research Institute
Amravati Road, Nagpur, Maharashtra

**Alemwati Pongener**
ICAR-National Research Centre on Litchi
Muzaffarpur, Bihar

**Anil Kumar**
Regional Fruit Research Station
Abohar, Punjab

**Awachare Chandrakant Madhav**
Division of Fruit Crops, ICAR- IIHR
Bengaluru, Karnataka

**A.P. Gaikwad**
Agricultural Research Station
Lonavala, Tal – Maval
Dist – Pune, Maharashtra

**B.N.S. Murthy**
Division of Fruit Crops, ICAR- IIHR
Bengaluru, Karnataka

**Devi Singh**
College of Agriculture (HAU)
Kaul, Haryana

**Deepa H. Dwivedi**
Department of Applied Plant Science
Babasaheb Bhimrao Ambedkar University
Rae Bareli Road, Lucknow, U.P.

**H.K. Bons**
Department of Fruit Science
Punjab Agricultural University
Ludhiana, Punjab

**J.S. Bal**
Former Professor of Horticulture
Department of Fruit Science
Punjab Agricultural University
Ludhiana, Punjab

**M. Tamilselvan**
ICAR-Krishi Vigyan Kendra,
Sikkal, Nagapattinam,Tamil Nadu

**Munni Gond**
Department of Applied Plant Science
Babasaheb Bhimrao Ambedkar University
Rae Bareli Road, Lucknow, U.P.

**M.I.S. Gill**
Department of Fruit Science
Punjab Agricultural University
Ludhiana, Punjab

**N.K. Arora**
Department of Fruit Science
Punjab Agricultural University
Ludhiana, Punjab

**Navjot Gupta**
Department of Fruit Science
Punjab Agricultural University
Ludhiana, Punjab

**Nirmaljit Kaur**
Department of Botany
Punjab Agricultural University
Ludhiana, Punjab

**Rajbir Singh Boora**
Regional Fruit Research Station (PAU)
Bahadurgarh, Patiala, Punjab

**S.D. Pandey**
ICAR-National Research Centre on Litchi
Muzaffarpur, Bihar

**S.R. Lohate**
AICRP on Arid Fruit Crops
(Custard apple and Fig)
N.A.R.P., Ganeshkhind
Pune, Maharashtra

**Vishal Nath**
ICAR-National Research Centre on Litchi
Muzaffarpur, Bihar

## SECTION - 1

## CULTURE AND TECHNOLOGY OF MAJOR SUB-TROPICAL FRUITS

# 1

# Citrus

*A.D. Huchche*

## INTRODUCTION

Citrus is the most economically important fruit crops widely grown in developed and developing countries of the world and is one of the main sources of vitamin C. There is also an increasing demand of "high quality fresh citrus" driven by World Health Organization recommendations. Citrus fruits contain the largest number of carotenoids found in any fruit and extensive array of secondary compounds with pivotal nutritional properties such as vitamin E, provitamin A, flavonoids, limonoids, polysaccharides, lignin, fiber, phenolic compounds, essential oils etc. These substances greatly contribute to the supply of anticancer agents and other nutraceutical compounds with anti-oxidant, inflammatory, cholesterol and allergic activities, all of them are essential to prevent cardiovascular and degenerative diseases, thrombosis, cancer, atherosclerosis and obesity. In spite of these beneficial traits there is still a major need to improve fruit quality to meet current consumer's demand.

## NUTRITIVE AND CULTURAL SIGNIFICANCE

Citrus fruits have long been valued as part of a nutritious and tasty diet. The flavours provided by citrus are among the most preferred in the world, and it is increasingly evident that citrus not only tastes good, but is also good for health of the populace. It is well established that citrus and citrus products are a rich source of vitamins, minerals and dietary fibre (non-starch polysaccharides) that are essential for normal growth and development and overall nutritional well-being. However, it is now beginning to be appreciated that these and other biologically active, non-nutrient compounds found in citrus and other plants (phytochemicals) can also help to reduce the risk of many chronic diseases. Where appropriate, dietary guidelines and recommendations that encourage the consumption of citrus fruit and their products can lead to widespread nutritional benefits across the population.

Although citrus fruits are known for their vitamin C, like most other whole foods, citrus fruits also contain an impressive list of other essential nutrients including both glycaemic and non-glycaemic carbohydrate (sugars and fibre) potassium, folate, calcium, thiamin, niacin, vitamin $B_6$, phosphorus, magnesium copper, riboflavin, pantothenic acid and a variety of phytochemicals. In addition citrus contains no fat or sodium and, being a plant food, no cholesterol. The average energy value of fresh citrus is also low , which can be very important for consumers concerned about putting on excess body weight. For example a medium orange contains 60 to 80 kcal, a grapefruit 90 kcal and a tablespoon (15 ml) of lemon juice only 4 kcal.

## ORIGIN, HISTORY AND DISTRIBUTION

Citrus fruits include oranges, lemons, limes, pummelo and grapefruit. Being a native of tropical and subtropical region of South East Asia, these fruits have been under cultivation from time immemorial in South China, Malaya and sub-Himalayan parts of Assam. From here, they spread to other tropical and subtropical parts of the world. Next to mango and banana, citrus represents the third most important group of fruits in India. Nagpur Santra is grown on a large scale in the Vidarbha region of central India. Similarly in Assam, Brahmaputra Valley and Dibrugarh district are famous for mandarin production. Khasi mandarin is an important cultivar of north-eastern regions of India. After mandarins, limes and lemons are also cultivated throughout India.

Citrus production takes place throughout the tropical and sub-tropical countries of the world.Citrus cultivation dates back to many centuries. This cultivation is said to be started in China as early as 2200 BC. South China and Assam are the origin of many citrus fruits. The citrus fruits include lime, lemons, and oranges. Bhattacharya and Dutta (1956) opined that limes, lemons and *Citrus reticulata* are indigenous to Assam. A broad analysis of the history and development of citrus is found in the book ‘The Citrus Industry’ Vol-I edited by Walter Ruther. As mentioned by Herbert John Webber in his article “History and Development of the Citrus Industry” in the said book the first member of the group of citrus to have known to European civilization was the ‘citron’ mentioned about 310 B.C. by Theophrastus. For several hundred years, this was the only citrus fruit known. The sour orange, the lemon and the sweet orange were discovered in a much later period. As far as preserved literature indicates, this last species i.e. sweet orange was not known in Europe until approximately 1400 A.D.

**Area and Production :** India ranks 3rd in world production of citrus with an area of 1.055 million hectares (16.25% of total fruit area) and production of 12.74 million tonnes (10% of total fruit production) with a productivity of 12.0 tonnes/ha (Table 1.1) (NHB, 2016-17).

**Table 1:** Area, production and productivity of Citrus fruit in India

| Citrus fruit | Area (000 HA) | Production (000 MT) | Productivity (MT/HA) |
|---|---|---|---|
| Limes/Lemons | 259 | 2789 | 11 |
| Mandarins | 429 | 4754 | 11 |
| Sweet oranges | 209 | 3497 | 17 |
| Others | 157 | 1706 | 11 |
| Total | 1055 | 12746 | 12 |

*Source:* Final estimate NHB 2016-17

## TAXONOMICAL AND BOTANICAL DESCRIPTION

### Taxonomical Description

**Genus:** *Citrus* L.

**Family:** Rutaceae (Rue family)

**Subfamily:** Aurantioideae (**Citroideae**)

**Subgenera:** The genus *Citrus* is further subdivided into subgenera: *Citrus* and *Papeda*, with the difference being the presence of acrid oil droplets in the pulp vesicles of *Papeda*. Of the species covered here, most belong to the subgenus *Citrus*, with *C. hystrix* and *C. macroptera* belonging to *Papeda* (Stone 1985).

While classifying citrus (*Citrus* spp.) Swingle has been considered to be a lumper (with 16 species only) and his student Tanaka a spliter (157 species). Hodgson (1967) classified citrus in to 36 species of commercial importance. This classification is the most accepted one. All the edible fruits of citrus come under subgenus *Eucitrus* which can be divided into 5 horticultural groups.

### Acid group

i. Acid lime : *Citrus aurantifolia*

ii. Tahiti or Persean lime : *Citrus latifolia*

iii. Rangpur lime : *C. limonia*

iv. Lemon : *Citrus limon*

v. Rough lemon : *C. jambhiri*

vi. Citron : *C. medica* (Kidarankai in Tamil, used for pickling)

vii. Sweet lime : *Citrus limettioides*

### Orange group

i. Sweet orange : *Citrus sinensis*

ii. Sour orange:*Citrus aurantium* (Narthankai in Tamil, used for pickling)

iii. Multiple leaf orange :*C. multifolia*

iv. Japanese summer grapefruit: *C.natsudaidai*

### Mandarin group : (loose jacket)

i. Coorg mandarin, Nagpur Santra : *C. reticulata* and Kodai orange

ii. Japanese Satsuma mandarin : *C. unshiu*

iii. Willow leaf mandarin : *C.deliciosa*

iv. King mandarin : *C. nobilis*

v. Kinnow mandarin : King x willow leaf

vi. Tangerine orange varDancy : *Citrus tangerina?* (trifoliate x mandarins)

### Pummelo and grape fruit group

i. Pummelo : *C. grandis*

ii. Grape fruit : *C. paradisi*

iii. Kumquat : *Fortunella* sp.

The fifth group consists of mainly hybrids of different citrus fruits with trifoliate orange (*Poncirus trifoliata*) and mainly used as rootstock. e.g.

i. Citrange (*Poncirus trifoliata* x *C. sinensis*) var. Troyer, var. Carrizo

ii. Citrangor (*Citrange* x *C. sinensis*)

iii. Tangelo (*Tangerine* x *grapefruit*)

iv. Citrangequat (*Citrange* x *kumquat*)

### Pomological Description

**Growth habit:** Citrus species ward the top of the tree, with upright medium to large, compact horizontal branches. Grapefruit produces large trunks (0.5–0.75 m [1.5–2.5 ft] in diameter) and a large conical head. Trees produced from seed tend to have more thorns and upright branch growth than trees produced from grafting.

**Flowering:** Flowers are 2–4 cm (0.8–1.6 in) in diameter, axillary, fragrant, single, few or cymose, and often perfect (having both functional stamens and pistils) or staminate. The calyx is 4–5 lobed and there are usually five petals with oil glands. Stamens number between 20 and 40. Petal colors range from white to pinkish in Kaffir lime to pinkish to purplish externally in citron and reddish in lemon varieties. The subglobose ovary is superior, with 8–18 locules (cavities), with 4–8 ovules per locule in two rows.

**Leaves and branches:** Leaves are entire, 4 to 8 cm (1.6–3.2 in) in length, unifoliate, fairly thick, with winged petioles. Leaves are ovate, oval or elliptical, with acute to obtuse tips, and glands containing oils in glands, which are released when crushed. Young twigs are angled in cross-section, green, and axillary single spined, while older twigs and branches are circular in cross section and spineless.

**Fruit:** The fruit is a hesperidium, a fleshy, indehiscent berry that ranges widely in size, color, shape, and juice quality. Citrus fruit range in size from 4 cm (1.6 in) for lime to over 25 cm (10 in) in diameter for pummelo. Fruits are globose to ovoid in shape. The fleshy endocarp is divided into 10–14 sections containing the stalked pulp and separated by thin septa. Each section contains pulp (juice vesicles) that contains a sour or sweetish watery juice. A whitish "rag" or mesocarp (also known as the albedo) covers the endocarp. In turn, the thin outer section of the leathery peel or exocarp containing many oil glands is known as the flavedo (Purseglove 1974).

**Seeds:** Seeds are pale whitish to greenish, flattened, and angular.The seeds are usually polyembryonic, meaning they have multiple embryos that can germinate. The embryos are either "zygotic" or "nucellar." The zygotic embryos are derived from pollination of the ovary, i.e., sexual reproduction, and therefore are not always similar in horticultural qualities to the parent tree. The nucellar embryos are derived wholly from the mother plant and display very similar characteristics to the parent plant.

**Rooting habit:** Over 70 per cent of citrus tree roots are in the top meter (3.3 ft) of soil. Citrus trees produce a taproot that can extend 2 m (6.6 ft) below the surface. Fibrous roots commonly extend well beyond the canopy.

## Similar looking species

All citrus species have dark green, waxy leaves with a characteristic citrus odor, and sweet-smelling flowers. Most species are easy to differentiate by their fruit. Kaffir lime and wild orange are often mistaken for each other. According to Walter and Sam (2002) two differ from each other on the basis of the fruit and the petiole shape. Wild orange fruit has a smooth skin, and the petiole wings are entire. Kaffir lime fruit has bumpy skin, and the petiole wings are crenulated (toothed). The leaves of sour orange have a petiole that is much larger than that of sweet orange (about the size of grapefruit petiole).

## CLIMATIC AND SOIL ADAPTABILITY

### Climatic Adaptability

Citrus trees are evergreen, grown in truly subtropical climates of the world although in tropical regions of the world they tend to produce cyclic growth flushes and hence regulating cropping in tropical areas for forcing them into concentrated bloom needs judicious management of water deficit stress according to soil type and growing season. 500-1500 m MSL elevation, a rainfall of about 150 cm to 250 cm is required. The winter should be mild and there should be no strong or hot wind during summer. Citrus fruits grow best between a temperature range of 13$^0$C to 37$^0$C. Temperatures below – 4$^0$C are harmful for the young plants. Soil temperature around 25$^0$C seems to be optimum for root growth. High humidity favours spread of many diseases. Frost is highly injurious. Hot wind during summer results in desiccation and drop of flowers and developing fruits. Barring these limitations citrus is grown in all subtropical and tropical areas of the world. The sub-tropical climate is best suited for citrus growth and development. Citrus trees are subtropical in origin and cannot tolerate severe frosts. Citrus production in South Africa is therefore confined to areas with mild and almost frost-free winters where temperatures seldom (not more than once in several years) drop below –2°C and almost never below –3°C. The average minimum temperature for the coldest month should not be below 2 to 3°C if no protection is provided.

**Temperature prior to flowering:** Citrus (except lemons) require shorter days and cooler temperatures in winter for a normal production rhythm. Flowering should occur almost exclusively in spring, and these spring flowers should produce a large fruit crop 7 to 12 months later, depending on the cultivar. In more tropical areas the flowering pattern is much less clearly defined and main-season crops tend to be considerably smaller. Navel trees, however, tend to set smaller crops in the warmer parts of the local citrus areas. The coldest month in most of the better navel areas tends to have a mean temperature of 12 to 13 °C. For the tree to become dormant, the mean temperature must be below 13 °C. Navels, therefore, need to approach dormancy in winter. The chances of producing consistently good crops diminishes progressively as the mean temperature for the coldest month rises above 14 °C.

### Soil Adaptability

Citrus requires deep, well drained soils, soil properties like soil reaction, soil fertility, drainage, free lime and salt concentrations, etc. are some important factors that determine the success of citrus plantation. Citrus fruits flourish well on light soils with a good drainage. Deep soils with pH range of 5.5 to 7.5 are considered good. However, by modifying cultural practices it can grow well from 5.5 to 8.5 pH. Presence of calcium carbonate concentration within feeding zone may adversely affect the growth. Light loam or heavier but well drained sub-soils appears to be ideal for citrus.Citrus is particularly sensitive to salt concentrations in the soil solution. High EC can show early decline symptoms in citrus species. Well drained loams are better than clay soils. However, it can be grown from hilly areas to semi-arid irrigated areas. If the soil pH is near 8.5 then EC of soil should be less than 0.5 mmhos/cm for the success of citrus orchards.The growth, development and production of a plant depend on the physical characteristics of the soil such as drainage, density, texture, water-holding capacity, structure, soil depth, the homogeneity of the profile, erodibility, and the degree to which water can infiltrate the soil. These characteristics differ in the various soil types. Physical soil properties determine the degree to which water is released for uptake by the plant roots, and the depth of the root system.

## RECOMMENDED AND POPULAR CULTIVARS

Large numbers of cultivars in genus citrus are being cultivated throughout India. Some of the commercial cultivars of North India are described here.

### Mandarins or Mandarin like Fruits

**Kinnow (King X Willow Leaf):** This is the most important cultivar of mandarin like citrus fruit. It is a cross between Willow leaf (*Citrus deliciosa*) X King mandarin (*Citrus nobilis*). The plants were planted First at Fruit Research Station Abohar. It performed well and now, it has become the number one commercial cultivar of North India. Tree is medium in growth with glossy and shining leaves. Leaves are lanceolate in shape.

Tree is precocious in bearing and comes to bearing as early as in the third-year of age. Fruits are of medium size and globose in shape. Rind is normally tight can become loose if kept upto February end on the tree. The colour of rind becomes golden yellow. Segments separate with little difficulty. Very sweet and juicy. TSS of juice varies from 12-15 per cent with 2 to 1 per cent acidity. Seeds per fruit vary from 10-25. Sometimes fruits may become seedless due seasonal change. Although Kinnow mandarin ripens in January, its fruits are used for juice purpose for blending with Musambi from November onwards.

**Local Santra (Desi):** The trees are erect growing with sparse foliage. Fruits are smaller than Kinnow, but have short neck. The rind is loose and easy to peel, yellow in colour. The segments get separated easily. Seeds are 4-5 sometimes seedless. The fruits ripen in December. T.S.S. varies from 8-10 per cent with 2-3 per cent acidity.

**Khasi mandarin:** It is commercially grown mandarin cultivar of Assam area and North Eastern states. There it is known as Sikkim or Kamala. Khasi name comes from the language Khasi of the Khasi tribes of the Meghalaya state. In fact all the mandarin cultivars grown in India (Santra, Nagpur santra, Coorg mandarin etc.) are the varietal strains of this original fruit. Fruits globose to ovate. Rind bright yellow to orange. The rind is smooth. The fruits may have short necks, very juicy. Juice is orange coloured. Seeds vary from 7-25.

**Coorg mandarin:** Most important cultivar of South India. It is grown on large scale in Coorg area of Karnataka. Trees are vigorous and upright with dense foliage. Fruits medium with rind developing a bright orange colour at maturity. The fruit shape varies from oblate to globose. Rind thin easily peeled. Very juicy with many seeds ranging from 10-30.

**Nagpur mandarin:** It is very popular mandarin cultivar of central India and famous in the world. The trees are vigorous with compact lanceolate foliage and usually erect. Fruits medium in size sub-globose. Rind loose develops cadmium colour at maturity. Sometimes the fruits have small neck. Very juicy. Seeds vary from 0-15 depending upon the season and climate. TSS varies from 10-12 per cent with upto 1 per cent acidity.

## Sweet Oranges (*Citrus sinensis*)

**Mosambi/Musambi** : This cultivar is named after the southern African country Mozambique from where it was introduced to India. It is erroneously associated with 'sweet lime' (*Citrus limettioides*) in the literature. Tree is of medium size with sparse foliage. Fruits medium in size and sub-globose in shape. The rind shows longitudinal furrows and a circular ring at the stylar end (bottom of the fruit). The colour of rind at ripening turns light yellow. The juice is acidless with 10-12 per cent TSS. Fruits ripen in November. Seeds vary from 20-25 per fruit. The plants raised on Rangpur lime or Pectinifera rootstock perform better than on rough lemon. Nucellar Mosambi, Phule Mosambi and Katol gold are the superior clonal selections of this cultivar.

**Satgudi :** Satgudi oranges are native to the south eastern part of India, particularly the region of Andhra Pradesh which is located on the Bay of Bengal.The fruits have a deep orange colored, thin outer skin with a very juicy inner fleshwith sweet flavour that is low in acid (but more than Mosambi) and mildly tart finish when fully ripened.

**Pineapple:** Tree is of medium size with spreading habit. Fruits medium-sized, round in shape. Rind colour yellow to orange, rind glossy and shining. Fruits are very juicy. TSS: Acid ratio well blended. Seeds vary from 10-20. Fruits ripen in December. Area under this cultivar is very negligible due to Kinnow mandarin expansion.

**Jaffa:** Tree is similar to pineapple. Fruits are bigger in size than Pineapple, round in shape. Skin yellowish orange. TSS: Acid well blended. Number of seeds per fruits varies from 9-12. Fruits ripen in December. Plants raised on Cleopatra rootstock perform better than rough lemon.

**Blood Red :** Trees are vigorous than Jaffa. Fruits are of medium-sized and round in shape. The skin develops deep orange colour at ripening in December. Flesh red when fully ripened TSS 10-13 per cent, acidity 1-2 per cent. The seed number per fruit ranges between 9-12. Juice reddish in colour. Plants budded on rough lemon show stem pitting and crease formation. Therefore, Cleopatra rootstock is being used to check decline in Blood Red orange.

**Valencia Late:** Tree is vigorous. Heavy bearer, fruits are of medium-size, slightly oval in shape. Skin turns golden yellow at ripening. Juice abundant with good blend of TSS and acidity. It is late ripening cultivar. The ripening time in North India is February. The seeds vary from 2-7 per fruit.

**Cutter Valencia:** Hodgson (1967) noted that: "This California nucellar seedling was derived about 1935 by H. S. Fawcett of the Citrus Research Center, Riverside, from an outstanding old Valencia tree in the J. C. Cutter orchard at Riverside. This seedling budline was released in 1957 and is currently popular in California.Primarily grown for processing and orange juice production, Valencia oranges have seeds, varying in number from zero to nine per fruit. Its excellent taste and internal color make it desirable for the fresh fruit markets, too. The fruit has an average diameter of 2.7 to 3 inches (70–76 mm).

## Grapefruits (*Citrus paradisi*)

**Red Blush:** The tree is dwarf in nature. The fruits are of medium size oblate in shape. Smooth rind, glossy and deep yellow with crimson patches, which become visible at ripening. Juice vesicles have deep crimson blush. Juice is less acidic with good TSS. Seeds usually get aborted. There can be 0-8 seeds per fruit. It ripens in end November.

**Marsh Seedless:** Trees are spreading and vigorous than Red blush. Fruits are medium to large in size oblate-roundish in shape skin light yellow and smooth. Juices have medium TSS and acidity. Mostly seedless due to seed abortion. It ripens in December.

**Duncan:** Tree is vigorous and spreading. It bears big- sized fruits of oblate shape. Skin pale light yellow or creamy. Juice acidic and sweet with well marked bitterness. It contains very high number of seeds, may be upto 50. It ripens in November.

**Foster:** Fruits are medium to large in size with oblate shape skin pale yellow. Flesh pink and juice have well blended acidity. Taste is bitter. Seeds 40-50 per fruit. It ripens in November.

**Flame grapefruit:** Flame grapefruit is reported to be from seed of a sport of Ruby Red grapefruit originating in the Houston, Texas grove of C. Henderson. C. J. Hearn made the final selection, and Flame was released as a new variety in 1987.Flame trees grow vigorously to a large size and are reported to be more cold-tolerant than Star Ruby. The fruit has a smooth yellow rind and usually has a pink blush. The flesh is tender and juicy and has an internal color almost as dark as Star Ruby.

## Lemons (*Citrus limon*)

**PAU Baramasi:** Tree is vigorous with dense foliage. The shoots drooping usually touch the ground. Fruits are lemon yellow, round in shape with rounded apex and tapering base. Skin is thin, fruit juicy and mostly seedless sometime containing few seeds. Bears throughout the year.

**Eureka:** Tree is semi-vigorous. Fruit oblong in shape with nipped apex. Skin-lemon yellow and sooth. Juice abundant and strongly acidic, with excellent flavour. Fruits ripen in August with few seeds rarely present.

**Punjab Galgal:** Tree is vigorous, foliage light green in colour. Fruit is medium is size and oval in shape. Peel yellow smooth and glossy. Juice is very acidic (5-6 % acidity), with 5-8 seeds per fruit. It ripens in November-December.

## Lime (*Citrus aurantifolia*))

**Kagzi:** Tree is spreading and spring with dense small leaves and twigs usually attacked by citrus canker. Fruits small round and thin skinned. Juice vesicles green in colour, juice is very acidic (7 % acidity).

**Sweet lime (Mitha) (*Citrus limettioides*):** Tree is spreading with light green foliage. Stout thorns present on twigs. Fruit medium globose to ellipsoid. Skin smooth with distinct aroma. Juice abundant, non-acidic and insipid. Seed 5-6 per fruit. Fruits ripens in September.

## PROPAGATION TECHNOLOGY AND ROOTSTOCKS

Most of the mandarin cultivars are propagated through seeds except Kinnow and Nagpur mandarins; usual practice in Coorg, Assam and North Eastern hills is to use seedlings as planting material. But with concerted efforts made to find out suitable rootstocks for different regions, orchardists in India shifted to vegetative methods, particularly 'T' budding because budded plants bear early and are tolerant to biotic and abiotic stress. The seedling trees not only bear late but also tend to become thorny and grow tall and slender.

## Propagation by Seed

For quality planting material, uniformly matured fruits from healthy, true to type and heavy bearing plants are selected to extract seeds. Freshly extracted seeds are mixed with ash and dried in shade otherwise, they may loose their liability seeds are sown at a distance of 2-3 cm. Germination may take place with in 3-4 weeks. Since the seeds are polyembryonic the zygotic (too fast growing, showing excessive vigour and some showing less vigour and stunted) seedlings are rouged out and the rest that are assumed to be produced from the cells of nucleus are allowed to grow. The seedlings thus selected are more or less uniform in growth and production.

## Vegetative Propagation

**T - Budding:** Budding is done using the buds on bud-wood taken from the disease free mother plants on Rangpur lime, Cleopatra, Jatti khatti, Karna katta and Troyer citrange. Rangpur lime is a vigorous, hardy rootstock with good adaptability to a wide range of soils particularly heavy soil, tolerant to tristeza and salt; it is susceptible to footrot, exocortis and xyloporosis. Cleopatra mandarin is the most salt tolerant rootstock with the ability to exclude sodium and chloride taken up by root system. It is tolerant to tristeza, exocortis and fairly tolerant to foot rot. Rough lemon is well adapted to a variety of soils ranging from high clayey to high sandy soils. However it is susceptible to foot rot and scab but tolerant to tristeza. Troyer citranges are used in areas where hardiness and resistance to tristeza are necessary. They are also resistant to foot rot but susceptible to exocortis.

**Shoot tip grafting:** The ICAR-Central Citrus Research, Nagpur has standardized shoot tip grafting technique for propagation of disease-free (virus-free) planting material of Nagpur mandarin.These shoot tip grafted plants are used as disease free mother plants for large scale production of bud-grafts for planting by the nursery growers in central India.

## Promising Rootstocks

Rough lemon (*Citrus jambhiri*) is ruling the citrus industry in India although Rangpur lime (*Citrus limonia*) is also evaluated as a promising alternative rootstock. In rough lemon, there are many strains like Jatti-khatti, Jalandhari khatti, jambhiri-I and jambhiri-II . Hence one should be sure enough to have Jatti khatti seeds for raising his rootstocks. Raising of quality planting material forms the basis of longevity and productivity of a citrus orchard.

**Rootstocks for Nagpur mandarin :** An exotic rootstock Alemow (*Citrus macrophylla* Wester) was found to be a most promising rootstock for Nagpur mandarin under black clay soils of central India conditions. The rootstock Alemow produced maximum fruit yield (21.t/ha) with medium canopy of Nagpur mandarin whereas conventional rootstock rough lemon and Rangpur lime yielded only 10 t/ha. The highest nutrient uptake of all the macro and micro(except Cu) was recorded with this rootstock. Alemow showed as the most potent rootstock for Nagpur mandarin having all the promising horticultural traits will go a long way in imparting not only production sustainability but improved orchard life as well in addition to fitting this rootstock under high density orcharding.

**Promising Rangpur lime and Rough lemon rootstock strains** A rootstock strain Rangpur lime (Brazillian) produced maximum fruit yield (15.61t/ha) than the conventional Rangpur lime strains whereas for rough lemon a strain 14-9-13 and Jullandhari khatti resulted higher fruit yield (14.48 &14.08t/ha resp.) with 100% tree survival after 18 years of tree life under limiting soil and water conditions at ICAR-Central Citrus Research Institute, Nagpur.

**Rootstock for Acid lime:** Alemow (*C. macrophylla* Wester, an old Philippine lemon/ pummelo hybrid) as a classical rootstock possessing outstanding traits as the most potent rootstock for acid lime with maximum macro and micro nutrient extraction capacity with 100% tree survival after 18 years of trial at ICAR-Central Citrus Research Institute, Nagpur. The highest fruit yield (13.4 t/ha) was recorded with Alemow whereas seedling produced only 5.88 t/ha on pooled yield basis.

**Rootstocks for citrus in Punjab**: Rough lemon (Jatti Khatti) is best rootstock for Kinnow, Jaffa, Grapefruit, Lime and Lemon .Similarly, Carrizo for Daisy and W. Marcott mandarins; Cleopatra for Blood Red and Pectinifera for Musambi are recommended rootstocks.

## Compatible and smooth bud union

Some of the rootstocks such as Alemow, Sun chusha, Willits citrange, Rangpur lime and rough lemon exhibited smooth bud union, showing the congenial relationship of stock and scion with Nagpur mandarin. As the incompatible rootstocks resulted over growth cup shape formation and bark splitting after 2-3 years. Mostly these rootstocks were affected by various diseases and declined at early age of 8-10 years. The bud union of all the strains of rough lemon and Rangpur lime were found smooth and showing congenial relationship with stock and scion of Nagpur mandarin.

## Disease tolerant rootstock

The rootstock X-639 (Cleopatra mandarin x *Poncirus trifoliata*) showed less disease susceptibility (1.25) on Nagpur mandarin with 100% tree survival after18 years of tree life. The least no. of gummosis lesions were also observed in sour orange Tirupati (1.11 lesions/tree) whereas conventional rootstock rough lemon and Rangpur lime resulted in higher disease susceptibility(2.17 and 2.06 lesion/ tree, respectively).

## Dwarfing Rootstocks

The exotic rootstock Flying dragon trifoliate orange(*Poncirus trifoliata*) and trifoliate orange (Chethalli) exhibited dwarfing growth with Nagpur mandarin in black cotton soil of Central India. The tree canopy of Nagpur mandarin on Flying dragon rootstock was only 27.2$m^3$ with better fruit quality whereas the plants on rough lemon and Rangpur lime canopy ranged from 80 to 100$m^3$ after 15 years of tree life. These rootstocks are useful for high density planting for Nagpur mandarin with double planting density.

## Flowering and Fruiting

Flowering in citrus is a least understood phenomenon across the citrus growing areas of the world. It is recurrent under tropical and subtropical conditions unless synchronised into a well-defined period of concentrated bloom by external conditions. In typical sub-tropical climates with cool winters, citrus tends to bloom profusely in the spring and set only one major crop which matures in late fall or winter or even the following spring depending on variety and seasonal temperature regime. On the contrary in the marginal tropical and semi tropical climates in the absence of sufficient low temperatures during winters water deficit stress is imposed which leads to induction of flowering in citrus plants. Hence flower formation in citrus species is promoted either by drought or low temperature, followed by restoration of climatic conditions favourable for growth. Apart from these a biotic stresses responsible for inducing flowering in citrus, biotic stresses like incidence of insect-pests e.g. bark eating caterpillar or stem borer which causes phloem inhibition of photosynthates to the root system as in the case of mechanical stem girdling and general debility of plants due to various fungal and viral diseases are the causal factors for flower bud formation in citrus.

Among the growth regulators studied the gibberellins have been found to have a definite role in flower bud formation in citrus plants. Flowering in citrus is inhibited by applied gibberellins prior to inductive stress conditions. That is how the anti-gibberellin chemical compounds such as CCC, SADH and benzothiazole and triazoles have been observed to induce flowering in citrus. Citrus plants tend flower and fruit profusely under optimum stress conditions followed by proper irrigation and fertilization.

There should be about 40-50 leaves for one fruit to reach maturity. Hence, thinning is necessary to improve fruit size and to check early decline. Thinning can be manually carried out during last week of May. It is very essential for the first two years of bearing of Kinnow mandarins. Chemical thinning can be effected by foliar application of 300-600 ppm of naphthalene acetic acid at full bloom stage when about 50 % flowers have set fruit (petal fall). Spraying at later stages results in to poor thinning response.

## PLANT AND FRUIT PHYSIOLOGY

Citrus species show a relatively long juvenility period (two to five years) before the trees reach the mature stage to produce flowers. The inflorescence developed in citrus may be either leafless or leafy and these may carry a single flower or several of them. The ratio of each kind of inflorescence in the tree varies with flowering intensity and cultivar. For example, high flowering intensities are generally related to high rates of leafless floral sprouts although some cultivars, such as Satsuma (C. *unshiu* Marc.) that tends to produce only vegetative shoots and single flower inflorescences, may show some particularities.

Citrus species usually produce a large number of flowers over the year. The floral load depends on the cultivar, tree age and environmental conditions. It has been reported, for example, that sweet oranges (*C. sinensis*) may develop 250,000 flowers per tree in a bloom season although only a small amount of these flowers (usually less than 1 %) becomes mature fruit. Thus, flowering represents a great input for citrus trees and to some extent even a waste of resources.

Citrus fruits are non-climacteric, hence their respiration rate and ethylene production do not exhibit remarkable increase along with changes related to maturity and ripening as in climacteric fruits. However, although they have a relatively long shelf life compared to other tropical and sub-tropical fruits, they may experience important physiological postharvest losses if they are not properly handled and stored. As with other horticultural products, major postharvest losses in citrus are caused by weight loss and physiological disorders.

## PLANNING AND PLANTING

Availability of quality planting material is of utmost importance in citrus cultivation. Citrus plants are very sensitive to various biotic and abiotic stresses. Therefore selection of an ideal rootstock is a continuing challenge for the citrus industry of India. Currently used rootstocks viz. rough lemon and Rangpur lime have gone through a lot of variation over the last five decades. Therefore ideal selections developed from the conventional rootstocks by ICAR – Central Citrus Research Institute (Formerly NRC for Citrus), Nagpur and at other places under State Agriculture Universities may be obtained for propagating quality planting material. For bud wood selection, disease free mother plants developed from the elite progeny of known pedigree through shoot tip grafting method available at ICAR-CCRI, Nagpur may only be used.

**Planting**: Generally, planting is done during monsoon in all mandarin growing areas i.e., June – December. In sub mountainous tracts, where planting is generally done on slopes, proper terraces are necessary, while in plains the land should be leveled properly. Pits of 45 cubic centimeters are dug at a spacing of 6 x 6 m and filled with FYM, sand and top soil and then basins are formed. The budlings are planted in the centre of the pits and irrigated.

In North-East parts of India, Khasi mandarins are very closely spaced. 6 x 4 m or 6x3 m is ideal for Kinnow budded on Jatti khatti for closed spacing plantations. Kinnow can be grown successfully under high density planting by using Troyer citrange as a rootstock and by spacing the plants 1.8 x 1.8 m, accommodating 3000 plants/ha. The optimum spacing for Nagpur mandarin is 6 x 6 m when budded on rough lemon. In Karnataka, Coorg mandarin on Trifoliate orange and Rangpur lime can be planted at a distance of 5 x 5, and 6 x 6 m, accommodating 400 and 275 trees / ha, respectively.

Land needs to be thoroughly ploughed and leveled. In hilly areas, planting is done on contour terraces against the slopes and on such lands, high density planting is possible as more aerial space is available than in flat lands. Since citrus trees are highly sensitive to water logging and water stagnation during rainy season providing drainage channels of 3-4 feet depth along the slopes around the orchard is essential.

## SOIL CULTURAL PRACTICES TECHNOLOGY

### Water Need

Water is essential for citrus trees (or for any plant) because it is an integral component of the biochemical reactions that occur within the plant. Water is also important because it is the carrier that moves plant nutrients and other substances throughout the tree. Water also helps maintain plant temperature through transpiration. Finally, water helps maintain leaf and fruit turgidity. There are several acceptable application methods. Basin irrigation is often the easiest for the homeowner. Then fill the basin as the tree needs water. Do not bank soil around the trunk. Water can touch the tree trunk safely if the trunk is not damaged, and if the tree is not planted too deeply.

Citrus trees require consistent soil moisture within available range during critical phenological stages especially when fruits are tender and their stem end vascular connections are not well set about one to one and half months after fruit set in the months of March and April in the northern hemisphere. Use of micro-irrigation systems has become indispensable in all the citrus growing regions of the India regardless of the abundance of irrigation water or scarcity of it. In drip irrigation, the quantity of water and interval can be adjusted according to the age of the plants, phenological stage and meteorological conditions of the region.

The goal of irrigation scheduling is to reduce tree stress to promote vegetative and/or reproductive growth while at the same time reducing leaching of fertilizers and other agrichemicals below the root zone. Water use efficiency of irrigation must also be considered in evaluating irrigation systems. Currently, low-volume drip and micro sprinkler irrigation systems have become the standard for Florida tree fruit crops. Compared with overhead sprinklers, low-volume systems can save water if they are properly managed. Because these systems usually operate at lower pressures than conventional overhead systems, there can also be appreciable savings from reduced energy costs.

**Irrigation scheduling in Nagpur mandarin and acid lime :** Irrigation scheduling based on soil water with 20% and 30% depletion of Available Water Content and based on pan evaporation with 0.8 of open pan evaporation resulted better growth, yield and fruit quality of Nagpur mandarin and acid lime besides improving the water use efficiency.

**Fertigation in Nagpur mandarin and acid lime :** The fertilizer dose of 500 : 140 :70 (N:P:K) is standardized as ideal for fertigation using water soluble fertilizers, which improved fertilizer use efficiency.

**Mulching in Nagpur mandarin and acid lime :** Black polyethylene of 100 μ and grass mulch @ 3 tonnes/ha. was found better in conserving the soil moisture, increasing the growth, yield and fruit quality in Nagpur mandarin and acid lime.

**Automatic drip irrigation scheduling :** The Nagpur mandarin yield was highest (30.91t/ha.) with irrigation on alternate day 120 minutes three times followed by irrigation scheduled with 90 minutes interval two times daily (30.11 t/ha.).

**Critical growth stages and water requirements:** The critical stage of water requirement in Nagpur mandarin was found from January to June (Stage I (Jan-Feb),II(Mar-Apr) and III(May-June). The stage wise water requirement in Nagpur mandarinwas optimum with 80 % ER in all the stages.

## Nutritional Need

Citrus plants are very demanding in their nutritional needs. Judicious use of manures and fertilizers not only helps to maintain productivity of citrus orchards, but also minimizes cost of production. Right choice of fertilizers in an appropriate dose in split applications at critical growth stages of the crop results into maximum advantage. The desirable fertilizer programme is one that provides maximum yield without affecting health of the trees. Orchards should be green manured after every 2 to 3 years. The fertilization should be adjusted on the basis of periodic soil and leaf analysis. Leaf analysis can indicate the quality of the effective absorption of nutrients and their distribution to the different plant parts.

Although fertilizer recommendations vary depending on type of citrus grown, soil, climate, water availability, crop phenology, crop load etc., the following Table 2 gives a general dose of fertilizer for a citrus plant.

**Table 2:** Manures and Fertilizers Schedule in Citrus

| Age of the Tree (Year) | F.YM.(Kg.) | N(g) | Superphosphate 16% P (g) | Muriate of Potash60% K (g) |
|---|---|---|---|---|
| 1-3 | 5-20 | 50-150 | 50-200 | 15-50 |
| 4-6 | 25-50 | 200-300 | 150-200 | 60-100 |
| 7-9 | 60-90 | 350-500 | 250-400 | 110-150 |
| 10 and above | 100 | 500 | 500 | 150 |

The cultivar and region-wise standardized and recommended doses of fertilizer with corresponding yield potential are given below in Table 3:

**Table 3:** Recommended Fertilizers Schedule in Different States

| Scion | Rootstock | Recommended fertilizer (g/tree) | | | Yield (Fruit/tree) |
|---|---|---|---|---|---|
| | | N | $P_2O_5$ | $K_2O$ | |
| **Punjab** | | | | | |
| Kinnow mandarin | Jatti Khatti | 400 | 200 | 400 | 713 |
| Jaffa orange | Jatti Khatti | 400 | 200 | 200 | 292 |
| **Assam** | | | | | |
| Khasi mandarin | Sohmyadong | 300 | 250 | 300 | - |
| **Maharashtra** | | | | | |
| Nagpur mandarin | Jambhiri | 600 | 200 | 400 | 684 |
| N.R.C.Citrus, Nagpur | | | | | |
| Nagpur mandarin | Jambhiri | 600 | 200 | 100 | 659 |
| Acid lime (Seedling) | | 800 | 200 | 100 | 67.5 (Kg/tree) |
| **Karnataka** | | | | | |
| Coorg mandarin | Rangpur lime | 400 | 200 | 200 | 500 |
| **Andhra Pradesh** | | | | | |
| Sathgudi | Rangpur lime | 400 | 200 | 200 | 500 |
| **Tamil Nadu** | | | | | |
| Acid lime (seedling) | | 600 | 200 | 300 | 115.7Kg/tree) |

In addition to these fertilizers, micronutrients especially $ZnSO_4$, $FeSO_4$ and $MnSO_4$ may be given alternate yearly to bearing plants @ 100 g of each of the salts/plant through soil. To non-bearing plants in the juvenile stage these doses may be reduced to 25 g each for plants below 3 years of age and to 50 g each for plants above 3 years of age. Apart from these if micronutrient deficiencies are seen on the foliage, foliar sprays of these salts @ 0.5 % may be given when the new flushes are at least 2-3 months old. Foliar spray should not be given after August since waxy cuticle layer of leaves may not allow absorption of nutrients.The fertilizers should be well mixed preferable with manures in the basins uniformly under the whole canopy of the trees.

## Weed Control

Both dicot and monocot weeds infest the citrus orchards. Frequent ploughings/ hoeings are expensive and hazardous to plant health. Deep ploughing injures the feeder root system which invites *Phytophthora* diseases. Hoeing should be done only once in a year to remove the weeds. Rotavator may be used soil decompaction and weeding during August-September.

**Use of Weedicides:** It is a cost-effective method of weed control in citrus orchards. Pre-emergence application either Diuron (3 kg/ha), terbacil (4.5 kg/ ha), Atrazine (5-6 kg/ha) or pendimethalin (3 kg/ha) keeps the weeds under control for whole of the monsoon period in India. Post-emergence, annual weeds can be controlled by spraying at their flowering stage with paraquat 24 WSC @ 6-7 ml/L of water. Glyphosate is a very potent post-emergence weedicide @ 8-10 ml/L of water. Parthenium may be killed at flowering stage by spraying common salt (NaCl) @ 20 percent.

## Inter Culture

Intercropping in citrus orchards can be done upto four years in grafted plants (juvenile period). Leguminous crops like soybeans, cluster beans, peas, moong beans, rajmash, grams should be grown leaving at least 50 cm radial distance around the plant stems or on either side of the plant rows. Other vegetable crops like leafy vegetables, melons, flower crops like marigold can also be grown as per the regional needs and preferences. Only care needs to be taken with respect to water, nutritional and plant protection measures such that they should not hamper the growth of the main citrus plants. If wheat is to be sown, separate irrigation system should be prepared for the plants to provide irrigation during March-April when wheat doesn't require irrigation. Crop like chillies and solanaceous crops should not be grown in citrus orchards.

## PLANT CULTURAL PRACTICES TECHNOLOGY

### Training and Pruning

The lowest scaffold should be selected at least 40-45 cm above the ground level. Regular visits to orchard during flushing should be made to select scaffolds on all directions on the main trunk. There should be 6-8 scaffolds at a distance of 20-30 cm. No scaffold should be selected one above the other at distance less than 15 cm. For getting high productivity strong tree with little pruning to open up the tree for proper sunshine and ventilation is necessary. The cut ends must be immediately treated with Bordeaux paint or Bordeaux paste.

The water shoots and rootstock sprouts should be periodically removed. Trees are trained to single stem with 4- 6 well-spaced branches for making the basic framework. Further no branches should be allowed from the trunk up to height of 45-50 cm from the ground level. Pruning of bearing trees consists of removal of dead, diseased, criss-cross and weak branches. Root pruning is also practiced in some parts of central and southern India to regulate flowering season. However, such prunings are not beneficial in the long run.Thinning out of branches in young nursery plants should be done atleast once a month during the year following planting and once in two to three months during the first three years just for avoiding very low headed trees. .

### Thinning of Fruits

Thinning is a good idea on trees that have very heavy crops one year and very little or no fruit the next year. Thinning will help balance the load so that the tree does not exhaust its energy and can bear a good crop every year. Thinning can help maximize the size and quality of your fruit and is worth the time invested.

Thinning fruit on citrus trees is a technique intended to produce better fruit. After thinning citrus fruits, each of the fruits that remain get more water, nutrients and elbow room.

The idea behind thinning fruit on citrus trees is to produce less but better fruit. Often, young citrus trees produce many more tiny fruit than the tree can bring to maturity. Removing some of these by fruit thinning in citrus trees gives the remaining fruits more room to develop.

### Merits of thinning

- Discourage overbearing and early fruit drop.
- Improve remaining fruit size, color and quality.
- Help to avoid limb damage from a heavy fruit load.
- Stimulate next year's crop and help avoid biennial bearing

Of the many chemicals used for fruit thinning in citrus fruits (Nagpur mandarin and Kinnow mandarin) use of naphthalene acetic acid at 300-600 ppm concentrations at full bloom stage has been reported to be effective in thinning the fruitlets by about 30 per cent of the potential fruit harvest. Earlier the time of spray after full bloom better are the results of thinning. At later stages thinning is less effective even at higher concentrations (400-600 ppm).

## Crop Regulation

In south and central India, mandarins bloom thrice a year. The February flowering is known as *ambe bahar* (since the flowering coincides with the flowering of Mango-*Amba* in Marathi) and June-July flowering as *mrig bahar* (since the flowering follows the onset of Monsoon Mrig Nakshtra) and October flowering as *hast bahar* (since the flowering coincides with the Hasta Nakshtra in October-November). Under such circumstances, plants give irregular and small crops at indefinite intervals. To overcome this problem and to get fruitful yield in any of the 3 flowering seasons, stressing mandarin trees has been practiced which is called resting or *bahar* (blossoming) treatment. Till the seventies in the twentieth century the practice of root exposure was followed in central and south India however, with the scientific researches that followed during the later part of the twentieth century, this practice has become obsolete due to its deleterious effects on tree health which ultimately affected the longevity of the trees. The current practice in vogue is water deficit stress wherein watering is withheld for a month or two before intended flowering. As a result of water deficit stress, leaves show wilting and fall on the ground. Subsequent irrigations are given at suitable intervals. Consequently, plants give new vegetative growth, profuse flowering and fruiting. It depends upon the choice of the grower as to which of the 3 *bahars* is to be taken to get maximum profit. As the availability of water is a problem in central India during April – May, the farmers prefer *mrig bahar* (June) so that the plants are forced to rest in April – May. Resting treatment is not feasible in North India (as for Kinnow mandarin, a major citrus fruit grown), as mandarin plants normally rest in winter and flower once a year

## Nature and Control of Fruit Drop in Citrus

Flowering, fruit set, subsequent fruit drop and ultimate fruit retention are affected by several environmental and physiological factors. Most commercially important citrus cultivars bloom prolifically producing as many as 100000-200000 flowers on a mature tree. However fewer than 1-2 per cent of these flowers will produce harvestable fruit. It was observed that out of total fruit drop of Nagpur mandarin, 70-83 per cent was pathological, 8-17 per cent entomological and 8-10 per cent was pathological in *ambia* crop (spring flowering and fruiting) whereas it was

73-78 per cent physiological, 11-12 per cent entomological and 10-13 per cent pathological in *mrig* crop (monsoon flowering and fruiting). All the fruits that fail to mature do not drop at one time but there are waves of light and heavy fruit drop in citrus.

Fruit drop in citrus occurs more or less in three distinct waves:

i) The first wave (post-setting drop) occurs soon after fruit set due to natural over production and is not of much concern to the grower.

ii) A second wave occurs from May to June (summer or June drop) in the northern hemisphere and November to December in the southern hemisphere. It is particularly severe in the hot dry climatic conditions.

iii) The third wave, called the pre-harvest fruit drop is the drop of mature fruits before the harvest. This drop is of economic importance to the grower as in this case nearly completely mature fruits are shed causing heavy losses to the grower.

The initial drop period involves the abscission of weak flowers, fruitlets with defective styles or ovaries, or flowers which do not receive sufficient pollination. During initial phases of abscission (6-8 weeks after bloom) most fruits have two abscission zones, one at the base of the pedicel and the other at the base of the ovary. Fruit drop occurs from the abscission zone at the base of the fruit leaving the pedicel attached to the tree temporarily.

Based on the causal factors involved the fruit drop can be classified broadly as (Plate1)

i) Physiological drop,

ii) Pathological drop,

iii) Enotomological drop.

## Physiological Drop

Although the initial fruit drop period in citrus apparently is primarily for physiological reasons, the term physiological drop generally has been reserved for the drop wave that occurs in May to June in subtropical regions in the northern hemisphere and in November to December in the southern hemisphere. Physiological drop is called June (or November) drop and describes the abscission of fruitlets as they approach 0.5-2.0 cm in diameter (Plate -1). Physiological drop is a disorder most probably related to competition among fruitlets for carbohydrates, water, hormones and other metabolites. The problem, however, is greatly aggravated by stress, especially high temperatures or water deficit. Consequently, physiological drop is usually most severe where leaf temperatures may reach 35-40°C and where water stress is a problem as in arid regions of north-central India.

One hypothesis is that high temperatures and severe water stress cause stomatal closure with a concomitant reduction in net $CO_2$ assimilation. Fruit abscission then results because the fruits maintain a negative carbon balance. Overhead sprinkling reduces temperature and has been used in some regions to reduce physiological drop. Differences in hormone or carbohydrate levels may also be involved.

## Pathological Drop

Fruit drop in citrus also occurs due to pathogenic fungi viz. *Botryodiplodia theobromae, Colletotrichum gloeosporioides* and in some cases due to *Alternaria citri* which causes the stem end rot of the fruits (Plate -1) which occurs predominantly on the mature fruits nearing ripening (pre-harvest drop). High inoculum of these fungi in the orchards builds up due to the dead twigs on the bearing trees and honey dew secreting insets like citrus black fly resulting in the development of sooty mould. This causes up to 22 per cent pre-harvest drop in the orchards having 10 per cent dead twigs of the canopy.

## Entomological Drop

Citrus bud mite and orange bug are some of the important pests which cause heavy drop of flowers and fruits in citrus species. Fruit fly *Dacus dorsalis* and fruits sucking moth O*theris fullonica* and O*theris materna* are the regular pests of ripening fruits. They suck the juice from fruits and cause rotting and drop of fruits (Plate -1). Such fruit drop is heavier during September-October in central India. This drop also contributes substantially to the pre-harvest fruit drop in citrus.

The continued retention of citrus fruits to the parent branch depends upon many factors, and the separation of the fruit from its stalk is preceded by a complex of physiological processes termed abscission.

**Control of Fruit Drop:** Some plants regulators (NAA, 2,4-D, 2,4,5-T, GA3, CIPA etc.) and related compounds seem to retard the formation of abscission layer by augmenting the endogenous auxin. Quite a variety of these compounds has been tried with various degree of success.

2, 4-D, a potent and time tested synthetic auxin compound when applied thrice at monthly interval in August, September and October @ 15 ppm with 1% urea gave a reduction in fruit drop up to 30%. For controlling fruit drop in early stages of fruit development after fruit set two sprays of growth regulators 2, 4-D (5 g) + Propiconazole 25 EC (500 ml) should be given in mid-April, August and

September. Spray $GA_3$ @ 20 mg/litre of water when broad leaf crops are cultivated in or around citrus orchards. It has been observed in three year trials that three sprays of either Benlate or Bavistin @1g/l at 15 days interval before harvest could reduce fruit drop to 9.94 % against 22.26 % in control. Thus more than (12%) yield was increased by saving the pre-harvest fruit drop.

The entomological drop can be controlled by poison baiting with 20g malathion W.P. or 50ml diazinon + 200g jaggery with some vinegar or fruit juice in 2 litres of water (two bottles containing poison bait per 25-30 trees) has been found quite effective. Light traps may also be used for collecting moths. The fallen fruits may be collected and disposed off from the orchard at once to avoid pest / disease development. For control of fruit flies also bait containing malathion or trichlorofon or fenthion 0.05 % + 1 % crude sugar, about two months before fruit ripening followed by 10 days interval. Male attracting traps containing 0.1% methyl eugenol and 0.05 % malathion reduces fruit fly population. Use 25 traps / ha starting from 60 days before fruit harvest and solution after every seven days.

## Physiological Disorders

**Fruit splitting:** Fruit splitting is a serious problem in citrus(Plate 2). Proper application of potassic fertilizers, maintaining regular moisture during summer around bearing trees and application of 2-3 sprays of gibberellic acid or 2,4-D(10 mg/litrewater)along with 1 % urea and 0.3 % boric acid in April and May are some preventive measures for this physiological disorder. In lime, the spray of 40 ppm NAA after fruit set of monsoon crop ($3^{rd}$week of May) reduces fruit cracking and improves weight and size of fruit. $GA_3$ 10-20 ppm can also reduce fruit cracking to some extent and improve fruit quality.

**Sun scorching:** This is a serious disorder which occurs in excessively high temperatures of more than $37^0$C coupled with humidity below 60 per cent (Plate 2). The fruits borne on the periphery of the canopy exposed to the sun are affected with yellowish red burn patches and deformity damaging the external appeal and internal quality of the fruits. Application of $CaSO_4$ or kaolin 4% when temperatures start rising after withdrawal of monsoon in September-October months protects the fruits from this disorder .

**Stylar end break down :**This disorder is caused due to the imbalance of nutrition to the developing fruits(Plate 2). Stylar end (blossom end) tissue breaks down due to the lack of nutrients and cracks are developed on the stylar end leading to drop of such fruits immaturely causing heavy losses. Application of $GA_3$10 ppm along with $CaNO_3$ 1.5% and boric acid 0.3% in the month of June immediately after the onset of monsoon prevents this disorder.

Fruit drop in Nagpur mandarin

Entomological fruit drop

Pathological fruit drop

Physiological fruit drop

**Plate 1:** Fruit drop in Nagpur mandarin (See colour version on page 311 )

Fruit cracking

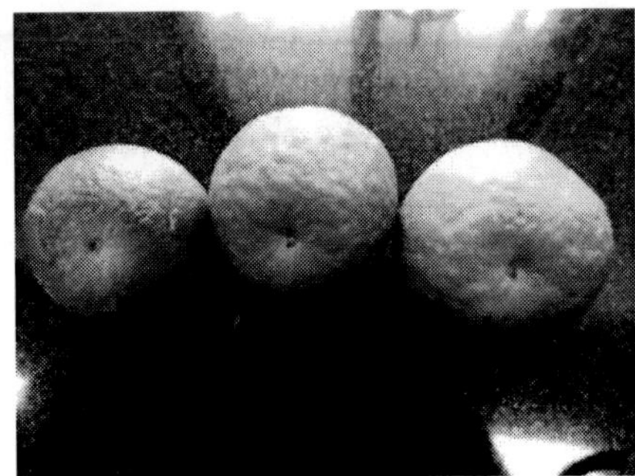

Stylar end breakdown

Sun burn

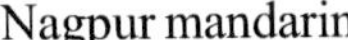

Nagpur mandarin

Kinnow mandarin

**Plate 2:** Fruit physiological disorders

*Waywar* affected fruits (*Note:* The greening of peel towards stylar end, symptoms characteristic of citrus greening disease) (See colour version on page 312)

***Waywar*a menacing fruit disorder:** *Waywar* (literal meaning in Marathi-wasteful), a local term for oblong fruit disorder, is prominently observed in the Paratwada, Morshi and Chandurbazar tehsils of Amravati district of Vidarbha region during last two decades and of late spreading in neighbouring areas of Nagpur in Vidarbha and Saunsar and Pandhurna tehsils of adjoining Chhindwara district of Madhya Pradesh. During the advanced fruit development, the stem attachment of the fruits develops pinkish/reddish tissues on the button albedo partially choking the supply of nutrients and photosynthates to the developing juice vesicles. Many of such fruits continue to remain stunted in growth assuming oblong shape but persistent, hence this disorder is also locally called as '*Cock bund*' (literally meaning 'tap closed'). Since this assumes pinkish/reddish colour at the button tissue, this disorder is also referred to as 'Red Nose' in China. The occurrence of the malady is being reported from distant places in Maharashtra and other states where Nagpur mandarin is being grown. Since the fruits did not drop down and only failed to develop in to normal shape/size/volume and internal quality, it seems that the movement of photosynthates is blocked and only xylem transport probably ensures water supply to the fruits thereby making them sour and insipid (Plate 2).

Preliminary studies have attributed this malady to a variety of abiotic stress factors viz. Nutritional, phenological and recently, also to a biotic factor, citrus greening disease.Since the leaf and affected fruit samples subjected to PCR for diagnosing citrus greening bacterium have become positive, there is conclusive evidence to attribute this disorder to citrus greening bacterium disease. Hence a comprehensive set of control measures involving vector control, spread of the greening bacterium and nutritional and hormonal supplements can overcome this malady.

## SPECIAL PROBLEMS

### Granulation

Granulation, crystallization or internal dryness of citrus fruit is the main concern for consumers. The juice sacs of granulated citrus are hardened, gelled or granular and are opaque white in colour. Juice levels are low. Soluble solids (sugars and acids) are at lower concentrations than normal fruit and granulated fruit as a consequence are tasteless.

The opaque white tissue in granulated vesicles appears to be due to thicker epidermises and/or due to thickened juice cell walls. The juice vesicles become hard, enlarged and turn opaque grayish in colour. The density of pulp is increased, juice contains increased minerals (Calcium, sodium,potassium) and decreased carbohydrate and organic acid. It results in lignification of juice cells that leads to formation of sclerenchyma.

**Causes:** Granulation usually occurs on rapidly growing fruit. Navel, Valencia, grapefruit and mandarin hybrids are most frequently affected. Judicious fertilization and watering practices and early seasonal harvesting alleviate this problem. The problem varies seasonally and is more of a problem on larger sizes or late blooming fruits. Premature fruit drying is also associated with young trees, a condition that is alleviated with tree maturity. Many of the factors mentioned above relating to granulation also result in reduced internal sugar and/or acid content (e.g. late bloom fruit, large fruit, high temperatures, increased fruit water content through irrigation or rain, shaded conditions etc.). Drying appears to be associated with over-maturity, lack of water, excessive tree vigor and extended warm, and/or dry fall weather. High humidity and fluctuation in temperature are also the major pre-disposing environmental factors affecting this malady.

**Management:** The incidence of granulation could be reduced to 50 per cent by applying two to three sprays of NAA (300 ppm) in the months of August, September and October. Spraying of GA 15 ppm followed by NAA 300 ppm in October and November also reduce granulation. Avoid excess moisture, spray lime @ 20kg in 450 l of water and also zinc and copper at 0.5% each.

**Plate 3:** Granulation in Sweet orange bright orange red coloured half is free others are granulated
(See colour version on page 312)

## HARVESTING AND PRODUCTION OF FRUITS

Citrus fruit is non-climacteric, hence should only be harvested at full maturity, and ripened stage when these have attained the desired TSS/acid ratio. Fruit has to be harvested according to maturity criteria for optimum taste and shelf life. The maturity indices for some important citrus cvs. such as Nagpur mandarin, acid lime and sweet orange 'Mosambi' have been already developed by ICAR-Central Citrus Research Institute, Nagpur. Thus for Nagpur mandarin indices for *Ambia* (spring blossom) and *Mrig* (Monsoon blossom) crop are set as TSS content 10%, TSS/acidity ratio 14 and juice content 40%. At least 1/3$^{rd}$ of fruit surface develops colour with desired taste and firmness. *Ambia* crop takes 270 days while *Mrig* crop takes 230-240 days to mature; Acid lime Cvs "Kagzi" (Central India) : minimum juice content 48% with 7-8% acidity. Summer crop matures in June- July after 160 days from fruit-set and Mosambi sweet orange: *Mrig* (monsoon blossom) crop matures after 230-240 days from fruit-set. Minimum juice content 40% and TSS: titratable acidity ratio of 13.90 or 14.

Kinnow needs to be harvested only when it gives brix value of 10 per cent and developed appropriate peel colour. Fully ripened Kinnows shows TSS/Acid ratio as 12:1 or 14:1. This situation develops in January. TSS further increases upto 15%. It will be beneficial for the growers, if they harvest the fruit and do marketing on their own.

**Yield:** On an average about 800-1000 fruits can be harvested from a tree per year amounting to 25 –30t/ha/year. The common practice of harvesting is to twist and pull the fruits from the branch, which may rupture the skin near the stem-end leading to fungal infection and rotting. Therefore, Kinnow fruits should neither be plucked nor torn off, but should be cut off with clippers, shears or secateurs. On the other hand Nagpur mandarin fruits can easily be twist plucked due to the button protection at the stem end. However, this practice is done by the experienced farm workers in the citrus orchards. Although mandarins may attain optimum internal maturity standards but the fruits may not be attractive at the time of harvesting due to lack of good yellow colour especially in central and south India. Accordingly, degreening of mandarins with the application of ethephon (150 ppm) one week before the harvesting develops golden yellow colour within 5-7 days of the treatment. Artificial degreening in ethylene chambers after harvesting is a deleterious practice and not being followed in this non-climacteric fruit.

## POST- HARVEST FRUIT TECHNOLOGY

### Spoilage control

Fruit is treated and packed as quickly as possible except unavoidable holding after its receipts in packing house. The imazalil and thiabendazole the principal fungicides used in citrus post-harvest decay control at present. Many countries have banned the use of benomyl and carbendazim on harvest fruit although they can be used 15-30 days before harvest. The residue limits of these fungicides have been set and fruit should not have residues of these compounds more than MRL.

### Mechanized Handling and Grading

The mechanized sorting, washing, wax coating, and size grading has been standardized for "Nagpur" mandarin, "Mosambi" orange and "Kagzi" acid lime. The process is now being used by growers and industry for domestic and export market in India. In case of 'Kinnow' mandarin grown in Punjab, Rajasthan and Haryana, mechanization of sorting, coating and grading is being followed on commercial scale. This practice has reduced losses as the handling and contamination is reduced considerably. In case of 'Nagpur' mandarin much needs to be done as the marketing is still in the hand traders on conventional line. This technology coupled with orchard management (timely fungicide sprays, sanitation), packaging and transportation would bring down losses from 25% to about 5%.

**Packing house machinery:** The development in packing house machinery witnessed a sea change in last 50 years and fully automatic citrus packing houses handling up to 300 tone fruits every day are common in industrially developed countries. In India, growth has taken place slowly in this area since 1990-91 and much needs to be done.

**Sizing and Sorting:** With developments in electronic sorting machine (vision technology), now industry has fully automatic packing houses. Hand sorters have been found to reject fairly large quality of good fruits. Now, with the help of electronic sorting consistently accurate sorting is ensured and rejection is reduced upto 8% or even less.

**Sizing on weight basis:** In place packing arrangement, fruit is packed in layers in the box. If height or diameter is not uniform and the fruits are not of same shape, the packed fruits do not appear uniform.

**Container for packing:** Corrugated cartons made of fiber boards are universally used in international and domestic trade all over the world. In Australia, 18 kg (30 lit volume) and 15 kg (24 lit) cartons are most common. Telescopic cartons are preferred. Similar is the case in most Mediterranean and South American countries. The cartons of 18-20 kg capacity are common for packing different types of citrus fruits in USA.

**Corrugated fiberboard cartons (50 x 30 x 30 cm size) for "Nagpur" mandarin:** The corrugated boxes (Universal and Telescopic type) are suitable for distant transport and have sufficient strength to withstand storage condition at 90-95% RH.

## Shelf life of Fruits

**Ambient Storage:** Nagpur mandarin fruit can be stored for 3 weeks at 30±5°C and "Mosambi" Orange and "Kagzi" acid lime fruits can be stored up to 30 days with mechanical wax coating and packing in vented polyethylene liner with 5-6hrs after harvest without curing.

**Refrigerated Storage:** Mandarins can be stored at 5-6°C temperature with RH of >90 per cent whereas sweet oranges, limes and lemons can be stored at 8-10°Cwith RH of >90 per cent for 60 days.

**Post-harvest losses:** The post-harvest handling losses of citrus fruits are 5-10 per cent in most developed countries, while in developing countries these losses are over 25-30 per cent which causes decay of citrus fruits produced under arid and semi-arid climate. The post-harvest losses of oranges are estimated about 8.3 to 30.7 per cent in Indian conditions including farm and wholesale market. This is due to faulty storage techniques, condensation of moisture and heat under high temperature condition which permit slow gas exchange leading to spoilage. Post-harvest dips to Nagpur mandarin in 1000 ppm fungicides (benzimidazole group) for 5 minutes provided effective control under storage conditions favourable for disease development for upto 21 days. Carbendazim 1000 ppm was quite effective which gives 70.26 per cent decay control. Storage temperature, relative humidity and storage life of fresh citrus fruits are given in table 4 below.

**Table 4:** Storage temperature, relative humidity and storage life of fresh citrus fruits.

| Sr. No. | Commodity | Temperature (°C) | Relative humidity (%) | Storage life (Weeks) |
|---|---|---|---|---|
| 1. | Mandarins & Tangerines | | | |
| | Satsuma mandarin | 1-3 | 80-85 | 12-18 |
| | Clementine mandarin | 3-4 | 85-90 | 4-6 |
| | Nagpur mandarin | 6-7 | 90-95 | 6-8 |
| | Dancy tangerine | 5-6 | 90-95 | 2-5 |
| | Kinnow and Coorg mandarin | 4-5 | 85-90 | 8-12 |
| 2. | Sweet oranges | 5-6 | 85-90 | 4-8 |
| | Valencia (Calif. and Arizona) | 3-9 | 85-90 | 4-8 |
| | Valencia (Fla. and Texas) | 0-1 | 85-90 | 8-12 |
| | Navel (Calif.) | 5-7 | 85-90 | 6 |
| | Shomouti | 5-6 | 85-90 | 8-12 |
| | Mosambi | 5-7 | 90-95 | 12 |
| | Sathgudi | 5-7 | 90-95 | 12 |
| 3. | Grapefruit | 10-12 | 85-90 | 6-8 |
| 4. | Pummelo | 8-9 | 85-90 | 10-12 |
| 5. | Lemons | | | |
| | (Dark green) | 13-14 | 85-90 | 16-24 |
| | (Light green) | 13-14 | 85-90 | 8-16 |
| | (Yellow) | 13-14 | 85-90 | 3-4 |
| 6. | Limes | | | |
| | Tahiti | 9-10 | 85-90 | 6-8 |
| | Maxican | | | |
| | (Dark green) | 9-10 | 90-95 | 10-12 |
| | (Yellow) | 8-9 | 90-95 | 8-12 |

Shelf life of mandarin and sweet orange fruits as influenced by various post-harvest treatments is given in table 5.

**Table 5:** Shelf life of mandarin and sweet orange fruits as influenced by various post-harvest treatments

| Cultivar | Treatment | Storage Temperature | Storage (Days) |
|---|---|---|---|
| **Mandarins** | | | |
| Nagpur | Wax 6%+2,4-D (100ppm) | 78±6$^0$ F | 24 |
| Nagpur | Film wrapping +carbendazim (2000ppm)+vented PE | 6-7$^0$ C; 90-95% RH | 62 |
| Kinnow | Stayfresh wax (6%) | AmbientCondition | 29 |
| Kinnow | Wax 6%+polyethylene | 1-3$^0$ C; 85-90% RH | 90 |
| Kinnow | Diphenyl paper wrapper | 2$^0$ C; 85-90%RH | 120 |
| | Wax 6%+diphenyl paper wrapping | 19±8$^0$ C | 90 |
| Kinnow | Curing+2,4-D (500ppm)+Carbendazim (1000ppm)+Wrap | Ambient condition | 84 |
| Coorg | 3% wax+2,4-D (100ppm)+PE | 20-26$^0$ C | 21 |
| Khasi | Wax 3%+Benlate (350ppm)+PE | 11-19$^0$ C | 24 |
| **Sweet Oranges** | | | |
| Sathgudi | Vitamin $K_3$ (Menadine) | 28-30$^0$ C | 16 |
| Mosambi | Perforated polyethylene | 10$^0$ C | 75 |
| | Wax (6%) | 28-30$^0$ C | 28 |
| Mosambi | Wax (8%)+2,4-D (100ppm) PE | Room temp. | 40 |
| Mosambi | Wax (9%)+2,4D(500ppm)+Bavistin (0.1%)+non perforated PE) | Ambient condition | 63 |
| Mosambi | Sta-fresh 451 (1:1 dilution)+PE | 25±5$^0$ C | 30 |
| | | 5-6$^0$ C; 90-95%RH | 90 |
| Blood Red | Wax (12%)+2,4-D (50ppm) | Room temp | 45 |
| Blood Red | Packing in Wooden box | Room Temperature | 60 |
| Pineapple | Wax (6%) | Room temp | 45 |

## Processing of Fruits

The scope for processing of citrus fruits in India is very limited. It is largely due to the fact that mandarins constitute about 45 % of area under cultivation which are highly seeded. The juice of the fruits develops bitterness during processing. Most of the citrus production in India is used as fresh fruits. Barely 2-3 % citrus produce is processed on small scale as ready to serve beverages like squashes, fizzy drinks and cordials. Limes and lemons are used processed for ready to serve beverages and pickles.

## INSECT-PESTS AND MANAGEMENT

Among the factors responsible for low production and also low fruit quality insect pests are of major concern. More than 250 insect pests have been recorded on various citrus species in India. However, about a dozen of them like psylla citrus leaf miner, blackfly, lemon butterfly, bark eating caterpillar, aphids, thrips fruit sucking moth and citrus mites attack citrus trees regularly right from nursery stage to the harvest causing cognizable damage thereby posing a serious threat to citrus cultivation and hence considered of major importance.

**Citrus psylla, *Diaphorina citri*** (Kuwayama) (Psyllidae : Homoptera)

Apart from causing massive deblossoming due to excessive de-sapping by numerous nymphs and adults consequently affecting the fruit set seriously (P.5). It also transmits the deadly Greening disease which accelerates 'citrus decline' syndrome.

**Chemical control:** Several insecticides like dimethoate, quinalphos, fenvalerate, oxydemeton methyl, malathion, fenitrothion, acephate, were found quite effective against psylla. Imidacloprid, and profenophos @ 0.025% are effective and persistent in action.

**Citrus leaf miner: *Phyllocnistis citrella*** Stainton (Phyllocnistidae : Lepidoptera)

It is a serious pest of nurseries, young plantations and tender flushes of citrus groves accounting for 30 per cent of the total damage incurred due to pest complex (P.6). Damage by this pest in acid lime predisposes the plant for development of canker disease and shelters thrips and mealy bug helping them to spread.

**Chemical control**: Insecticides having systemic and fumigant action are more effective to control this pest. Acephate (0.075%), quinalphos (0.05%) have been found effective in giving good control upto 21 days after spraying. Fenvalerate (0.01%) foliar spray once in 45 days offers effective check of the pest.

**Citrus black fly: *Aleurocanthus woglumi*** Ashby (Aleurodidae:Homoptera)and white *Dialeurodes citri* (Ashmed) (Aleurodidae : Homoptera)

Citrus black and white fly is one of the economically important insect pests of citrus orchards particularly in central, north-west and north eastern parts of India, causing heavy yield loss and also affecting the quality of the fruit.It is an endemic pest. Both nymphs and adults suck cell sap and the nymphs excrete voluminous honey-dew on which sooty mould (*Capnodium* sp.) grows widely leading to black layer manifestation, locally called as 'Kolshi' in central India, covering entire plant parts due to which photosynthesis is affected. Plants are devitalized due to excessive de-sapping and the fruit bearing capacity of the tree is also affected.

**Control:** Application of insecticides acephate @ 1.25 g/l or imidacloprid @ 0.5 ml/l alternatively at the appropriate developmental stage of the pest is important. Pesticides are to be applied twice at 50% egg hatching stage when 1st and 2nd larval instars predominate at 15 days interval.

**Lemon butterfly: *Papilio* spp**. (Papilionidae : Lepidoptera)

Caterpillar is the damaging stage of the pest which feeds on foliage voraciously causing severe defoliation.

**Control:** Hand picking of various stages of the pest and their destruction is suggested as it is very easy to do in both nurseries and young orchards.In case of severe infestation spray with quinalphos (0.05%) at 10 days interval is recommended

**Scales: *Aonidiell aaurantii, A. Citrina,*** Diaspididae, and *Coccus viridis (Green), Coccidae, Homoptera)*and **Mealy bugs** *Planococcus citri* Risso (Pseudococcidae :Homoptera)

Both these pests suck the cell sap and devitalize the plants. While sucking the sap they also excrete honeydew due to which black sooty mould layer is formed over the leaves and the stems.

Mealybugs cover their body with wax ovisac where they lay eggs. The pest attacks fruit stalks and joints of branches and stems which lead to fruit drop. Growth of the attacked plants is arrested in case of severe infestation

**Control:** Soft scales and mealybugs can be controlled by spraying insecticide like nuvan (0.01%). Pre-bloom spray of cypermethrin or nuvan 0.01% can control the scales effectively. Dimethoate (0.07%), chlorpyrifos(0.1%) or neem oil (2%) also give good control of scales.

**Fruit sucking moth: *Ophideres* (*Otheris*) *fullonica, O. materna*;***Castor Semi looper,* ***Achaea janata* L.** (Noctuidae, Lepidoptera)

The adults come out in the late evening dusk and puncture the ripening fruits (P.8). Such fruits drop prematurely as a result of rottening due to fungal and bacterial infections introduced through punctures resulting into considerable fruit loss.

**Control :** Poison baiting with 20 g malathion w.p. or 50 ml diazinon + 200 g gur with some vinegar or fruit juice in 2 litres of water (two bottles containing poison bait per 25-30 trees) has been found quite effective.In case of fruit fly male attracting trap containing 0.1% methyl eugenol and 0.05% malathion at 25 traps per hectare (change the solution every day) before 60 days of first harvest.

**Bark eating caterpillar:** *Inderbela quadrinotata* More: (Metarbelidae: Lepidoptera)

Dirty grey coloured caterpillar having dark robust head makes the tunnel near the union of branches (P. 10). It rests in these tunnels during the day time while comes out in night through the way covered with the fross and feaces and start eating the bark. Thus, xylem tissue supplying water and nutrients to the tree are eaten away. Attacked trees lose their vigour. Branches above the attacked portion start yellowing and drying and splitting of the trunk occurs near the union of branches.

In case of trunk borer, the larva makes entry hole into the trunk at ground level horizontally in sapwood whereas the right angled exit hole to vertical tunnel is intermittent 1 or 2 holes are also made , perhaps for airation. Sawdust/ frass at the holes is indicative of the borer.

**Control :** Nuvan 0.01% should be applied into the larvel tunnel with the help of plastic syringe or its cotton swab inserted into the tunnel is quite effective due to its contact and fumigant action.

**Citrus aphid: *Toxoptera citricida*** and ***T. aurantii*** (Aphididae :Homoptera)

Brown citrus aphid (BCA) *Toxoptera citricida* and another, black citrus aphid *Toxoptera aurantii* are the most efficient of all the aphid vectors of citrus tristeza virus (CTV) causing stem splitting and debilitation of grapefruits and oranges with least hope of eradication. Heavy infested trees may suffer significant loss of sap, build up of sooty mould from copious honeydew excretion and leaf distoration.

**Control:** The pest can be managed by two-three sprayings either with any of the systemic organophosphatic insecticides *viz.,*quinalphos, dimethoate or malathion.

**Citrus thrips: *Scirtothrip*** sp. ***Heliothrips haemorrhaeodalis*** Bouche (*Thripidae, Thysanoptera*)

The nymphs and adults rasp and suck the sap from fully developed flower and leaf buds, young and grown-up fruits and also the leaves (P. 7). The leaves become cup shaped and leathery.

**Control :** Dimethoate or acephate @ 1 ml/1iter of water should be sprayed at bud burst stage and on berries and the surrounding vegetation should also be sprayed as the pest thrives on it.

**Citrus mite: *Eutetranychus orientalis*** Klien (Tetranychidae; Acarina)

Mite infestation produces a multitude of grey spots on the upper leaf surface which gives a chlorotic appearance to leaves (P. 9). The leaf surface appears to have been ash spread or gives mottling appearance, becomes weak and finally drops.

**Management :** Acaricides like dicofol 1.5ml water should be sprayed to get effective control. Spraying should be carried out on the upper surface of the leaves where the pest is present.Kelthane @ 1.5 ml/litre or wettable sulphur @ 3 g / litre are recommended against mite.

## DISEASES AND MANAGEMENT

Apart from insect-pest and nematodes, the fungi, bacteria, virus, fastidious prokaryotes and viroids are the other biotic factors affecting the health of citrus trees. The most important among these are elaborated here.

### *Phytophthora* diseases

*Phytophthora* species cause foot rot, root rot, crown rot, gummosis, leaf-fall and brown rot diseases in citrus. Foot rot lesions develop as high as 60 cm from the ground level on the trunk and may extend below the soil on crown roots as crown rot (P.4). In severe cases, when foot and significant portion of root system is damaged, the large branches of the same side of the affected plant are killed due to the rot of conducting tissues near the bark. Usually the disease is confined to feeder roots. The diseased plants thus have comparatively fewer fibrous roots than healthy plants. In severe cases, where regeneration of feeder root does not cope with the rate of destruction, the affected plant will show starvation, less canopy volume with naked branches, die back and slow decline symptoms.

**Plate. 4:** *Phytophthora* in citrus (See colour version on page 313)

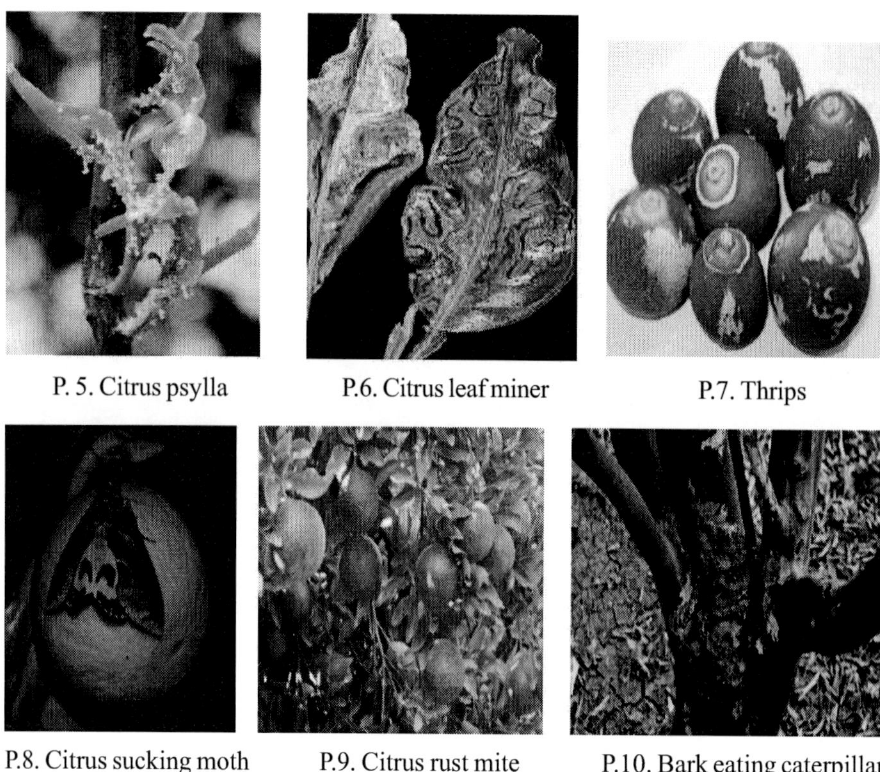

P. 5. Citrus psylla P.6. Citrus leaf miner P.7. Thrips

P.8. Citrus sucking moth P.9. Citrus rust mite P.10. Bark eating caterpillar

(See colour version on page 313 )

**Citrus canker (*Xanthomonas campestris* pv.*Citri*)**

Darkbrown, rough, raised, pimple-like corky spots appear on leaves, twigs, branches, petioles, fruit stalks, fruits and thorns. On leaves, spots first appear on lower surface and then on upper surface. Spots are surrounded by a yellow halo lesions on twigs are elongated.

## Citrus Greening Disease (CGD)

It is an important disease which is prevalent in the relatively cooler, high-lying areas (above 600 m). Typical symptoms are yellowing of the leaves and malformed fruit. One side of the fruit along the central axis does not develop normally and remains smaller, resulting in asymmetrical fruit. The smaller side remains greenish while the rest of the fruit turns orange. If the entire tree is affected especially in pre-bearing juvenile period, it would be better to remove and replace it. It is transmitted by psylla (see Citrus psylla). As greening is usually localised within one or two branches of the tree, it is advisable to cut out such branches. In Nagpur mandarin it causes damage to the maturing fruits limiting the movement of photosynthates to the fruits which generally show oblongation and lack in internal juice qualities. This disorder is locally called as *Waywar* (literal meaning in Marathi-wasteful) in central India.

**Tristeza:** Tristeza is probably the most destructive disease of citrus. It is caused by the largest known for a plant RNA virus. Tristeza can be transmitted by grafting, budding, dodder, and by aphid vectors. The citrus can cause diverse field symptoms based on citrus cultivars, environmental conditions and virus strain involved. Sudden collapse or abrupt wilting that justifies the name 'quick decline' followed by defoliation in sweet orange , grapefruit and mandarin on sour orange rootstocks are indicative of tristeza infection .Typical symptoms on Kagzi lime develops vein flecking of leaves , leaf cupping and stem pitting. Strains of tristeza also induce stem pitting on trunk and branches of grapes fruit, pummelo, tangelo, lime and sweet orange, but rarely in lemon, sour orange, trifoliate orange and mandarin.

## Management of Disease

There are several aspects that must be included in a successful budwood certification programme such as regulation, cultural practices, sanitation, quarantine, vector control and cross protection. An integrated programme with all the above aspects is essential for the success of any certification scheme.

**Cultural practices :** All pruning and grafting tools should be adequately disinfected with 1-2% sodium hypochlorite or other detergents, prior to any fruit picking, grafting or cutting of any tree or nursery plant to avoid spread of mechanically transmitted diseases.

**Vector control:** Citrus tristeza virus is transmitted efficiently by aphids. The non-biological methods include the use of suitable insecticidal sprays, insect traps, reflective mulches etc. Different biocontrol agents include parasitoids, predators and microbes. Parasitoids are host (insect) specific.

**Use of resistant rootstocks:** The primary reason for shifting citriculture from seedling to budded plants was the appearance of *Phytophthora* foot rot.

**Improved cultural practices:** Plant should be selected from *Phytophthora* free certified nurseries and with high budding (above 9" height). While planting care should be taken to keep bud union as high as possible so that irrigation water should not touch the rootstock-scion union.

**Chemical control:** Copper fungicides are used as foliar spray, drenching of basins and as trunk paste to control *Phytophthora* diseases.

**Biocontrol**: *Trichoderma harzianum* shown antagonistic action against *Phytophthora* root rot of Coorg mandarin when applied along with Coffee waste, poultry manure and FYM.

### *Citrus Decline*

In different parts of the citrus growing areas of the world, citrus decline is attributed to multiplicity of causes. Some times or the other, most of the citrus industries of the world were threatened by this decline syndrome and hence, it deserves utmost attention. Decline as such is not a specific disease but is a symptomatic expression of many disorders in the plant.

**Symptoms:** The symptoms comprise of retarded growth and sickly appearance of the tree. Trees show sparse foliage with mottling leaves, stunted growth and sickly appearance. There is an excessive flowering, but the fruit set is either poor or the fruits are of small size. The yield could be bumper in the initial years of decline leading the trees to exhaustion thereby aggravating their susceptibility to decline. In the advanced stages of decline the yields drop down drastically owing to the reduced canopy. The quality of fruits is adversely affected.

**Causal factors:** 'Citrus decline' is attributed to a number of factors. They include general neglect of the orchard, stress for induction of flowering, scarcity of water for irrigation during later part of summer, poor drainage, non–availability of disease free planting material, susceptibility of conventional rootstocks/scions to various insect-pests and diseases, malnutrition, alternate bearing, improper spacing, injudicious water management, neglect of pruning and disposal of dead wood.

**Rootstocks :** Lack of an ideal rootstock for sustainable commercial citrus production is one of the serious constraints in management of citrus decline. Although rough lemon is tolerant to CTV, it is highly susceptible to *Phytophthora.* The other rootstock, Rangpur lime is also susceptible to *phytophthora* as well as the quality of fruits on Rangpur lime does not match to those on rough lemon. All the conventional rootstocks in India are found susceptible to citrus nematodes.

**Soil texture :**The soil texture indicating the relative proportion of coarse and fine particle in soil plays an important role in proper root activity. The hard pan in the subsurface affects root activity and age of citrus orchards.Some of the mandarin growing soils of Nagpur district having higher clay content (60-80%) mainly concentrated in subsurface, low coarse fragments, sand and $CaCO_3$ upto 150 cm depth were not suitable for Nagpur mandarin for regular flowering and fruiting.

**Soil drainage:** The high soil moisture, poor drainage, aeration and permeability are associated with high clay contents of soil whereas light soils are not able to retain sufficient moisture. Citrus trees are sensitive to excessively moist soil conditions within their root zone and thrive best on soils that are well drained.

**Nutritional factors:** Deficiencies of major nutrients, namely nitrogen, calcium, magnesium and potash have been considered to be responsible for the decadence of citrus trees. Nutrient imbalance in citrus plants can lead to adverse consequences and ultimately the decline of trees gradually takes place.

**Irrigation:** Poor management of irrigation sources is one of the main factors associated with citrus decline in India. The large scale drying of citrus plants is also due to scanty rainfall, gradual deepening ground water table, absence of measures for ground water recharging and absence of dams.

**Insect- pests:** Citrus trees in India are attacked by number of insect-pests. However, only some of them lead to citrus decline in varying intensities. Two species of bark eating caterpillar (*Inderbela quadrinotata* Walker and *Inderbela tetraonis* Moore) found to be quite serious in the neglected orchards in Central India. Similarly citrus trunk borer (*Monohanmus versteegi* Nitzema) is a serious insect pest of mandarin orange in North Eastern region and Sikkim.

**Nematodes:** The citrus nematodes, *Tylenchulus semipenetrans* is the most important pathogen found associated with trees showing decline symptoms. In addition to this, *Radopholus similes* and other nematode fauna have been seen to predispose the plants for declining stage.

**Rejuvenation of Declining Citrus Orchards**

**Improved cultural practices:** Improved cultural practices such as pruning, top and frame working must be implemented. Injuries to roots should be avoided during ploughing and harrowing.

**Removal of nutritional disorders:** The nutritional problems have to be identified based on the available leaf nutrient standards and adjusting the fertilizers schedules accordingly.

**Clean cultivation:** Weed control is of foremost importance in citrus cultivation. The pre and post emergence herbicides have been identified for different citrus growing regions for an effective control of weeds up to 300 days in a year.

**Efficient use of water:** Flood irrigation is conventionally followed in most of the citrus growing areas in plains. It acts as pre disposing factors for foot and root rot. Hence, regulation of irrigation through drip is of utmost importance since the adopted in has. The adoption of fertigation technique will meet the nutritional as well as water requirements of the plant at different growth stages to ensure maximum water use efficiency.

**Crop regulation:** Only optimum load of crop should be allowed in a season and necessary measures should be taken at flowering/ berry stage for thinning etc. In case of Nagpur mandarin, two commercial crops are taken in a year that is not advisable for the normal health and productivity of the tree hence double cropping should be discouraged. The bearing should be regulated in such a way that only one crop of optimum loads e.g. 700-900 fruit/tree is taken in a year.

**Management of insect-pests and diseases:** Citrus plantation all over India have been found susceptible to fungal, bacterial and viral diseases and insect-pests like blackfly, psylla, whiteflies, leaf folders, trunk borer, scales etc. Regular monitoring of pest and disease complex is must by taking prophylactic and curative measures at a right time.

**Removal of dry shoots:** After harvest, the dried twigs should be removed along with one inch live stem portion followed by the Carbendazim spray @ 1g/lit water in order to arrest further sickness of the plant.

**Inarching :** Declining trees can be rejuvenated by inarching with the help of three-four seedlings around the tree trunk to provide new root system to the plant.

## MARKETING OF FRUITS AND EXPORT POTENTIAL

There are many constraints hindering the true potential of the country's fruit production and exports. Market information received by producers are always partial and sketchy. Resource poor farmers under-invest in farming inputs like pesticides and fertilizers that leads to lower yields and poor quality products. Advance sales are also a root cause of financial constraints amongst farmers. Scarcity in storage and transportation infrastructure resulted in 25-40 per cent post-harvest losses that shrinks supply and put pressure on prices. The prevention of such losses would further improve exportable surplus and their international competitiveness. In order to lower the shares of middlemen in consumer's rupee, access to credit and market information, control over the output losses, improvements in market infrastructure and cheaper availability of transport and packing material is needed. Fruit markets are not perfectly competitive. There is a need to improve efficiency and effectiveness to promote export of fruits. A product-specific market development strategy needs to be initiated with the active participation from the production and marketing systems.

### Export of Citus Fruits

During 2014-15, India earned Rs 31.94 crores from 17231 tonnes of export of orange. The major share of oranges export of 76.9 per cent was made to Bangladesh. However,oranges were also exported in small consignment to many countries in the world.

**Fruit size requirements for export markets:** The minimum size requirements for export markets of major citrus fruits grown in India are given in table 6 below:

**Table 6:** Minimum size requirements of citrus fruits

| Sr. No. | Citrus fruits | Minimum Size (mm) |
|---|---|---|
| 1. | Sweet oranges | 53 |
| 2. | Mandarins | 45 |
| 3. | Acid lime | 33 |

Lack of access to institutional credit for market participants may reduce the efficiency of the marketing system by constraining investment in improved practices. The financial institutions are desired to reformulate the laws and regulations to ease out the loaning process in order to alleviate the financial constraints of producers, contractors, exporters and other market traders. Exporters are also suggested to launch development schemes for their contract farmers by providing credit and technical guidelines to produce high quality fruits for export. Appropriate incentives should be extended by the policy makers in establishing export zones where all necessary infrastructures like cold storages, refrigerated transport facilities, financial institutions, SPS certifying laboratories, marketing information analysis department, etc. are available.

## FUTURE STRATEGIES

- Early maturing, seedless and uniform colour development through clonal selection programme
- Development of crop regulation technologies for extended fruit harvest to ensure round the year availability of citrus fruits
- Integrated nutrient and water management practices using micro-irrigation system and rain water harvesting to restore the ground water recharge
- Low cost technologies of citrus fruit production for wider adoption
- IPM strategy for insect pests and diseases to improve quality and enhance fruit yield. Rejuvenation programmes to improve declining orchards
- Implementation of climate change oriented fruit crop insurance schemes in citrus orchards
- Strengthening of the post-harvest management practices and processing including value addition
- Mechanical hoeing tools for under tree basin weeding

- Mechanical tree pruners for canopy management in declining & senile orchards.
- Mechanised pit makers
- Clipping tools for smooth plucking of mandarins and sweet oranges.
- Suitable long handle tool for plucking and collection of lime fruits from thorny lime trees.
- Microprocessor based sorting, washing, waxing and grading machine according to fruit size and colour for citrus fruits.

## LITERATURE CONSULTED

Aulakh, P.S., Mehan,V.K., Chahil,B.S. and J.S.Bal. 2011. Prospects and Problems of citrus in Punjab. PAU,Ludhiana.

Bal,J.S. 2014. Citrus. In: Fruit Growing. 3rd Edition. Kalyani Publishers, New Delhi.

Bal,J.S., Gill,S.S. and Sandhu, A.S. 2016. Raising Fruit Nursey. 2nd Edition. Kalyani Publishers,New Delhi.

Bhattacharya,S.C.and Dutta,S. 1956. Classification of Citrus fruit of Assam.ICAR Monograph No.25.

Bose, T. K., Mitra,S.K. and Sanyal, D. 2001. Fruits Tropical & Subtropical Vol 1, NayaUdyog , Calcutta.

Davies, F. S. and Larry Keith Jackson, 2009. Citrus Growing in Florida 5thn edition, University Press of Florida.

Ghosh, S.P.1985. "Citrus" Fruit of India Tropical and Subtropical. Ed. T.K.Bose: Naya Prokash. Kolkata,Pp: 159-213.

Hodgson,R.W.1967. Horticultural varieties of citrus. In: The Citrus Industry vol 1.Uni.California,Berkley. Edited by W.Reuther.H.J.Webber and L.D. Batchelor.

Ladaniya, M. 2010. Citrus Fruit: Biology, Technology and Evaluation, Elsevier Science Publishing Co Inc.

Mankad, N. R. 1994. Citrus in India. New Delhi: Wiley Eastern Ltd.

Mukhopadhyay, S. 2004. Citrus Production, Postharvest, Disease and Pest Management, Jain Book Agency, Connaught Place, C-9, Connaught Place New Delhi – 110001.

Purseglove, J.W. 1974. Tropical Crops:dicotyledons. John Wiley and Sons,New York.

Shrivastava, A. K. and Singh,S.2003. Citrus Nutrition, International Book Distributing Co.

Shrivastava, A. K. and Singh,S.2002. Citrus Climate and Soil, International Book Distributing Co.

Singh,A., Naqvi,S.A.M.H. and Singh,S. 2002. Citru Germplasm. Kalyani Publishers, New Delhi.

Singh,S., Shivankar,V.J., Ladaniya,M.S., Huchche,A.D., Shirgure,P.S., Marathe,R.A., Das,A. and Wanjari, V.2001. Citrus: 1001 Questions-Answers -Kalyani Publishers, Ludhiana..

Steve, Dreistadt, 2012. Integrated Pest Management for Citrus, Univ of California Agriculture & Natural Resources; Third edition.

Stone,B.C.1985.Rutaceae.In: A Revised Handbook of the Flora of Ceylon,vol V.(Ed.Dassanyake,M.D.& Fosberg,F.R).Smithsonian Institution and the National Science Foundation,Washington,DC.

Walter,A. and Sam, C. 2002. Fruits of Oceania. ACIAR Monograph no.85. Australian Centre for International Agricultural Research Canberra, Australia.

Wood ford, R. C.2005. Citrus Classification. Delhi: Biotech Books.

# 2

# Grapes

*N.K. Arora, Navjot Gupta and M.I.S. Gill*

## INTRODUCTION

Grape (*Vitis vinifera* L.) is one of the important fruit crop in world grown commercially in temperate, tropical and sub-tropical regions. At present, it is one of the major horticultural industries over an area of 7.5 million hectares. Grape berries are attractive for their unique flavour and are utilized in many different ways, so its cultivation is becoming more popular. Approximately 71% of world grape production is used in wine industry, 27% as fresh fruits and only 2% as dried fruit. China is the largest producer of grapes followed by Italy, USA, France and Spain. In world, India hold 9$^{th}$ position in grapes production with an annual production of 2822.78 thousand MT.Peninsular India comprise of accounts for 90 % area under grapes. India has the distinction of having highest productivity (21.4 MT/ha) of grapes in the world

## NUTRITIVE AND CULTURAL SIGNIFICANCE

Grapes have good amount of polyphenolic antioxidants particularly resveratrol, vitamins like $B_1$ and $B_2$, and minerals like calcium, phosphorus and iron, (Table 1). Especially the coloured grapes are very rich in antioxidants and its use has proven beneficial for high cholesterol patients. Grapes are consumed for table purpose/raisins or processed into wines and juices. Grape juice acts as laxative and stimulant to Kidney. The ripe table grapes are good source of sugars like glucose, fructose and sucrose. The Perlette a commercial variety of north India has TSS: acid ratio in the range of 23-30. The varieties like Thompson Seedless and Anab-e-Shahi grown in other grape growing regions have TSS: acid ratio in the range of 28-35 and 22-28, respectively.

In Chinese culture, grapes are symbol of wealth, abundance, fertility, many descendants and family harmony. The mentioning of grapes is available in Holy Bible

**Table 1:** Nutritive value per 100 g of ripe grapes (European type red or green e.g Thompson Seedless).

| Constituents | Nutrient Value | Constituents | Nutrient Value |
|---|---|---|---|
| Energy | 69 Kcal | **Electrolytes** | |
| Carbohydrates | 18 g | Sodium | 0% |
| Protein | 0.72 g | Potassium | 191 mg |
| Total Fat | 0.16 g | **Minerals** | |
| Cholesterol | 0 mg | Calcium | 10 mg |
| Dietary Fiber | 0.9 g | Copper | 0.127 mg |
| **Vitamins** | | Iron | 0.36 mg |
| Folates | 2 µg | Magnesium | 7 mg |
| Niacin | 0.188 mg | Manganese | 0.071 mg |
| Pantothenic acid | 0.050 mg | Zinc | 0.07 mg |
| Pyridoxine | 0.086 mg | Phyto-nutrients | |
| Riboflavin | 0.070 mg | Carotene-a | 1 µg |
| Thiamin | 0.069 mg | Carotene-ß | 39 µg |
| Vitamin A | 66 IU | Crypto-xanthin-ß | 0 µg |
| Vitamin C | 10.8 mg | Lutein-zeaxanthin | 72 µg |
| Vitamin E | 0.19 mg | | |
| Vitamin K | 14.6 µg | | |

(*Source:* USDA National Nutrient data base. www.ndb.nal.usda.gov)

## ORIGIN, HISTORY AND DISTRIBUTION

The grape (*Vitis vinifera*) one of oldest flora on earth is native to the area across the southeast coast of the Black Sea (near the south of the Caspian Sea) to Afghanistan. The presence of imprints of vines and leaves in createaceous chalk deposits about 90-95 million year back at time of dinosaurs lead to the evidence that grape is amongst the oldest flora on earth. However, new world species are believed to be native to South America and North Eastern America. Initially, cultivation of *vinifera* species in North America was not successful due to occasional low temperature of this region.However, later on the viticulture got successful in this region after growing of varieties selected from native species or from hybrids of native species with *vinifera*.From centre of origin grapes cultivation spread to both west (Europe) and east (Iran and Afghanistan). Approximately, before 600 B.C., Phoenicians took wine grape varieties to Greece, Rome and to Southern France. During second century A.D the Romans introduced the vines in Germany. Probably, table and raisin grapes spread from eastern end of Mediterranean Sea to the countries of North Africa. The distribution of table varieties is different from processing grape varieties. It is due to variation in cultural and religious values among the people of northern and southern area of Mediterranean.

Grape is the most important fruit grown in 6.98 million hectares with total production of 68.4 million mt in the world. It ranks first in occupying the area among other fruits. The average productivity is 9.8 mt/ha. China is the leading grape producing country with a share of 14.00 per cent of total grape production in the world. The total grape production in China is 9.6 million mt. USA is second in rank with the share of 9.7 per cent of total grape production. The other important grape producing countries are Italy, France, Spain and Turkey.

In India, the cultivation of grapes is known to be 1356 B.C. during the period of Susruta and Charakha. In ancient time, the Aryans knew the grape cultivation and preparation of wine from grapes. Muslim invaders brought grape from Iran and Afghanistan by end of 12$^{th}$ century. Mohamed Bin Tuglak took grapes to south India when he shifted his capital to Daultabad in 1338. During 16$^{th}$ century Akbar spread grape cultivation in India. Later Mughal ruler also encouraged grape cultivation and introduced many more varieties during 1758. Further grape cultivation in India got a boost after introduction of Anab-e-Shahi variety from Middle East during 1890 by Abdul Banquer Khan. In 1928, S. Bahadur Lal Singh brought 116 grape varieties at Layallpur. During 1958 a handsome collection of grape varieties was made at IARI. In 1962, the first Chief Minister of Punjab, Late S. Partap Singh Kairon brought one lac cuttings of Perlette from California and boosted its cultivation in North India.

The major grape growing states in India are Maharashtra, Karnataka, Andhra Pradesh and Tamil Nadu and in small pockets of northern Indian states viz. Punjab, Haryana and Uttar Pradesh. However, Out of total production in India, 78 % is used as table grapes, 20 % is dried for raisins production and the remaining 2 % is used for processing into juice and wine. India exports grapes to Middle East, European, Netherland, Hongkong, Singapore and Japan markets. Grapes occupy 1.23 lakh hectares and produced 28.22 lakh metric tonnes of fruits annually. Maharashtra is the leading state in grape production.

## TAXONOMICAL AND BOTANICAL DESCRIPTION

Grape belongs to family *Vitaceae*, has 14 genera and 1000 species extensively distributed in the tropical, subtropical and temperate climate of world. The genus *Vitis* which only contains edible species got further subdivided into two subgenus *Euvitis* and *Muscadinia*. The *Muscadinia* subgenera can be differentiated from *Euvitis*.It has chromosomes number (2n = 40), tight bark, simple tendrils, nodes without a diaphragm and small clusters. It has small detachable berries at maturity, seeds oblong without beak. While, the subgenera *Euvitis* (2n = 38) has forked tendrils, bark that sheds, a diaphragm at the nodes and elongated clusters with berries that adhere to the pedicles at maturity. The leaves are shiny, smooth with 3, 5 or 7 lobes. Berries are round or oval with edible skin, seeds pyriform with beak. Fruit is berry, epicarp and mesocarp is edible portion. The *Euvitis* contains 60 species, while *Muscadinia* has only three species. The most important species of *Euvitis* is *Vitis vinifera* Linnaeus which has more than 90 per cent of cultivated grape varieties.Varieties which belong to *Vitis vinifera* species are referred 'old world grapes' or 'European grapes'. Other important species of Euvitis is *Vitis labrusca* Linnaeus. The grape varieties of new world grape or American grapes belong to *labrusca* or its hybrids having *labrusca* blood.

The species are classified under Middle Asian and Mediterranean, North American, Caribbean, Asiatic and Muscadine grapes as Table 2 follows:

**Table 2:** *Vitis* Species Under Different Groups

| | |
|---|---|
| 1. Middle Asian and Mediterranean | *Vitis vinifera*, Linnaeus. |
| 2. North American | |
| *V. labrusca*, Linnaeus. 'Fox grape' | *V. lincecumii*, Buckley. 'Post Oak grape' |
| *V. aestivalis*, Michaux 'Summer grape' | *V. longii*, Prince. ' Bush grape' |
| *V. argenifolia*, Munson. 'Silverleaf grape' | *V. monticola*, Buckley. 'Sweet Mountain grape' |
| *V. arizonica*, Englemann. 'Canyon grape' | *V. movae-angliae*, Fernald. 'Pilgrim grape' |
| *V. berlandieri*, Planchon. 'Spanish grape' | *V. palmata* (rubra) Vahl. 'Cat grape' |
| *V. baileyana*, Munson. 'Possum grape' | *V. riparia*, Michaux. (V. vulpina, Linn.) 'Frost grape' |
| *V. californica*, Bentham. 'Pacific grape' | *V. rufotomentosa*, Small. 'Redshank grape' |
| *V. candicans*, Englemann. 'Mustang grape' | *V. rupestris*, Scheele. 'Sand grape' |
| *V. champini*, Planchon. 'Calcarie grape' | *V. shuttleworthii*, House. 'Calloosa grape' |
| *V. cinerea*, Englemann. 'Grayback grape' | *V. smalliana*, Bailey. 'Fig Leaf grape' |
| *V. cordifolia*, Lamarck. 'Winter grape' | *V. simpsoni*, Munson. 'Current grape' |
| *V. gigas*, Fennel. 'Florida Blue grape' | *V. sola*, Bailey. 'Curtiss grape' |
| *V. helleri*, Small. 'Round Leaf grape' | *V. treleasei*, Munson. 'Gulch grape' |
| *V.illex*, Bailey. 'Manatee grape' | |
| 3. Caribbean | |
| *V. indica* (tiliaefolia, caribeaea) | |
| 4. Asiatic | |
| *V. coignetiae,* Pulliat | |
| *V. flexuosa*, Thunberg | |
| *V. pentagona*, Diels and Gilg | |
| *V. amurensis*, Ruprecht. | |
| *V. embergeri*, Golet | |
| *V. betulifolia*, Diels and Gilg | |
| *V. eticulate*, Pampanini | |
| *V. armata*, Diels and Gilg | |
| *V. davidii*, Romanet du Caillaud | |
| *V. lanata*, Roxburgh | |
| *V. pedicellata*, Lawson | |
| 5. Muscadinia | |
| *M. rotundifolia*, Michaux.'Muscadine grape' | |
| *M. munsoniana*, Simpson. 'Little Muscadine grape' | |
| *M. popenoei*, Fennel. 'Mexican Muscadine grape' | |

## CLIMATIC AND SOIL ADAPTABILITY

Grapes are one of the rare fruit plants grown all over the world in climate ranging from temperate, subtropical and tropical. Grapes can be grown successfully in long, hot, dry and rainless summer with sufficient sunshine followed by winter cold to induce dormancy in the vines. The native place of grapes i.e warm temperate zone between 34 degree North and 45 degree South latitude is considered most ideal area for grape cultivation. However, grapes are grown outside this zone in both hemisphere as special areas. In tropical zone, vine do not shed leaves hence behave like as an evergreen plant. Temperature is considered as a main factor affecting quality of grapes. Each variety requires certain specific heat units for ripening. For example the early ripening varieties need 1600 heat units for ripening. The late varieties need upto 3500 heat units for ripening. In general hot and dry climate at time of ripening is most suitable for production of quality table grapes. The moderate cool weather at the time of ripening, slow down the ripening process and is rather helpful for production of grapes suitable for preparation of dry table wine.

In dormancy,vines can tolerant low temperate due to falling of leaves. Occurrences of late frost in spring just after sprouting injure the newly sprouting buds. Heavy and early rains are not good for cultivation of late maturing varieties. The high humidity during rainy season also favour the fungus diseases. The area having rainfall during ripening is not fit for cultivation of grapes. Heavy wind storm and hail storm during fruit development period also damage the fruit vis-à-vis cause economic loss to the growers. The grape species which belong to *Labrusca* group are considered better under humid summers and cold winter as compared to grapes which belong to *vinifera* species.

Grapes are adopted to a wide range of soil types and are grown on wide variety of soil in different regions. Grapes have a strong root system and most of the soil types are considered ideal for cultivation. The ideal soil should be sandy loam, well drained and fairly fertile with good amount of organic matter. However, heavy clays, shallow and ill drained soils should be avoided for cultivation of grapes. As compared to others fruit plants, grapes can tolerant soil having high salt or alkalinity. But higher lime concentration (>0.3 %) can damage the vine and have adversely effect on growth. Poorly drained soil do not allow the vine to enter into dormancy thus should be avoided for cultivation of grapes. The well drained, clay loam, black clay loam and even shallow soils produce successful grapes. Salt affected, nematode infested and water-logged soils should be avoided. Soils having EC up to 1.5mmhos/ cm, calcium carbonate up to 10 per cent, lime concentration upto 20 percent and pH upto 8.7 are suitable for successful grape cultivation.

## RECOMMENDED AND POPULAR CULTIVARS

Owning to wide cultivation of grapes in tropical, sub-tropical and temperate regions of the world, more than 8000 varieties have been described and named in different regions. However, the commercial varieties are less than one fifth of the known varieties. In the developed countries like USA, France, Italy, Australia, the grape production is primarily aimed at to produce processed products like wine, juice and raisins. Accordingly, the varieties suitable for processing are being grown. On a limited scale, table purpose varieties are grown. The reverse is true for India, where more than 90 per cent production of grapes is aimed at to produce table fruit, while less than 10 per cent is being grown for production of processed products. The growing of a particular cultivar in a region depends upon several parameters, the most important being soil characteristics, climatic region and the purpose e.g table/processing for which the grapes are to be grown. Primarily in India grapes are being cultivated for table purposes and the table varieties are classified into four primary groups viz., coloured seeded, coloured seedless, green/amber seeded, green/ amber seedless. Further the grouping has been done on the basis of maturity i.e early, mid or late season varieties which are based upon the time taken (days) from flowering to ripening. India has a vast climate ranging from E-W-N-S and thus a wide range of varietal wealth is available in India. The vine morphology as well as grape quality also varies with the agro-climatic zone, for instance vines go dormant under north Indian conditions, while in south the grapevines remain evergreen. Thus as a result the varieties also respond differently to the yield potential and its performance. Single pruning and one crop is practiced in North India which has primarily sub-tropical region, while, double pruning and one crop is a general practice in vine growing regions of Maharashtra. In Karnataka, Andhra Pradesh and Tamil Nadu two crops are harvested in a year.

**Table 3.** List of Important grape varieties grown in India.

| S.No | Characteristics | Varieties |
|---|---|---|
| 1. | Coloured seedless | Flame Seedless, Shard Seedless, Fantasy Seedless, Beauty Seedless, Crimson Seedless, Arka Krishna, |
| 2. | Amber/green seedless | Perlette, Thompson Seedless, Pusa Seedless, Arka Shweta, Arkavati, Pusa Urvashi |
| 3. | Coloured seeded | Bangalore Blue, Muscat Hamburg/ Gulabi, Red Globe, Cardinal, Arka Majestic, Arka Shyam |
| 4. | Amber/ green seeded | Anab-e-Shahi, Bhokri, Italia, Seeded Hybrid, Arka Kanchan, |
| 5. | Wine (red) varieties | Punjab MACS Purple, Zinfandel, Cabernet Sauvignon, Grenache, Merlot, Pinot Noir, Shiraz, Tempranillo, Pusa Navrang, Rubi Red, |

| S.No | Characteristics | Varieties |
|---|---|---|
| 6. | Wine (white) varieties | Chardonny, Chenin Blanc, Clairette, Riesling, Sauvignon Blanc, Ugni Blanc, Viognier |
| 7. | Juice purpose varieties | Bangalore Blue, Punjab MACS Purple, Pusa Navrang |
| 8. | Raisin purpose varieties | Thompson Seedless, Tas-A-Ganesh, Sonaka, Manik Chaman, Arkavati, A 17-3, Manik Chaman, A 18-3 |

*Source:* Tropical and Sub-tropical Fruit Crops (2014): Crop Improvement and Wealth.

## Brief Descriptions of Important Varieties

### Northern Region

**Perlette:**It is a hybrid of Scolokertek Hiralynoje 26 x Sultanina Marble evolved at the University of California, Davis by Dr. H.P. Olmo in 1936.Being earliest maturing seedless variety, it was introduced and released for commercial cultivation in the Punjab State in 1964 and has now spread in whole of the North-West India. This variety has medium to large sized, long conical, shouldered, compact to very compact bunches, borne on 3$^{rd}$ and 4$^{th}$ nodes. Berries are attractive, yellowish-green, glossy, spherical to slightly ellipsoidal, seedless, a few rudimentary seeds often found. The TSS 16-18%, acidity 0.5-0.1%, juice 60%. This variety is ripens during first week of June and have short harvesting span which often cause glut in the market.

**Beauty Seedless:**It is an early ripening, coloured, seedless, sweet and prolific bearing grape variety introduced from California, U.S.A. Its bunches are medium to large, long, conical to cylindrical, well filled to compact, borne on 3$^{rd}$ to 5$^{th}$ nodes. Berries bluish black, spherical, small, seedless. This variety has T.S.S 18-21%, acidity 0.6-1.0 % and juice 70-75 %. It ripens in the first week of June.

**Flame Seedless:** An introduction from California (USA). This variety is a descendant of three cross breeds, the Cardinal and Red Seedless, Red Malaga and Tifahfi, and the Muscat of Alexandria and the Thompson Seedless. The clusters are medium in size with small, bright red, crisp seedless berries. It ripens in 2$^{nd}$ to 3$^{rd}$ week of June and yields upto 27 kg per vine. Its juice is light yellow in colour having 18.0 per cent T.S.S. and 0.7 cent acidity. For obtaining uniform colour of berries, retain 75 % crop load and treat bunches with 400 ppm (a.i) ethephon. This is excellent table grape variety and holds promise for North-West Indian plains.

**Punjab MACS Purple (H-516)**: This variety is suitable for processing into juice, nectar and RTS. It has medium and loose bunches. The berry is medium in size, seeded and purple colour at maturity. It contains 60-65 per cent juice with total soluble solids 17-18 per cent and acidity 0.50 per cent.The juice is rich in anthocyanins (a source of antioxidants). This variety ripens in first week of June under northern India.

**Pusa Seedless:** It is a clonal selection from Thompson Seedless at IARI, New Delhi. It ripens from 2$^{nd}$ to 3$^{rd}$ week of June under northern India and has total soluble solids (TSS) 20-22 %. Berries are more elongated than Thompson Seedless. It is susceptible to rust, downy mildew and highly susceptible to anthracnose.

**Pusa Navrang:** This is teinturier hybrid (Madeleine Angevine x Ruby Red) developed at IARI, New Delhi. Bunches are loose and medium in size, berries medium, round and seeded. Ripening is uniform during first week of June in northern India. This variety is relatively resistant to anthracnose and is considered good for processing into juice and wine.

**Pusa Urvashi:** This hybrid has been developed at IARI from a cross between Hur x Beauty Seedless. This is basal bearer variety having loose bunches which are medium in size. The berries are greenish yellow, seedless medium in size with high TSS 20-22 %. This variety ripens during 1$^{st}$ week of June. Suitable for table and raisin purpose and is reletively tolerant to diseases anthracnose and powdery mildew.

**Himrod:** It is evolved from a cross between Ontario (American type) x Sultanina (Thompson Seedless). The bunches of this variety are attractive, medium to large, conical to cylindrical. Berries are yellow-green, dull amber when fully ripe, slightly ellipsoidal, and medium-large in size. Seeds very small or abortive. The TSS at time of ripening varies from 20-25%, acidity 0.5-0.6%, juice 60-65%. It ripen early i.e end May in northern India It has been reported to be moderately susceptible to anthracnose.

**Superior Seedless:** This is an early ripen variety, suitable for table purpose.Vines are medium in vigour, bunches loose, medium to large in size. Berries are seedless, large, amber coloured and crisp in texture.At time of ripening TSS (17.0 %), acidity (0.51 %).

## Southern Region

**Anab-e-Shahi:**It was introduced from the Middle East around 1890 in Hyderabad. Bunches of this variety are very attractive, medium to large, moderately compact, and borne on 5$^{th}$ to 7$^{th}$ node. Berries medium-large, yellowish green when raw and amber when fully ripe, TSS 14-16 %, acidity 0.5% and juice 55-75 per cent. It is highly susceptible to downy mildew and moderately susceptible to anthracnose. However, it can withstand high alkalinity. It is mid-season to late variety i.e. ripens during last week of June to mid-July in North India.

**Thompson Seedless:** Its bunch is medium to large in size, conical to cylindrical, shouldered, and well filled to compact. The berries are yellowish green to golden yellow when fully ripe, small and seedless. The TSS ranges from 18 to 22 %, acid content 0.5 to 0.6 % and juice 70-75 %. It is a multipurpose grape variety being used for table, wine and raisins. This variety has good keeping quality. Average yield is 20 – 25 t/ha.

**Tas-A-Ganesh:** This is a clonal selection from Thompson Seedless by late Sh. Vasant Rao in his vineyard in Sangli district. It has comparatively loose medium to large size bunch. TSS range from 18 to 20 %, acidity 0.8 to 1.0 % and juice 60 %.

**Bangalore Blue:** This is cross between *vinifera* and *labrusca*, resistant to anthracnose,cercospora leaf spot and susceptible to downy mildew. Bunches are small, well filled to compact, berries dark purple, small sized with skin thick, juicy, TSS 15-18 %, acidity 0.8- 1.0 %. This variety is ripen uniformly in south. But in northern part bunches do not ripen even kept on vine upto September. This variety is suitable for processing into juice and wine.

**Bhokri:**This is an oldest variety introduced in India (Deccan) during 1338 by Mohammaden rulers. Its bunches are large, conical to cylindrical, well filled, compact, borne on 3-4 buds. Berries seeded, greenish yellow, brownish when over mature, adherence weak, TSS 16-18 %, acidity 0.5- 0.8 %, juice 60-75 %. Highly susceptible to rust and downy mildew and moderately susceptible to anthracnose. It is heavy yielder and seeded variety. This variety is popularly grown in India before the introduction of Thompson Seedless.

**Cheema Sahebi:** Its bunches are medium to large, conical, well filled, borne on 1-4$^{th}$ bud. Berries pale yellow, oval, cylindrical, small, seeds 2-3, TSS 20-21 %, acidity 0.5 %, juice 75 %. Susceptible to rust and highly susceptible to downy mildew and anthracnose. High yielding variety but its attachment of berries to pedicel is not firm.

**Delight:** Bunches are medium to long, conical, well filled to compact, borne on 2$^{nd}$ to 5$^{th}$ node. Berries seedless, green, small, TSS 18-21 %, acidity 0.5 %, juice 70 %. Susceptible to rust, downy mildew and highly susceptible to anthracnose. It is an early maturating variety.

**Muscat Hamburg/ Gulabi/ Karachi Gulabi:** Bunches of this variety are attractive, medium large, short conical or cylindrical, loose to well filled borne on 2$^{nd}$ to 5$^{th}$ node. Berries deep purple, spherical, small-medium, skin medium thick, seeds 2-3. TSS 18-20 %, acidity 0.5-0.6 %, juice 60-75 %. Highly susceptible to rust and downy mildew. It has musky flavour and can withstand rain to certain extent.

**Cardinal:** This is an introduction from U.S.A., where it was evolved as a cross between Flame Tokay and Ribier. The clusters are small to medium in size and mostly loose. The berries are very large and some what round in shape, dark red to reddish black in colour and seeded. It ripens in the first and second week of June. The TSS ranges from 16 to 18 %, acidity 0.4 to 0.6 % and juice about 70.0 %. Because the berry is thin skinned and the stem attachment is week, the clusters must be handled carefully to avoid injury to the berries.

**Kishmish Beli:** Bunches are large, cylindrical to conical, winged, well-filled to compact, borne on $3^{rd}$ to $5^{th}$ nodes. TSS 20-22%, acidity 0.6-0.7% and juice 80%. It is susceptible to rust and downy mildew. It is suitable both for table and raisin making.

**Kishmish Chorney:** This variety has large, conical, shoulder, well-filled to compact bunches, borne on $3^{rd}$ to $5^{th}$ node. Quality good, TSS 20-22%, TSS goes upto to 24% acidity 0.5-0.8, juice 70-75%. It is susceptible to rust and downy mildew and moderately susceptible to anthracnose and *Cercospora* leaf spot.

**Shard Seedless:** It is clonal selection from Kishmish Chorney. Berries are bold, seedless, crisp and brilliantly black coloured with high TSS (22-24$^{o}$ Brix). It is moderately susceptible to anthracnose and *Cercospora* leaf spot. It is a prolific bearer and is one of the choicest table varieties in India at present.

**Arkavati:** Bunches are medium-sized, well filled, long conical to cylindrical in shape.Berries are medium in size, spherical to ellipsoidal, yellowish green, seedless (rudimentary soft seeds often found); TSS 22-25%, acidity 0.6-0.7%, juice 70-74% and pH 3.5-3.7. it is moderately susceptible to anthracnose and *cercospora* leaf spot. It has high yield potential, since almost all the buds are fruitful. This variety is good for raisin making.

**Arka Hans:** It is a cross between Bangalore Blue x Anab-e-Shahi. Bunches are medium-sized, well filled and cylindrical shaped. The berries are medium to large in size, yellowish-green, spherical to ellipsoidal, seeded with pleasant foxy flavour; TSS 18-21 %; acidity 0.5 %, juice 68-69 % and pH 3.6-3.7. Moderately susceptible to anthracnose and *cercospora* leaf spot.

**Arka Kanchan:** Late maturing variety evolved by cross between Anab-e-Shahi x Queen of vineyards. Bunches are medium to large, well filled, long conical. The berries are large, golden-yellow, ellipsoidal to ovoid, seeded with pleasant Muscat flavour, TSS 19-22 %; acidity 0.5-0.6 per cent; juice 60-65 % and pH 3.5-3.8.

**Arka Shyam:** It is crossed between Bangalore Blue and Black Champa. This variety has medium-sized, well filled to compact, cylindrical bunch. The berries are medium large, shining black, spherical to ovoid, seeded with mild foxy flavour, TSS 22 - 25%, acidity 0.5-0.6 %, juice 60-72 % and pH 3.7-3.8. Moderately susceptible to anthracnose.

**Arka Shweta:**Evolved by cross between Anab-e-Shahi x Thompson Seedless and released in 2006. Bunches are well filled, medium in size. The berries are greenish yellow, large in size, TSS 18-19 %. It is moderately susceptible to anthracnose. Good for table purpose and for export.

**Arka Chitra:**This is hybrid variety (Angur Kalan x Anab-e-Shahi). Bunches are well filled, medium in size, berries golden yellow with pink blush, slightly elongated and large in size, TSS ranged from 20-21 %. Tolerant to powdery mildew. All buds are fruitful therefore no specific pruning requirement.

**Arka Krishna:**It is crossed between Black Champ and Thompson Seedless.Bunches are well filled, medium in size. Berries are dark coloured, seedless, large in size, sweet in taste, TSS 20-21 %. Moderately susceptible to anthracnose. It is suitable for juice making.

**Red Globe:**This is one of the famous table variety in most part of world grown in green house. However, can be grown outside in warm area with long growing season such as Australia and California. Large part of this variety grown in USA and Australia exported to Asian countries. Bunches are large with bold seeded berries, have crisp pulp with firm thick skin, often peel able. TSS 18 %. Moderately susceptible to anthracnose, downy mildew and moderately tolerant to powdery mildew.

**Manjri Naveen:** It is clonal selection from Centennial Seedless recently released by National Research Centre for grapes, Pune for commercial cultivation in Maharashtra. This variety has self-thinned medium to large size bunch. The berries are seedless bold, oval to elliptical shaped crisp in texture. The ripe fruit has TSS 17-18 % with 0.45- 0.55 % acidity. Susceptible to powdery mildew. Less prone to downy mildew attack. This is an early maturing variety ripen 20-25 days early than Thompson Seedless in south India.

**Table 4:** List of Exotic Wine Grape Varieties (White and Red) commercially grown for Wine Production.

| Wine varieties (White) | | |
|---|---|---|
| Variety | Origin | Important characteristics |
| Sauvignon Blanc | France | Clusters and berries are small. Dry wines are very pleasant, fine, balanced and typical. Acidity is quite prominent making a refreshing wine. |
| Chenin Blanc | France | Clusters are medium to large with small to medium berries. It is suitable for dry, sweet or sparkling wines. Acidity remains high. Wines have honey flavor. |
| Ugni Blanc | France | Clusters are very large and berries are small to medium. Dry wines are balanced, but relatively neutral. It gives brandies of very high quality. |
| Chardonnay | France | Clusters and berries are small. It gives dry, sparkling or sweet wines. Must sugar content is often high and acidity remains high. Wines are well balanced, strong, ample, soft and full. |
| Clairette | Rhone | Produces musky and perfumed wines which appear rich on the palate with high levels of alcohol. |
| Riesling | Germany | This classic grape yields light to medium wight wines characterized by citrus and floral scents backed by steely acidity. The wines are usually dry. |
| **Wine varieties (Red)** | | |
| Cabernet Sauvignon | France | Clusters and berries are small. Wines are dark and rich in tannins when harvested at full maturity. Aromas are subtle and complex. |
| Shiraz | France | Clusters are small to medium and berries are small. Wines are of good quality with high alcohol level. They are dark and strong with complex tannins and delicate aromas. |
| Merlot | France | It has winged clusters with medium sized berries. Wines are strong, alcoholic and coloured, but slightly acidic. |
| Zinfandel | Italy | Produces rich and voluptuous (aroma of Christmas cake) wine. It gives full bodied wine with a slight earthy and spicy nose. The wines are juicy with red fruit on the palate. |
| Pinot Noir | France | Clusters and berries are small. Wines have long keeping quality. Must sugar content is high, acidity is medium and the colour is weakly intense but usually lasting. Used for making sparkling wines. |
| Grenache | Spain | Produces fairly high alcohol wines with raspberry flavours. However, it is usually made into meaty, spicy wines of medium to full body. |

## PROPAGATION TECHNOLOGY AND ROOTSTOCKS

Propagation through cuttings is the most commonly used method in grapes, particularly in the region where own rooted plants are grown. However, in the recent past, due to abiotic stresses, the rootstocks have been used instead of own rooted plants. In such cases, grafting is the method of choice for propagation. In northern India, most of the commercial plantations are own rooted, and propagation through hard wood cuttings is the only method of propagation. While in states like Maharashtra, Karnataka most of new grapes plantations comprise of grafted plants.

## Preparation of cuttings

The plants, which are healthy, free from diseases and earmarked for quality bunches should preferably be used for preparation of cuttings. The cuttings of pencil thickness and having atleast 4 buds are prepared from one-year-old mature shoots (canes). The terminal shoots are avoided for preparation of cuttings, as they are very thin and may be infected with anthracnose disease, which starts from tips towards middle of shoot. The upper cut of cuttings should be away from the bud, while lower cut should be near the bud, which helps to promote better, rooting. The main criteria of the time of preparation of cuttings are the period before sprouting of dormant shoots. Keeping this in view, cuttings must be prepared before mid-February. The cuttings after preparation are tied in bundles and are buried in moist soil for 1-2 weeks. The process allows callusing on cut ends which subsequently help in better rooting of these cuttings.

After callusing the cuttings are planted in nursery beds at a distance of 15 cm from cutting to cutting and 30 cm from line to line. While planting cuttings, 2/3$^{rd}$ portion should be buried in soil, and 1/3$^{rd}$ above the soil. At the time of planting, chlorpyriphos treatment is must to avoid termite attack. Grape cuttings are highly susceptible to termites.

## Rootstocks

In Europe, U.S.A., Australia and Russia to tackle the problems of soil borne diseases, insect-pests (phylloxera) grapes are grafted on compatible rootstocks. Recently, in commercial grape growing area of India, due to problems of soil salinity, drought, nematode and poor fruitfulness, use of rootstock tolerant to these problems is gaining popularity over own rooted plantation. Grape rootstocks have been found to impart resistance to soil borne fungi, drought and vigour to the scion. Characteristics of some important rootstocks are given as below.

**Dogridge**: It is seedling selection from *Vitis champini,* which is recommended for use with wine and raisin varieties on light sandy soils. It is highly resistant to nematode infestations and produce most vigorous grafted vines. Cuttings of this rootstock root with great difficulty and once rooted they bud and graft readily.

**St. George**: It is also known as Rupestris St. George. This rootstock is a pure seedling selection chosen from the wild type species *Vitis rupestris.* It has been planted more extensively than any other rootstock in most of the grape growing countries. It is vigorous, roots-well from cuttings, grafts well and is highly resistant to phylloxera and drought. This rootstock produces only staminate flowers.

**Teleki 5-A:** It is a hybrid between *Vitis berlandieri* and *Vitis riparia* which show moderate degree of resistance to nematodes and highly resistant to phylloxera.

**1616** (Syn. Couderc 1616, Solonis Riparia 1616): It is hybrid between *Vitis solonis x Vitis riparia,* having moderate resistance to phylloxera and is highly resistant to nematodes. In sandy loam soils, it produces scions of moderate vigour that yield good crops. Cuttings of this variety easing toroot and graft readily.

**1613** (Syn. Couderc 1613, Solonis x Othello 1613, Solonis-othello): This rootstock is a hybrid between the species *Vitis solonis* and the fruiting variety Othello. It produces moderately vigorous scion and is resistant to nematodes and moderately resistant to phylloxera.

**Salt Creek** *(Vitis champini)*: This rootstock imparts great vigour to its scion variety and is resistant to nematodes.It has performed well with wine and raisin varieties of California in light sandy soil of low fertility. Rooting of cutting is poor but bud take is higher than Dogridge.

**110 R** (Richter 110): It is reported to be moderately vigorous to vigorous. The initial vine growth is slow because it first develops the root system. It is not appropriate for varieties with irregular fruit set. It is reported to have a very long vegetative cycle, which delays maturity. Its roots are not as deep growing as those of Richter 99 or Rupestris du Lot. Richter 110 is well suited to all kinds of soils, including acid soils. It is an excellent rootstock in warm grape-growing areas with an arid climate.

## PLANT AND FRUIT PHYSIOLOGY

Flowering appears on one-year-old mature shoots called canes. The temperature 10 to 15°C is most congenial for bud swelling and growth. The swelling of buds leads to differentiation of leaf or flower primordial. Flowering takes place at temperature range of 18-21°C. Induction and differentiation of inflorescence primordia for the next year's crop begins soon after bud break of the current season and is completed between veraison and harvest stage. During this period, commercial table or wine grape vineyards undergo several changes that determine canopy micro-climate. Too much shade during this period result in low bud fruitfulness during the following season. However, it is not known how much sunlight (quantity/ length of exposure) is necessary to achieve maximum bud fruitfulness. Light and temperature are the most important climatic factors for inflorescence induction and bud differentiation. The results of various studies showed a direct correlation of shoot light exposure and temperature with fruitfulness. High temperature has been found to promote fruitfulness in developing grapevine buds. Fruitfulness of individual shoots is positively correlated to light exposure. Higher fruitfulness has been observed under conditions of 16 hours day and at temperature of 25°C.

Following pollination and fruit set, the grape flower ovary develops into a fleshy berry. The grape berry is a simple fruit, consisting of two locules surrounded by an ovary wall (pericarp). In seeded varieties, there may be as many as four seeds. In the case of stenospermocarpic varieties, the locules contain seed traces resulting from the abortion of the ovules early in their development.

Three stages of grape berry development have been identified. During stage I, starting at fruit set, berries grow undergo cell division. Stage II, called the Lag Phase, starts with a pause in berry growth, while seed embryos start to form and grow. Cell division stops, and further growth is through enlargement of cells. Stage III starts at veraison, when berries change colour, softens, accumulate sugars, and metabolize acids. Acids and tannins that accumulate before veraison ensure that they remain unpalatable. After veraison, changes occur (colour change, softening, sugar accumulation, and acid reduction). The different phase of berry development is as follow:

*Stage 1- Rapid berry growth phase:* This stage comes immediately after bloom, when berries grow both through cell division and cell enlargement. Berry texture is firm, while the peel colour is green due to the presence of chlorophyll. The sugar content of the berry remains low, while organic acids accumulate. This stage lasts between 3 and 4 weeks for most of the seedless grape varieties.

*Stage II- Lag phase:* During this phase, the berry growth markedly slow down, whereas the organic acids concentration reaches to their highest level. Berries remain firm, but begin to lose chlorophyll. The lag stage normally lasts between 2 and 3 weeks for most varieties.

*Stage III-Rapid berry growth, and Fruit ripening phase:* In this phase a rapid berry growth occurred and ripening of berries starts. The term veraison or berry softening is characterized by change in berry colour nearing maturity. Berries soften and lose chlorophyll, while in coloured varieties red pigments begin to accumulate in the peel. Sugar also begins to accumulate, and the concentration of organic acids declines, though aroma and flavour components accumulate. Berry growth during this stage is limited to cell enlargement, and normally lasts between 6 and 8 weeks.

Vines undergoing water stress during stage I normally produce smaller berries than non-stressed vines. Since the effects of water stress during stage I on berry growth cannot be reversed by subsequent watering, decreased growth probably indicates a reduced number of cells per berry or a permanent reduction in the size or volume of the cells. Water stress during phases II and III may also decrease berry weight, but in that case the reduction is related to reduced cell volume or lesser solutes (sugar) in the cells. Nutrient deficiencies and other disorders that reduce photosynthesis may also reduce berry growth or slow ripening by decreasing the accumulation of sugars.

## PLANNING AND PLANTING

In grapes plant density depends upon training system and type of variety. In head system, the planting distance of 2.0 x 1.5 m is kept, which accommodates 3300 vines per hectare. In case of kniffen system, the vines are planted at the distance of 3 x 3 m which accommodates 1100 vines per hectare. When the vines are trained on Bower system, the planting distance is maintained at 3.0 x 3.0 m which adjusts 1100 vines per hectare. Spacing of vines on Y-trellis should be kept at 1.5 x 4m, which thus accommodates 1650 vines per hectare.

The direction of rows will matter only in the case of Kniffin and Telephone system and not in head or bower systems. In deciding the direction of rows, the usual direction of the wind should be taken into consideration. Planting the grapevine rows right angle to the wind need to be avoided, as while blowing across it is likely to damage the young shoots and the fruit bearing canes, especially in the areas which are affected by windstorms of strong velocity. In plains of north India, where there is intense heat at the time of ripening of grapes, the rows should preferably run from East to West, so that the ripening bunches are saved from the direct rays of the sun. In areas where the climate is comparatively cool, it would be advisable to set the rows from north to south so that the sunshine is well distributed on both the sides of the vine to facilitate proper maturation and colour development of the fruit.

The planting of vines under north Indian conditions is done in January to first fortnight of February when vines are still dormant. The size of the pit should be kept 1 x 1 x 1 m. If the soil is light, the size of the pits can be reduced. The refilled pits should be watered a few days before planting the vines. In each pit, add 5 ml of chlorpyriphos 20EC mixed in about 2 kg soil against white ants. At the time of planting, it is usual practice to prune the top three to four buds, keeping only one cane. The straggling roots are also trimmed; broken and damaged roots should also be removed. Vigorous and healthy, one-year-old rooted cuttings are considered best for planting

## SOIL CULTURAL PRACTICES TECHNOLOGY

### Water Need

The growth of vines and berries on the clusters depends directly on the supply of irrigation throughout the growing period of vines. The grape vines have fibrous root system which mostly spread in the top 20 cm of soil. A little depletion of soil moisture can cause diffusion pressure deficit within the vines, leading to growth inhibition. During dormancy vines can tolerate soil moisture stress for a sufficient, long time as the soil moisture depletes to a very low level. As such, vines would not resume normal growth in spring unless irrigation is applied to replenish the soil moisture. Any water stress condition hampers growth and productivity of the vine viz; the shoot growth decreases and the internodes become shorter, the tendrils drop, the leaf margins curl and the older leaves turn yellow. The irrigation or watering of plants is need based. The climatic factors such as temperature, humidity and wind influence greatly the water needs of vineyards. Young vines need more frequent irrigation than grown up vines. The vines trained on bower system need more water than those on kniffin or telephone or head system. There are different phases of plant growth viz. dormant season, sprouting or initiation of shoot growth, flowering, fruiting, maturity and ripening, and the period after fruit harvest. The requirement of irrigation varies with different phases of plant growth.

- The irrigation during dormant season should be restricted, as it results in early sprouting of vines. The inappropriate irrigation during dormant season also affects the pruning process – it leads to oozing of cell sap from cut portion. In general, dormancy is a natural process to overcome winter, which is disturbed by inappropriate irrigation.
- The irrigation during sprouting or initiation of new growth is must to have uniform vine growth and initiation of flowering.
- The irrigation during flowering should again be restricted, as the excessive irrigation during flowering diverts food energy towards vegetative growth and thereby leading to imbalanced flowering and vegetative growth.
- Period after fruit set and during fruit development is most critical. The restricted irrigation during this period has direct co-relation with the berry size, as the berries are constituent of more than 90% of water.
- During maturity and ripening, the irrigation should again be restricted. The excessive irrigation during this period affects the fruit quality. The

berries loose T.S.S. and thus have flat taste. In the contrary, if the moisture stress is created during ripening of berries, it leads to increase in T.S.S/ Acid ratio and results in better fruit quality.

- The irrigation after harvesting should again be restricted, as after harvest due to onset of rainy season in couple of months, do not necessitate irrigating the vines. For adequate growth of vines and good fruit yield and quality, the recommended irrigation schedule should be followed. The irrigation through drip save water and improve fruit quality.

Under Punjab conditions, one irrigation should be done after pruning in the first fortnight of February. The second irrigation is applied in first week of March. After fruit set in April till first week of May, irrigation is recommended at 10 days intervals. During the rest of May, at weekly interval, during June at 3 or 4 days interval and during July to October, irrigate when prolonged dry spell prevails or rainfall is insufficient. One irrigation is recommended from November to January, if soil gets extremely dry.

In southern India, the grapevine needs regular irrigation except at the time of pruning and monsoon. Irrigation is given at 6-8 days interval during fruiting period. In Madurai, the vines are irrigated four times a week during summer and two times a week during winter.

## Nutritional Need

Vineyard fertilization practices aim to ameliorate the supply of available soil nutrients to the levels required for optimum grapevine growth and yield. Most soils will contain adequate amounts of micronutrients. However, nitrogen, phosphorus, potassium, and magnesium are the nutrients that usually limit grape production. It is basically essential to know the role of different nutrients on growth, fruit yield and quality in order to include them in various proportion in vineyards manuring in different stages of growth and fruit development.

**Nitrogen (N):** Nitrogen is the mineral element that grapevines require in the greatest amount. When grapevines become deficient of N, vegetative growth slows and the foliage becomes chlorotic. In contrast, vines with an abundant supply of N have dark green foliage, growth is vigorous and canopies are dense, making canopy management difficult and may also contribute to other problems such as poor bud fruitfulness, poor coloration of red grapes, excessive shatter and increased levels of bunch rot and bunch stem necrosis. The timing of N fertilizers, like other nutrients, should occur when demand is high and uptake is rapid. Nitrogen is needed most during the period of rapid vegetative growth, which occurs during the spring, from bud break to early berry development. Nitrogen absorption is most rapid between bloom and version, with the developing clusters being the largest sink for N during this time.

**Phosphorus (P):** Phosphorus promotes floral bud initiation. Application of phosphorus has been found to promote fruitfulness in Anab-e- Shahi vines. Increasing levels of applied phosphorus was associated with increased fruit yield in Anab-e-Shahi and Thompson Seedless.

**Potassium (K):** Potassium is essential for grapevine growth and yield and serves an important purpose in several different plant functions. Potassium is readily translocated throughout the grapevine and may be involved in carbohydrate transport and metabolism. Potassium also neutralizes organic acids and plays a role in controlling acidity and pH of the fruit juice. Potassium deficiency is generally not widespread in the vineyard and is often observed in areas with sandy soils with low native K fertility or where topsoil was removed for leveling. Compact and poorly drained soils, water stress and vines with weak root systems due to damage by soils pests (nematodes) may also contribute to K deficiency due to poor K uptake. The need for K is most critical during berry development and ripening, and it is during this time that the fruit becomes the strongest sink for available K. This period also corresponds with the time at which root uptake for K is most rapid. The developing fruit is such a strong sink for K, timing of K fertilizers should be applied preferably during the early spring (a few weeks after budbreak) up to veraison.

**Magnesium (Mg):** Magnesium in grapevines plays two main roles. First, magnesium is an essential component of the chlorophyll molecule and is vital for photosynthesis. Magnesium also activates enzymes required for plant growth. Magnesium deficiency is more prevalent where soils have become acidic (pH d" 5.5) after years of repeated use of urea and/or ammonical fertilizers. This can be corrected with the application and incorporation of limc, thus neutralizing the acid and adding calcium and Mg to the base exchange site. Furthermore, calcium, potassium and Mg interact on the soil's exchange site and compete for entry into plants. Seasonal uptake and partitioning of Mg within the grapevine begins at budbreak and from the period of budbreak to bloom. Reserve Mg (mainly from roots) contributes 18% toward the requirement of new vine growth. Leaves and shoots account for the greatest portion of total vine Mg throughout the season. The greatest amount of absorbed Mg partitioned to the permanent vine structures occurs about 4 weeks after harvest.

**Zinc (Zn):** Zinc deficiency in grapevines is observed on sandy soils of low Zn content and calcareous (high lime) soils where the high pH reduces Zn availability. Vines grafted on rootstocks of *Vitis champinii* parentage such as, 'Freedom' and 'Harmony' is also prone to Zn deficiency. Zinc deficiency in grapevines, depending on the severity, may affect both fruit and foliage. Fruit symptoms include reduced fruit set and the formation of shot berries.

## Micronutrients

While much lower levels of micronutrients are needed to satisfy grape crop production, the correct balance of these trace elements is essential, especially for high yielding crops.

More iron and zinc is taken up than any other micronutrient. A lack of iron reduces leaf growth and resultant berry size and yield. Zinc deficiencies can be a serious problem causing poor fruit set and stunted shoots with small, misshapen leaves. Foliar application or fertigation helps minimize in-season micronutrient deficiencies by quickly correcting the problem.

As vineyard soils are either sandy loams or heavy clays, the usage of organic manure has assumed high importance in India. A standard dose of 500:500:1000 kg of N, $P_2O_5$ and $K_2O$ per hectare is followed in light sandy soils, while 660:880:660 kg are applied for heavy clay soils. The annual dose is fixed based on the petiole analysis carried out at 45 days after spur pruning. While 40 percent of the annual dose is given through organic sources, 60 percent is given as inorganic fertilizer. Calcium ammonium nitrate is usually not used. Sulphate of potash is the only source of potash used in place of muriate, particularly in heavy clay soils. Recently application of soluble fertilizers through drip irrigation is picking up. 40 percent of N, 50 percent of $P_2O_5$ and 33 percent of $K_2O$ of the annual dose is given during the growth season and the rest in the fruiting season.

## Weed Control

Maintain a weed-free environment because during the growing season, weeds suppress the grape crops and this leads to competition for moisture and nutrients, especially if the plants are still young. The occurrence of the weeds in vineyards also varies with the weather and climatic conditions. For instance in rainy season, the weed infestation is at the peak. During this period, the occurrence of weeds if left uncontrolled leads to increase in disease problems especially anthracnose, which has its own negative impacts on crop yield of the subsequent year. Vineyards trained on kniffin and head system the weeds get ample irrigation, nutrition and light to flourish. For better growth, make sure that the yard is clean from early spring until midsummer. The commonly found weeds in vineyards include *Bathu, BilliButi, Halon, Jangli Sarson, Maina, Pitpapra, Bhang* and *Chulai*. Mulching is a valuable way of eliminating many problem weeds. The method of weeding depends on the weed pressure and weeding can be done by three methods, i.e. mechanically, chemically and biologically.

**Mechanical method:** Weeds can be controlled by hand hoe. The hoe is used to control and remove weeds that surround the bark/trunk. To control weeds, grapevines should be cultivated, using a flat cultivator, a disk or a roller. Avoid cultivation when roots begin to spread out.

**Chemical method:** Weeds can be controlled through the application of herbicide, depending on species. Follow the label instructions carefully. The use of herbicides is a risk to soil that has a low organic matter content but it remains acceptable although it is harmful to the soil structure. This method must be repeated each year. There are two groups of weedicides, one which are used before emergence of weeds and other after weed emergence. Hexuron 80 WP (Diuron) at a rate of 1.2 kg per acre should be used before emergence of weeds. While, glycel 41 SL (Glyphosate) or gramoxone 24 WSC (Paraquat) 1.6 litre per acre should be sprayed after weed emergence (15-20 cm tall). The weedicides should be dissolved in 200 litres of water for an acre. These weedicides should be sprayed on a calm day to avoid drift to the foliage of vines.

**Biological control:** Grasses can be used for mulching. They must be left on the soil surface to suffocate and repress emergent summer weeds. Any disturbance of the soil will destroy the weed-controlling effect of the mulch. Mulching reduces tillage during the growing season.

### Inter Culture

No intercrop should be grown in vineyards as growing of intercrop shall be at the cost of training of vines. More over juvenile period of grape vine is short and vines are planted very close where no implement can move freely. However, in the first year some vegetable crops like peas, potato, turnip, radish or carrots can be grown. Cucurbits can be grown for first two years. To improve the organic matter content of the soil, leguminous crops like guars, cowpeas and senji can be grown for green manuring. No rabi season crop should be sown in vineyards as these crops do not require water to maturity in April when vines need it urgently.

## PLANT CULTURAL PRACTICES TECHNOLOGY

### Training and Pruning

Grapevines have a week stem which cannot stand of its own especially in the initial years of growth; hence there is a need of artificial support. During the first two to three years of growth, pruning is done to shape and train the vine. This is achieved by tying the growing vine along with a strong support to develop a strong trunk with well-placed laterals to form a permanent framework. The grape bears on growth of one-year-old shoots called cane. Pruning is done to regulate yield, size and quality of fruit. Regular and right pruned vines bear the good quality fruit for longer time. The technique of training and pruning is explained below:

## Training

An ideal support / training system is one on which quality fruits/ bunches can be produced, involves ease in carrying out cultural operations, there is less disease incidence, can accommodate more number of plants per unit area and is durable over the years. The choice of training system depends upon the bearing habit and vigour of a cultivar. For instance, Anab--e-Shahi cultivar of grape which is very vigorous should be trained on Bower system. Some cultivars with medium vigour (Perlette, Flame Seedless, Beauty Seedless, Thompson seedless and Punjab MACS Purple) may be trained on head, kniffin, bower, Y- trellis or telephone system.

## Training System of Grapes

1. **Head System:**This system is suited to those cultivars which produce fruitful shoots close to the base of the canes. The vines are planted at the distance of 2.5 to 3.0 meters. After planting the rooted cutting, vine grows luxuriantly. The side branches should be removed as soon as they appear and the main trunk should be cut back at a height of about one metre in July. This would encourage the growth of a few laterals below the cut. During the next January, when the vines are dormant, these lateral branches should be shortened to 1 or 2 bud spurs. When growth starts in spring these spurs will produce a number of shoots which are retained as framework arm. During the third dormant season, 8-10 arms growing around the head with one or two spurs are retained. This training system is the cheapest and very easy to adopt and maintain. Although the yield per vine is low, however, it is compensated by having more number of vines per unit area. This system is suitable for spur bearing variety like Flame Seedless, Perlette, and Beauty Seedless.

2. **Kniffin System:** This system is named after William Kniffin of New York who first developed it in 1850. This system of training of grape vines is good for cultivars bearing on spurs as well as on canes. Trellising of kniffin system should be completed before planting. The angle iron posts are securely fixed in the ground at a distance of 6 m from row to row and 3 m from plant to plant. Two or three wires (10 gauges) are stretched along these posts; the lower wire should not be less than 75 cm from the ground and second wire about 50 cm above the first and so on. In the beginning, the vine is trained on this system just as it is done for head system. A strong trunk is developed which is then cut-back at the level of the top wire of the trellis. This would firstly encourage the growth of canes, the two of which are selected and tied firmly to the top wire. Similarly, two canes growing nearest to the lower wire are tied to it, one

on each side of the trunk removing all other canes. Two renewal buds are kept near the base to produce canes for the next years' crop. This system is comparatively more expensive to establish than head system.

3. **Telephone System:** This system is constructed on a horizontal plane like that of wires on telephone poles. This system combines most of the advantages of kniffin and bower system while eliminating many of their disadvantages. Angle-iron posts, standing two metres above the ground, are fixed in the soil at a distance of 6m from each other in the row and 3m from plant to plant. Two angle-iron arms, about 13m long, one on each side of the post are shouldered on the post at 45° angle at the height of one metre. Three wires (l0mm gauge) are then stretched, one through the main post and one through each of the side arms. The side arms are strongly braced.

4. **Bower System:** This system is suited for training very vigorously growing cultivars like Anab-e-Shahi and others. The 'Perlette' commercial cultivar of grapes in Punjab is also trained on this system. Bower may be constructed with angle-iron, concrete or brick pillers. Brick pillers are best as they withstand the stress. The brick pillars 2.10m tall, 37cm thick are constructed at a distance of 3-5 metres in the row and from row to row. Angle-irons are fitted at the top of the pillars and 10-12 gauge wires are stretched both ways at a distance of 50cm through these angle-irons.

   The vines are planted at distance of 3.0m in the row and double the distance between the rows in case of cane pruning cultivars and at distance of 3m X 3m in case of spur pruning cultivars like Perlette. The vines are trained during thc first year just to develop a good root system. When the vines start growth during the second year, they are tied in a fan shaped manner to the front wires of both sides. The arbour should be covered as early as possible and developing shoots should be regularly tied loosely to the wires. This system is costly to establish than the kniffin and Telephone system. Other disadvantages are (1) Difficulty in spraying the vines against insect, pests and diseases (ii) pruning is very tedious and time consuming. However, this system gives higher return than the other systems of training and is suitable for medium as well as vigorously growing cultivars of grapes.

5. **Y-trellis System:** Conventionally bower/ pandal system is used for training vines. The bower system has several disadvantages like difficult and labour intensive cultural operations, high disease incidence and poor fruit quality. Due to shade under the bower, there is enhanced incidence of fungal diseases like anthracnose and powdery mildew. The manual sprays of

fungicide and insecticides for disease and insect pests control virtually fall on the field worker, thus making it a health hazard. As an attractive alternative a new training system (Y-trellis system) has been evaluated and recommended by PAU, which has several advantages over conventional bower system.

## Advantages of Y-trellis system over bower system

*Early fruit maturity*: The fruit on the vines trained on Y-system mature 2-3 days earlier than on the bower system.

*Better fruit quality*: The TSS of the fruits on Y-system is 17-18% as compared to 16-17 % TSS obtained on bower system.

*Low incidence of diseases*: As the bunches are exposed to sun, there is lower humidity around the bunches as prevalent under bower system, due to shade of bower itself, the incidence of fungal diseases like anthracnose and powdery mildew is relatively low in Y-system.

*Ease in cultural operations*: As the vines are at the average height of field worker, it is easy and economical to carry out fruit quality improvement technologies like flower bud thinning, $GA_3$ dips and girdling.

*Less health hazard*: The health hazards while using fungicides and insecticides are relatively less as compared to conventional bower system.

## Training of vines on Y-trellis

*First year*: The planting should be done in the months of Jan- Feb at a distance of 5 x 12 feet, plant-to-plant and row-to-row, respectively. The side growing shoots should be regularly pinched off, so as to maintain a single main shoot (a) upto 4 feet. Then the main shoot should also be pinched off, so that side shoots emerge. Two side shoots (primary arms) (b and c) in each opposite direction should be developed, upto the distance of lowest wire.

*Second year*: The side shoots (b and c) should again be pinched, so that the two secondary shoots (b1, b2 and c1, c2) in each horizontal direction of a wire are developed. A sort of H is thus formed after this training *i.e* one main shoot, two primary shoots and four secondary shoots.

*Third year*: The shoots developed on secondary shoots should be allowed to grow vertically on both the sides of Y system, so that the maximum canes (one year old shoots) are developed on secondary shoots.

The pruning in the 4$^{th}$ and the subsequent years should be aimed at to maintain 40-50 equally distributed canes per plant and 4 buds per cane. In general, the shoots crossing the upper most wire of a Y should be pinched off after fruit harvest.

The Y system of training is superior as compared to conventional bower system in respect of early fruit maturity, better fruit quality and low cost on cultural operations and has minimal health hazards.

**Flat roof gable system**: This system is particularly followed for vigorous vines (vines grafted on rootstocks) and has the advantage of both bower and the extended Y systems. In this system an inter-connected Y trellis forming a flat roof gable is being adopted. This system has been developed at IIHR, Bangalore and is gaining popularity among the growers due to certain advantages like ease in management; clusters are protected from direct sunlight and well exposed to sprays of pesticides and the clusters hang within the reach of the worker of an average height.

## Pruning

Grapevine branches shows acropetal growth habit and continue climbing with tendril. This behaviour is more pronounced in tropical conditions and if branches left as such proximal buds become unsprout or bear small size cluster. Hence, grapevine needs regular pruning every year. During the process, almost 50% of the shoots developed in a year are removed. The aim of pruning is to evenly distribute the bearing branches all over the vine and to regulate the so that it bear crop for longer period.

The one-year-old mature shoots are called canes. Pruning severity depends upon the bearing habit of a cultivar. The number of canes and renewal spurs to be kept on each vine at the time of pruning depends upon vigour of the cultivars, training system and cultural practices. There are two systems of pruning:

1. **Spur Pruning:**Spur pruning is followed in the cultivars which bear at the basal portion of the cane and each spur possesses upto 4 buds e.g., Beauty Seedless, Perlette, Flame Seedless and Punjab MACS Purple etc.

2. **Cane Pruning:**This type of pruning is followed in cultivars which bear on their distal buds, i.e 5-10 buds e.g. Thompson Seedless and Anab-e-Shahi.

**Table 5:** Pruning schedule of important grapes cultivars grown in India.

| Sr.No. | Cultivar | Distance (m) | Training System | No. of cane per vine | No. of buds per cane |
|---|---|---|---|---|---|
| 1. | Perlette, Beauty Seedless, Flame Seedless, Punjab MACS Purple | 3x3 | Bower | 60-80 | 4 |
| 2. | Thompson Seedless | 3x3 | Bower | 40-50 | 5-12 |
| 3. | Anab-e-Shahi | 3x6 | Bower | 80-100 | 6-8 |

**Pruning Time and Intensity:**The pruning time and intensity is distinctly different in northern and southern India as below:

**North India:** In north India, pruning is generally done during dormant season before sprouting i.e from second fortnight of January to second fortnight of February. Severity of pruning depends upon variety and training system. At present maximum area under grapes in this region is occupied by Perlette, Flame Seedless or other spur bearing varieties. Only one crop is taken in these varieties under northern India. The spur pruning is done, retaining 3-4 buds/ canes. These are termed as fruiting canes. About 60-80 canes on bower system are retained per plant, every year after pruning. It is important to keep canes uniformly placed on the vine. Along with fruiting canes, almost an equal number of shoots having only 1-2 buds (spurs) are also retained on a vine. In a particular growing season, the fruiting canes will bear bunches and the spurs will develop into vegetative shoots, which in the next season act as fruiting canes. In this way, the vine potential to yield uniform fruit over the years is retained. Furthermore, pruning is necessary to keep the fruiting wood near the center of vine, otherwise if pruning is not done regularly every year, the central portion of vine near main stem becomes barren. It is also easy to manage the vineyards by practicing necessary regular pruning.

**South India:** In peninsular India region comprising of Telangana, Andhra Pardesh, north Karnataka and Maharashtra, grapevines do not shed leaves and continue growth throughout the year. In this region trend of two prunings and single crop is prevalent as follow:

1. **Foundation/Back pruning:** During April-May at the time of pruning, canes (one-year-old shoots) is pruned to one bud and process is known as back or foundation pruning. This pruning is done in all varieties.

2. **Forward Pruning/ Fruit pruning:** During October, the canes are pruned by retaining number of buds which depends upon variety and thickness of cane. For e.g. in Anab-e-Shahi variety 4-6 buds per cane is retained at time of forward or fruit pruning. While in Thompson Seedless and Bangalore Blue 5-12 and 3-4 buds per cane, respectively is retained at time of second pruning.

In the grape grown area of Tamil Nadu and Bangalore and Mysore districts of Karnataka double pruning and double cropping practice is followed. The 'Gulabi' and Bangalore Blue are the famous varieties of these regions. The five crops can be harvested in two year in Gulabi variety in Tamil Nadu by continuous pruning. In Bangalore Blue variety fruit can be harvesting throughout the years by staggering pruning.

In tropical region to promote fruitfulness and to manage the current growth when main shoot attains 7-8 leaf stage shoot pinching is done. The tip of the mature shoot is removed after fifth node. Shoot pinching promote growth of lateral buds called sub canes. 1-3 buds from base on sub cane bear 2-3 clusters per cane.

## Fruit Thinning

Fruit thinning in terms of flower bud thinning or removal of some clusters are important practices in grape cultivation helpful for quality improvement. The management of optimum load on vine also maintain the health of vineyard for longer time. Different grape varieties do not set the fruit equally well with normal thinning. Some tend to set straggly clusters, others nearly perfect clusters while some like Perlette over compact clusters. To obtain maximum improvement in fruit quality in these varieties, a set of different methods of thinning have been recommended.

**Flower cluster thinning:** This practice consists of reducing the number of clusters without reducing the number of leaves. With an increase in leaf cluster ratio, the flowers on the retained clusters are better supplied with food materials. The flower parts develop more perfectly, fruit set will be more and there will be a large percentage of normal berries. Flower cluster thinning is followed in cultivars which bear more number of clusters than desired. Flower cluster thinning can also be practised to rejuvenate weakened vines. The flower cluster thinning in Flame Seedless cultivar under northern India beneficial for uniform colour development of cluster.

**Flower bud thinning:** In this process, almost $1/3^{rd}$ of unopened flowers are retained and the rest are removed using a special brush having distantly placed bristles. Flower buds are thinned carefully by holding a bunch with one hand and brushing slowly from base of bunch towards the tip (narrow side of bunch). After brushing, uniform flower buds should be seen on the bunch. It is necessary to irrigate the field after brushing, so as to achieve cent percent retention of unbrushed flower buds. Flower bud thinning is done one week before flowering, when flower buds are still closed. Flower bud thinning after flowering damages the ovary and leads to deformities in developing berries at later stages. This methods is very famous in cultivar Perlette which tend to set heavily and bear compact cluster.

**Cluster thinning:** It consists of removal of entire clusters after the fruit set. This method has no effect on fruit set or cluster length. In this practice misshapen, small sized and oversized clusters are removed. This provide more favourable conditions of growth for the remaining clusters, giving larger berries. Cluster thinning is easiest and best means of reducing crop on overload vines of highly productive wine grape or raisin grape vineyards. Cluster thinning, however, is a time consuming and costly affair. In case of Perlette, thinning of bunches is necessary so that the remaining bunches develops properly. It is recommended that not more than 100 bunches be left on each vine (trained on bower), planted at a distance of $3 \times 3$ m. The thinning should be done soon after the berry set and all small and under developed clusters should be removed.

**Berry thinning:** Berry thinning consists of removing parts of clusters or individual berries after fruit set. The rachis of the cluster can be cut back to retain the derived number of berries. In addition to its removal of some of the laterals it provides more space for berry development and avoids compactness of the clusters. Berry thinning should be done soon after fruit set especially in seeded cultivars. The removal of individual berries is very time consuming and uneconomic. Berry thinning is not as effective in increasing the berry size in seedless cultivars. It is, however, very important in preventing overcompactness of the clusters when berry size is either increased by girdling or use of growth regulators.

## Quality Improvement Practices in Grapes

In northern India, Perlette occupy around 90 per cent area under grapes. This cultivar has compact clusters and bear low quality small berry size fruit, if kept as such. Hence, in Perlette grape and others cultivars bearing compact clusters, the following practices should be adopted to get optimum yield and quality:

1. Thin flower buds one week before flowering by leaving 100-120 flower buds per panicle.
2. When berry size is 4 mm, girdle the vine by removing a 2 mm wide ring of bark from the main stern and dip the clusters in 40 ppm $GA_3$. While girdling, there should be no injury to the wood and no piece of bark should remain attached with the wood.
3. One week after the first $GA_3$ treatment, give a second dipping in 40 ppm $GA_3$.
4. The field should have sufficient moisture during girdling and at least 3 weeks after that.

5. Harvest the crop when attained the requisite TSS content.
6. Two foliar sprays of potassium sulphate @ 1.5 per cent, first one week after fruit set and second at colour break stage, improve the quality and colour of Perlette grapes.

## Girdling

Girdling consists of removing a narrow ring of bark entirely around some member of the vine like arm or cane. The common width is 4-5 mm. It was introduced into Greece accidentally as a means of improving the set of Block Corinth in 1833. It is essential that the ring be completely removed and even if a small portion of the bark is left, there is no effect. The immediate effect of a complete girdle is to interrupt the supply of carbohydrate and hormones so that their level increases in the parts above the girdle. The width of the girdle should be such that the wound heals in a short time. It is especially true when the trunk is girdled otherwise restricted food supply to the roots results in death of the vine. Similarly removal of wood below the bark injures the vine. Vines are girdled to accomplish one or more of the three objectives viz. to improve fruit set, to increase berry size and to advance maturation.

Fruit set in variety like Perlette can be improved by girdling the vine before blooming. Sizing of the Seedless cultivar like Thompson Seedless and Perlette can be hastened and their ultimate size can be appreciably increased by girdling timed to be effective during the period of rapid berry enlargement. If girdling is done to improve berry colour or enhance ripening, the girdle must be open and effective during the early part of the ripening period.

## Use of Plant Growth Regulators

Plant growth regulators are organic compounds other than nutrients which in small quantities promote, inhibit or otherwise modify any plant physiological process. These plant growth regulators play important role in viticulture production. They can control specific plant processes that cannot be conveniently or economically controlled by any other means. In grapes growth regulators now can be used either for root initiation, bud burst, fruit set, cluster and berry size, loosening of clusters, induction of seedless berries in seeded grapes, colour improvement in coloured varieties and post-harvest keeping quality of the berries etc.

**Root initiation:** The growth regulator IAA (Indole-3-acetic acid) was identified as the first naturally occurring compound having auxin activity, which takes part in inducing rooting in the cuttings. For initiation of rooting in cuttings, a continuous supply of auxins is required for first 3-4 days. The IBA (Indole butyric acid) and NAA (α- Napthaleneacetic acid) are found to be more effective in inducing rooting in cuttings.The most commonly used auxins for rooting of grapevine cuttings are IBA and IAA. Soaking of the basal ends of the cuttings in IBA @ 2000 ppm to 2500 ppm solution for 10-15 seconds ensures better rooting of the cuttings. However, the lower concentration of IBA @ 500 ppm can also be used by increasing the time of dipping to overnight.

**Bud burst and differentiation:** The hydrogen cynamide commercially known as Dormex (49 % aqueous solutions) is widely accepted for early bud break and ripening all over the world. Application of Dormex @ 1.5-2.0 per cent in first week of January immediately after pruning hastened the bud burst and proves helpful in early ripening upto one week as compared to untreated vines in northern India. The thiourea can also be used just after pruning to break the bud dormancy. In Maharashtra just after October pruning practice of swabbing of buds once with hydrogen cynamide (1.5 ppm to 4.0 ppm) is followed for early and uniform bud break. Likewise, after April pruning CCC @ 500-1000 ppm is used at 5 leaf stage to restrict shoot vigour and enhance fruit bud differentiation. BAP 6 (10 ppm) 50 days after back pruning or Uracil (100 ppm) 45 days after back pruning are used for increase in fruit bud differentiation and bud fruitfulness, respectively. CCC treatment significantly increased the number of bunches per vine in Thompson Seedless grapes.

**Cluster and berry size:** Growth regulators like $GA_3$, cytokinins and BA are widely used for increasing cluster size and berry elongation in India and abroad. $GA_3$@ 75 ppm 2-3 days after full bloom on Pusa Seedless variety was used to increase bunch length, berry size and weight. In Maharashtra 6-BA @ 10 ppm and CCC @ 250 ppm is commonly used for increasing the cluster size and thinning of berry. $GA_3$ @ 10 ppm to 40 ppm at different pre bloom or flowering stages is also used for elongation of rachis and increase of cluster and berry size.

**Ripening and uniform colour development:** Maturity of grapes can be hastened by 6 days by spraying the clusters with ethephon@ 500 ppm at the time of colour break stage. Ethrel application at 250 ppm two months after set reduced the percentage of green berries to as low as 6.3 and 3.3 in Bangalore Blue and in Muscat, respectively as against 30.9 and 25.9 in control. Ethephon at 600 ppm in Perlette at colour break stage advanced the ripening. For uniform colour development in Flame Seedless grape, retain 80-90 bunches per vine (75% crop load) immediately after bunch emergence and spray 400 ppm ethephon at colour break stage.

**Storage and post-harvest life:** Pre-harvest application of calcium chloride (0.5 to 1.0 %) or NAA (20 ppm) one week before harvesting improves the storage life of grapes. Post-harvest berry shattering inAnab-e-Shahi grape can also be controlled by pre-harvest application NAA @ 100 ppm. Shattering of berries during storage can also be minimized with pre- harvest sprays of cycocel (4000 ppm) and kinetin (100 ppm).

## SPECIAL PROBLEMS

### Barrenness

It has become the major problem of vineyards of North Indian plains. The vines develop unproductive wood. At the time of pruning most of the shoots are found to be dry. A few fruiting canes per vine are available, hence reduced productivity. The canes produce bunches with few berries. It has been noticed that excessive vegetative growth of vines leads to over shading or immature elongated shoot growth resulting into failure of production of floral primordia development.

There may be many reasons for this physiological state. Over irrigation, excessive nitrogen application, defective training, wrong pruning and keeping of high number of fruiting canes in early years of bearing. Both Perlette and Anab-e-Shahi, cultivars are prone to bareness. Some workers feel that it is due to lack of flower bud formation due to increased diameter of canes. There is direct relationship with vigour of shoots and dryness.

Barrenness of vines can be prevented by following proper pruning practices and plant protection measures. Avoid taking heavy yield from the young vines. The excessive growth can be checked by spraying cycocel (CCC) @ 1000 ppm to 2000 ppm just after fruit harvest. Spray of Bordeaux mixture 2:2: 250 during July-August to control fungal diseases can help in the reduction of barrenness.

### Water Berries

Development of water berries in grape clusters is very common. Water berry is associated with fruit ripening and most often begins to develop shortly after berry softening. These berries do not develop into full size, remain off coloured, lack normal sugar and have more acidity. The berries look like small cellophane bags half filled with sap and remain hanging on the clusters.

The berries do not have firm pulp and shrivel or may dry up by the harvest time. Such berries mostly confine to the tip of the main rachis or its branches.The main reason is overcrowding of berries in bunch. Excessive vegetative growth at the cost of developing of berries in a cluster due to more nitrogenous fertilizer application and over irrigation in heavy soils is also responsible for the water berries formation. Water berry formation can be reduced by applying balanced dose of fertilizers and checking flooding of vines during irrigation.

## Cluster-Tip Wilting

The wilting of apex of the cluster is a big problem in Perlette and Thompson Seedless cultivars.The apical portions of the clusters contain wilted berries at maturity. Light brown lesions on the apical end of the rachis affect the conductivity of the rachis. This results in shriveling and drying of the rachis at the tip of the bunch. In severe cases the tip of the bunch up to 30-40% dries up completely leaving hard small and light brown berries at the tip. This disorder is usually associated with compact bunches and low moisture in soil at development stage of the berries. Cluster pinching or berry thinning is recommended to reduce excessive crop load on the vines. Ensuring adequate irrigation during the berry development and protection of bunches from direct sunlight also help in reducing the incidence of cluster-tip wilting. These berries remain acidic in taste.

## Shot Berries

The presence of very small berries in the clusters of grape cultivars particularly Perlette is very common. These berries are usually seedless and are called millerandage. Shot berries are smaller, sweeter, round and seedless as compared to normal berries. They are formed due to delay in pollination and fertilization of a few flowers or due to inadequate flow of carbohydrates into the set berries. Generally small berries drop due to embryo abortion or lack of nourishment. Those which do not drop become shot berries. Boron deficiency, incorrect stages of GA application and girdling are the known reasons for shot-berry formation. Boron or Zinc deficiencies should be corrected. The treatment of clusters with $GA_2$ and thinning of the clusters help in checking the formation of shot berries.

## Pink Berry

This is a serious problem of Thompson Seedless in Maharashtra. As the bunch approaches maturity some berries in the bunch develop pink colour at random. The pink colour changes to dull red colour rendering the bunch unattractive. Incidence of pink berries is low in the early season crop and increases with the rise in temperature late in the season. Indiscriminate use of Ethrel for berry colouration can also cause this disorder.

### Bud and Flower Drop

This physiological disorder is known as coulure or shelling. This phenomenon has been reported from North India in the states of Punjab, Haryana and Rajasthan. Flowers drop from the clusters just before and after opening. The buds drop on shaking the panicle. Excessive bud and flower drop results in reduction of yield. Association of a number of factors such as atmospheric temperature, high phosphorus and total salt contents of the soil has been reported as the factors causing this malady. Therefore, judicious irrigation practices and canopy management practices to improve ventilation during the flower development helps to minimize the flower bud and young berries drop.

### Poor Cane Maturity

Poor cane maturity is a common phenomenon observed in peninsular India. In this type of disorder shoots fail to mature and their barks remain green until late in autumn. Such shoots turn pink-red due to low temperature in winter. It is more serious in vineyards, where the shoot growth is vigorous and dense; vines are planted closely and excess nitrogen and irrigation are provided. Previous season's crop load also has been found to affect the shoot maturity. Judicious shoot pinching to check excessive vegetative growth; shoot thinning 30 days after summer pruning to prevent mutual shading of the shoots and promote light interception are some of the suggested remedial measures. Avoiding excess irrigation and nitrogenous fertilizers during 40-70 days after back pruning helps to overcome cane immaturity.

## HARVESTING AND PRODUCTION OF GRAPES

The grape is a non-climacteric fruit – a fruit which do not ripen after harvesting from the vine. In other words, grapes if harvested unripe (sour in taste) will remain soar even after couple of days/weeks in storage. This is because in non-climacteric fruits like grapes, the carbohydrates are not converted to sugars, once the fruit is harvested.Grapes should be harvested only when all the berries have developed the unique colour and desired TSS of the cultivar.The best harvest index or the indication of harvesting is the time when the berries at the tip of bunch tastes sweet or are good to eat. This indicates that the whole bunch is ready for harvesting, as the bunches start ripening from base (broader side of bunch) towards the tip. The other more reliable and accurate indication of harvesting is when berries develop adequate TSS/acid ratio. This ratio varies from one variety to another. As the fruit ripens, the colour changes, the sugars increase and the acids decrease. The timing will vary from year to year. Berries should be sampled daily as the harvest approaches to determine sugar, acid and pH levels. In Perlette grape TSS should be around 16-17%, while in Flame

Seedless and Punjab MACS Purple the TSS should be between 17-18% at time of harvesting.

Grapes are harvested by repeated pickings since the bunches do not ripen at one time. Taste is the most valuable indicator of the ripeness of the bunch. The berries at the shoulders ripe first followed by centre and tip of the clusters. There may be some bunches having unripe berries at the tip which may be harvested and unripe portion clipped for marketing. The bunches should be plucked with secateurs very close to the canes. The bunches should never be held from berries, rather bunches should be held from the peduncle (attachment part of a bunch which holds bunches on stem). Holding of bunches from berries removes natural bloom from them, which helps to protect, bunches from attack of microorganisms. Do not jerk or pull clusters from the vine as this may crush some of the grapes. During harvesting, bunches should be placed in the baskets very gently. After harvesting in the evening the bunches should be kept in shade for packing.

As grapes ripen during hot months of May-June, we should harvest grapes, early in the morning when air temperature is much lower than temperature during day, especially at noontime. Harvesting time affects the shelf life of harvested produce. Moreover, the changing weather conditions should also be considered, while the produce is harvested.

**Yield:** Approximately one million tons of grapes are harvested annually in India. Grape is harvested almost all the year round. If not all the varieties, one or more varieties are always available at any given time of the year.

**Table 6:** Yield and period of harvest in commercial grapes varieties.

| Variety | Yield (t/ha) | | Period of Harvest |
|---|---|---|---|
| | Average | Potential | |
| Anab-e-Shahi | 45 | 90 | February-May, July, Nov.-Dec. |
| Bangalore Blue | 40 | 60 | January-March, June-December |
| Bhokri | 30 | 50 | November-December, June-July |
| Gulabi | 30 | 50 | January-March, June-December |
| Perlette, Flame Seedless | 40 | 50 | June |
| Thompson Seedless & other seedless varieties | 25 | 50 | January-April |

However, the major proportion of produce, mainly of Anab-e-Shahi, Thompson Seedless and its clones, is harvested during March-April from the hot tropical region, which contributes more than 70 percent of the total harvest.

The productivity of grapes in India is very high, particularly in the Hyderabad region. Yields as high as 100 t/ha in Anab-e-Shahi and 75 t/ha in Thompson

Seedless were recorded in this region. However, quality of grapes is usually poor as a result of high yields.

## POST-HARVEST FRUIT TECHNOLOGY

### Grading of Fruits

Grading is an important pre-requisite to profitable marketing of the produce. Grading in grapes is based on bunch shape, bunch size, berry size, uniformity of berries in the bunch, berry colour, uniformity of bloom especially visible in coloured bunches. The grading restrictions are much more stringent, if the grapes are to exported. Each country has its own criteria for acceptability of grapes.The unripe, misshapen, rotten, immature / green, over-ripe berries, small sunburnt, soft and damaged berries in the bunch should be trimmed from bunches, with help of sharp pointed scissors. It is important for the workers to wear rubber gloves. Then bunches of uniform shape and size are separated. In coloured bunches, uniformity in colour is also considered. The remaining lot is again sub-divided into medium sized to small sized bunches. Thus three grades of bunches can be separately packed for sale.

### Packaging of Fruits

Packaging of grapes depends on the destination of its disposal. If grapes are to be sold in local markets, the expensive packaging in CFB boxes may not be required. For local marketing mulberry or bamboo baskets can be used. However, for distant markets packaging should be proper, so as to avoid minimal losses during transit. For distant markets, CFB boxes of 2-4 kg capacity should be used. The boxes are lined with fine shredded paper, which is spread at the bottom and the top of the box for protection (cushioning). The open flaps of the box are secured firmly by means of adhesive tape.On the top of bunches, in the box, grape guard (a paper coated with sodium metabisulphide) is placed. These chemically coated papers absorb the ethylene released during respiration of grapes and thus help to extend shelf life and avoid rotting during storage and transit of the grapes.

Grapes that are sent to foreign markets are packed in five ply corrugated boxes, 500 × 300 mm in size to accommodate 5 kg of grapes. The graded bunches are weighted into 5 kg lots of plastic trays. One or two bunches weighing between 350 g and 650 g are placed in small, thin polythene pouches. Before the pouches are placed into the carton, a sheet of bubble wrap is spread with its rough surface facing toward the base of the box. A white, soft polythene liner is spread over the top of the bubble sheet. These pouches are arranged in a single layer before precooling of the grapes. After precooling, dual purpose $SO_2$ release pads are placed over the pouches and the polythene liners are folded in. In the cold storage maintain the temperature of – 2 to 0°C with relative humidity of 85 to 90 percent where it can be kept for 40 days. Shelf-life can be improved if heat of berries is immediately removed before actual storage by forcing the air through the boxes at 2°C less than cold storage temperature.

**Use of Grapeguard:** It is a paper, chemically treated with Sodium meta-bisulphate or Potassium meta-bisulphate, which when comes in contact with moisture releases sulphur dioxide, which helps in retaining freshness of fruits and checks fungal infection in transit and storage. Fast release grapeguard is used for transport and slow release grapeguard is used for storage; while dual release paper is used to take care of both. To eliminate discolouration and to avoid direct contact with fruits it should be wrapped in suitable material before keeping in cartons meant for export or domestic trade.

**Pre-cooling**: No phase is more critical in the postharvest handling of grapes than that of cooling – removal of the sensible or field heat from the fruit after harvest. Cooling is necessary to reduce the rate of fruit respiration, retard the development of decay and most importantly to minimize water loss from the fruit. As grape is a non-climacteric fruit with a very short post-harvest life, pre-cooling is essential to minimize water loss, avoid decay and reduce the physiological and metabolic activities of grape by lowering the temperature. Harvested grapes will deteriorate more in 1 hour at 32ºC than they will in 1 day at a temperature of 4ºC or in 1 week at a temperature of 0ºC. In fact, the amount of time between fruit harvest and cooling is key for final fruit quality, as is also that of cooling to optimum temperatures. Assuming that harvest is carried out early in the morning and that harvested fruit is always kept in the shade after picking, the amount of cooling needed to decrease fruit temperature will be minimized. In pre-cooling, the temperature is quickly brought down by fast and prompt cooling to check drying and browning of the stems and maintaining the firmness of berries. The moisturized air is directly brought in contact by parallel flow system. The temperature of grape is brought down within 2 to 6 hours after the harvest by pre-cooling. In pre-cooling, temperature is maintained at 0-4 ºC and relative humidity at 90% and above. The voyage transit period to the European markets ranges between 20 to 30 days. Therefore, unless the shelf life of grapes is extended to 60 days, it may not be possible to maintain quality during the post shipment transit period. The grapes are either pre-cooled in naked form or after packing into individual paper bags or in ventilated CFB cartons. The pre-cooling service has facilitated the export diversification of grapes to Gulf, Europe and South Asian countries. In order to control decay of table grapes, fumigation with sulfur dioxide is a common practice. However it is important to consider that grapes treated with sulfur dioxide will necessarily call for storage under low temperatures.

## Storage of Grapes

Under favorable storage conditions, and if cooling has been carried out appropriately, different varieties can be held for varying periods in order to obtain better prices or sale conditions. The best temperature to attain the full storage potential of table grapes is -1°C. The relative humidity of the air should be as high as 95 to 98% if possible. Frequent and close inspection of the grapes during storage is necessary to detect any disorders early and dispose of fruit before market quality is impaired significantly.

**Post-Harvest Losses:** All fresh fruits including grapes are inherently perishable. During the process of distribution and marketing, substantial losses are incurred which ranges from a slight loss of quality to total spoilage. Post-harvest losses may occur at any point in the marketing process, from the initial harvest through assembling and distribution to the final consumer. The causes of losses are many: physical damage during handling and transport, physiological decay, water loss, or sometimes due to glut in the market and there are no buyers.

## INSECT-PESTS AND MANAGEMENT

**Grapevine Thrips (*Rhipiphorothrips cruentatus*):** Thrips are the most troublesome insect pest of grapes. It is highly polyphagous. The nymphs and adults attack tender leaves and flower-stalks and suck the oozing cell sap. As a result, the leaves develop silvery white scorchy patches with curly tips, which gradually get deformed and ultimately fall down. Attack on flower-stalks results in shedding of flowers resulting in poor fruit set. It is also responsible for scab formation on the berries. The peak infestation is during hot weather from March to October in various parts of India. The pest breed throughout the year, except in winter, when it is found as a pupae in the soil at a depth of 8-18 cm under host plants. The control measures targeted at early stage (February-March) are more effective compared to late stage.Expose the hibernating pupae to sun heat and natural enemies by digging the soil near vine roots. However, if thrips and chaffer beetle attack occur simultaneously in September then the insecticide application may be done. To control thrips spray 500 ml of malathion 50EC in 500 litres of water per 100 vines once before flowering and again after the fruit set.

**Jassid (*Arboridia viniferata*):** Nymphs and adults of hoppers usually suck the cell sap from the ventral surface of leaves. The feeding spots on the leaves become pale. In case of severe infestation, the affected leaves turn yellow, which gradually start curling, become brown and ultimately fall down. It is serious after rainy season. It is also responsible for an indirect loss by producing honeydew, which serves as a substrate for the growth of sooty mould fungus on foliage and fruits. It affects the production of fruit and also depreciates the quality of grapes. To damage can be reduced by adopting clean cultivation practice in the vineyard

**Leaf Roller (*Sylepta lunalis*).** The eggs are laid on ventral surface of leaves. On hatching, the young caterpillars feed on epidermis of leaves and make skeletons of the same. Later these caterpillars roll the leaves and feed within. Pupation take splace within rolled leave. The pest is active during monsoon. To control leaf roller spray 500 ml Malathion 50EC in 500 litres of water as soon as the attack begins.

**Defoliating Beetles (*Adoretus* spp).** Adult beetles appear with break of monsoon and feed on leaves during night and hide during day. In case of severe attack, fruits are also scrapped near the apical end. Eggs are laid in the soil, grubs feed on roots and other organic matter and sometimes the grubs feeding on roots cause the death of tree. To control defoliating beetles adopt clean cultivation practice in the vineyard.

**Yellow and Red Wasps (*Polistes hebraeus* and *Vespa orientalis*).** These cause much more damage by feeding on ripe berries having thin skin and high sugar content. To contol wasps, burn or smoke the wasp nests in hedges on trees etc. at sunset. On a small scale the damage by wasps can be avoided by covering bunches with muslin cloth.

**Mealybugs:** (*Nipaecoccus viridis* and *Maconellicoccus hirsutus*). These are active in grapevine orchards during July-October, *N. viridis* cause damage to twigs branches, leaves and fruits while *M. hirsutus* is active on tender shoots. To control these, spray 1875 ml Durmet/Dursban (chlorpyriphos) or 750 ml curcaron 50 EC (propenophos or 500g Asataf 75 SP (acephate) in 500 litres of water after the harvest of crop in July to avoid residue of insecticides.

## DISEASES AND MANAGEMENT

**Anthracnose or Dieback:** Caused by *Elsinoe ampelina.* It is a serious disease and commonly found in all grape growing regions. Small light brown spots appear on young leaves, which later enlarge, turn dark brown and give shot-hole appearance. In severe attack, early defoliation occurs. Dark brown sunken spots with raised margins develop on new shoots/canes leading to their death from tip backwards. Similar spots appear on laterals of clusters. The fungus thrives well in warm wet weather. Berries may drop if there is infection on the stalks and subsequently girdling takes place. Under conditions favorable for the disease dark brown depressed spots appear on berries also. This disease on the berries is referred to as "birds eye spot" To control anthracnose, an integrated approach is needed as follows:

i) Prune the shoots and canes during January-February and give one dormant spray of Bordeaux mixture (2:2:250) after pruning using 125 litres of water/acre.

ii) Spray with Bordeaux mixture (2:2:250) in the last week of March using 250 litres of water/acre.

iii) Spray Bavistin 50 WP @500g/acre in last week of April using 500 litres of water.

iv) Spray Bordeaux mixture (2:2:250) in the last week of May in 500 litres of water/acre.

v) Spray Bavistin 50 WP @500g/acre in mid-July using 500 litres of water.

vi) Spray Bordeaux mixture (2:2:250) in mid-August in 500 litres of water/acre.

vii) Spray Bavistin 50 WP @500g/acre in first week of September using 500 litres of water.

**Cercospora Leaf Spot:** Caused by *Cercospora spp.* It manifests as necrotic small area on leaves with straw-coloured center and reddish brown margins. To control cercospora leaf spot, spray the fungicides as recommended for Anthracnose. Proper fertilization of the vines results into reduced attack of leaf spot.

**Downy Mildew:** Caused by *Plasmopara viticola*: Downy mildew is considered as the most destructive fungal disease of grape, which has been recorded from 91 grape-growing countries of the world, ranging from temperate to tropical climatic conditions. In India, it causes heavy losses to grapevine orchards in south India. Northern parts are, however, relatively free from this malady. This disease has been reported to be associated with the accidental discovery of Bordeaux mixture during 1885. In infested vines light yellow oily spots appear on upper surface of leaves which on the lower surface are covered with white downy growth of the pathogen. Later the spots become brown and brittle. Leaves with many active spots drop pre-maturely. The disease starts appearing in nursery and on grown up vines in March-April. After rainy season *i.e.* August-September it may assume serious proportions and continue to appear in humid weather. It affects tendrils and fresh growth of the shoots also.The inflorescence is also attacked and destroyed. If the berries are infested, they turn reddish brown in colour and fail to develop and drop down. To control downy mildew, all the pruned canes should be destroyed as the buds may contain dormant mycelium. Old fallen leaves should be collected from the orchard soil and destroyed.The vines should be properly spaced and trained in a way that leaves don't come in contact with the soil surface. Spray the fungicides as recommended for Anthracnose. Also give one additional spray of Bordeaux mixture (2:2:250) in mid-September using 500 litres of water/acre.

**Foot rot or Collar rot:** Caused by *Rhizoctonia sp.* In case of young vines the roots and the collar region turn brownish with the shredding of bark and internal brown discolouration. Leaves turn yellow and ultimately the vines wilt and die. To control foot or collar rot following measures should be taken:

i) The cuttings before planting should be dipped into 0.2% Ziram suspension (2g per litre of water)

ii) The soil in the pit should be drenched with 0.4% Captan (400 g in 100 litres of water) thoroughly before planting the cuttings.

**Powdery Mildew:** Caused by *Uncinula necator*. In case of attack, white powdery patches may be seen on the upper surface of leaves, stems, tendrils, flowers and berries. Fruit set is reduced. Mature berries develop cracks. On leaves, white powdery patches appear which enlarge, coalesce and become dirty white at a later stage. In southwestern Punjab, it appears in the form of yellowish diffused spots on the upper surface of the leaves and its presence is felt when it appears on the barriers as dirty white growth. To control powdery mildew, spray the vines with 0.25% wettable sulphur (1.25 kg in 500 litres of water) or spray Bayleton @ 200 g or Topas 10EC @200 ml/500 litres of water in mid-March, last week of April and first week of May.

**Rotting of berries:** (Black mould rot, blue mould rot etc.). It is caused by various kinds of air borne fungi, such as *Botrytis, Rhizopus, Aspergillus, Penicillium* spp, yeast. Grape berries are attacked when still on vines. The wasps injure the berries and releasing the juice, which serves as substrate for the growth of fungi. In Perlette, which has compact bunches the growth pressure ruptures some berries and the released juice flows into other berries where the fungi grow.

i) For Perlette, practice the thinning of bunches as recommended under quality improvement.

ii) Use insecticides or repellants to guard against wasps and other insects causing injuries to berries, as given under recommendation for insect-pests.

iii) During June spray grape vines with 0.2% Ziram (1 kg/500 litres of water) at 7 day interval. Stop spraying a week before harvesting bunches

## MARKETING OF FRUITS AND EXPORT POTENTIAL

More than 80 percent of the total production is consumed as table grapes in India, and more than 70 percent of the total production is harvested in March-April, but the cold storage facilities are inadequate. Therefore, market gluts and fall of prices of grapes in March-April are common. The producers sell the fruit either to the pre-harvest contractor or to the wholesaler through an agent with middlemen sharing profit. The responsibility of harvesting, packing, transportation and marketing vests with the contractor to whom the produce is sold on the basis of price agreed for unit weight of the produce or without weighing for a mutually agreed price. Co-operative grape marketing societies are in existence in many grape producing states of India. The advantage of marketing by producers' cooperative are:

- Reduction in the price gap by avoiding the commission agent and wholesaler
- Regulate supplies to different markets; and
- Minimize marketing problems arising out of unhealthy competition among producers.

Approximately, 2.5 percent (22,000 t) of fresh grapes are exported to the Middle East and European countries. The rest of the produce is marketed within the country. Grapes are exported through three different agencies. These agencies have established their own facilities for pre-cooling and cold storage in the vicinity of major production sites. The channels of selling grapes exist in the international markets viz., Grower Exporters, Growers' Cooperatives who collect, pack, cool, transport, market it abroad and share the profit with the growers and the Trader exporters who purchase, pack, pre-cool, store and then ship these in refrigerated containers to overseas markets.

To boost exports regular guidance is being given to the farmers and their co-operative societies on different aspects e.g. pre-harvest, proper use of pesticides, post-harvest, packaging, pre-cooling, cold storage and transportation. There is a need to increase the share of Indian grape in the imports of European countries like U.K., Germany, France and Netherlands where higher prices can be fetched. There is also potential for increase in the export of Indian grape to Asian countries like Japan, Singapore, Hong Kong and China in which prices are very high.

Grape is the important fruit of the India occupying 1.21 lakh hectares area, then covering 1.70 of the total areas. India is a major exporter of fresh grapes to the world and exported 1.56 lakh mt of grapes for the worth of Rs 1557.32 crores. Grapes were mainly exported to Netherland, UK, Russia, UAE and Saudi Arabia.

## FUTURE STRATEGIES

India has the distinction of achieving the highest productivity in grapes in the world, with an average yield of 30 t/ha. Production currently is much higher than demand in the domestic market. There is need to promote export of grape and its products to sustain present production trend. Otherwise growers can incur heavy financial losses.

Till recently export of grapes from India was mostly confined to neighbouring countries due to inadequate pre-cooling facilities and consciousness about quality as well as residues of pesticides by countries like U.K., USA, Germany, Canada and Switzerland.

There is a need to diversify the uses of grapes. Currently more than 80 percent of the produce is used for table purposes. The major bulk of the produce is harvested in March-April, but as cold storage facilities are currently inadequate there are frequent market gluts. Diversification of uses as wine/juice and export of table grapes can ease the marketing problems. Maintenance of quality of table grapes by crop regulation is the priority consideration to increase exports. For the survival of the grape industry in India, the produce should be of quality and cost competitive. The Government of India is supporting the grape industry of the country in the following ways:

a) Encourage and support the farmers for establishing the vineyards and installing drip irrigation systems by providing soft loans and subsidies.

b) Provide research support to sustain the productivity of grapes under adverse situations.

c) Promote and support the export of fresh grapes by training the growers and providing soft loans and subsidies for pre-cooling and cold storage facilities.

## LITERATURE CONSULTED

Adsule, P.G., D.S., Yadav, Anuradha Upadhyay, J. Satisha and A.K. Sharma 2013. Good agricultural practices for production of grapes. Published by P.G. Adsule, Director NRC, Pune ICAR.

Anonymous ,1985. Grape Bulletin. Maharashtra Grape Grower Association, Pune, pp.16.

Anonymous ,2009. Manjri Naveen. National Research Centre for Grapes, Pune, India. Extension Folder No. 35.

Anonymous, 2011. Report on the wine sector: Indian update. Published by Indian Grape Processing Board, Ministry of Food Processing Industries, GOI, Pune pp 15-17.

Anonymous, 2012. Manual on good agricultural marketing practices for grapes Ministry of agriculture (department of agriculture & cooperation) Directorate of marketing & inspection branch, Nagpur (agmarknet.nic.in/grapes.pdf).

Arora, N.K.,Navjot and Gill, M.I.S. 2011. Effect of hydrogen cyanamide on enhancing bud burst, maturity and improving fruit quality of Perlette grapes. *Indian J. Plant Physiol.*, 16(2): 171-74.

Bal, J.S. 2014. Grapes.In : Fruit Growing. Kalyani Publishers; 3rd revised edition. Pp 213-242.

Chadha, K.L. 1999. The Grape-Improvement, production and post-harvest management. Malhotra Publishing house New, Delhi.

FAO stat 2015. FAO statistical database, Food and Agricultural Organization of United Nations, http://www.faostat.org/site/340/ Date sited : 2015-9-7.

Gill, M.I.S. and Arora, N.K. 2009. Performance of different grape varieties under North Indian conditions. *Indian J. Ecol.*, 36 (1): 15-17.

Gill, M.I.S., Arora, N.K., Kumar, K., Karibasappa, G.S., Tetali, S., Karkamkar, S.P., Misra S.C., Ghosh, S.N.2014. .Gprape. In: S. N. Ghosh (ed.) Tropical and Sub-tropical Fruit Crops: Crop Improvement and Wealth. Jaya Publishing House, New Delhi. Pp 293-334.

Keller, M. 2010. The Science of Grapevines: Anatomy and Physiology. Academic Press, Burlington, Massachusetts.

Pearson, R.C. and Goheen, A.C eds. (1988) Compendium of Grape Diseases. APS Press, St. Paul MN.

Pandey, R.M. and Pandey, S.N. 1990. The grape in India. ICAR, New Delhi, pp115.

Ray, P.K. 2002. Grape. Breeding Tropical and Subtropical Fruits. Narosa Publishing House, New Delhi.pp 172-200.

Shanmugavelu, K.G. 2003. Grape cultivation and Processing. ISBN No. 81-7754-168-4. Published by Agrobios (India).

Singh, R. and Murthy, B.N.S. 1993. Improvement of Grape. In: Advances in Horticulture Vol. 1 – Fruit crops Part 1 (Chadha KL and Pareek OP eds.), Malhotra Publishing House, New Delhi, pp. 349-381.

Singh, S.K. and Dhillon, W.S. 2013. Grape In: Dhillon WS (ed.) Fruit Production in India. Narendra Publishing House, New Delhi. Pp 235-261.

Thakur, A., Arora, N. K., and Singh, Som Pal.2008., Evaluation of some grape varieties in arid irrigated region of North-West India. *Acta Hor.*,(ISHS)785: 79-84.

Winkler A.J .1965. General Viticulture, Univ of California Press, Berkeley and Los Angeles.

## SECTION - 2

## CULTURE AND TECHNOLOGY OF COMMON SUB-TROPICAL FRUITS

# 3

# Pomegranate

*B.N.S. Murthy and Awachare Chandrakant Madhav*

## INTRODUCTION

The pomegranate(*Punica granatum* L.) is one of the oldest known edible fruits, occupies the eighteenth place among the main world-fruit cultures.The fruit was naturalized throughout the Mediterranean region through ancient times and today the main areas of world production are Turkey, Spain and California. Around the world pomegranates are generally grown in Mediterranean climates, often with very warm dry summers (Wetzstein *et al.*, 2011). During recent past, this ancient fruit has emerged as commercially important fruit, owing to its enormous medicinal and nutritional properties, built-in ability to tolerate heat and drought, low resource input demanding nature and high returns on investment (Singh *et al.*,2012). The fruit has growing consumer demand both for fresh use as well as processing into juice, syrup, squash and *anardana* (an acidulant product), owing to its attractive, juicy, and refreshing arils (Pruthi and Saxena, 1984).

## NUTRITIVE AND CULTURAL SIGNIFICANCE

### Nutritive

Pomegranate is believed to be one of the oldest fruit crop, domesticated by mankind since the dawn of civilization.Due to its immense potential for health benefits, pomegranate has acclaimed as 'superfood'. It has been studied that 100 g arils provides 72 kcal of energy, 1.0 g protein, 16.6 g carbohydrate, 1 mg sodium, 379 mg potassium, 13 mg calcium, 12 mg magnesium, 0.7 mg iron, 0.17 mg copper, 0.3 mg niacin and 7 mg vitamin C (Grove and Grove, 2008; Silva *et al.*, 2013). Further, its peel extracts have also been found to be suitable for applications in the food industry as they are an important source of phenolics, flavonoids and tannins.

## Medicinal and pharmaceutical

Pomegranate has a long history of medicinal uses, having been used as a herbal cure for cancer, diarrhea, diabetes, blood pressure, leprosy, dysentery, hemorrhages, bronchitis, dyspepsia and inflammation (Wang *et al.*, 2010), anti-microbial (Al-Zoreky, 2009), anti-plasmodial (Dell'Agli *et al.*, 2009), anti-diabetic (Julie, 2008; McFarlin *et al.*, 2009), and anti-carcinogenic (Khan, 2009).Several studies indicated that it has potentially active phytochemicals like lignins,sterols and terpenoids in the seeds, bark and leaves; alkaloids in the bark and leaves; fatty acids and triglycerides in seed oil(Newman *et al.*, 2007); simple gallyol derivatives in the leaves;organic acids in the juice (Ender *et al.*, 2002), anthocyanins and anthocyanidins, catechin and procyanidins in the juice and rind (Kashiwada *et al.*, 1992). It has also been observed that pomegranate juice is more effective in treatment of depression and bone loss in menopausal syndrome in women, owing to its high phenolic and antioxidant properties (Mori-Okamoto *et al.*, 2004). Recently, Ismail *et al.*(2012) reviewed the anti-inflammatory and anti-infective effects of pomegranate peel and fruit extracts.

These are some of the compelling and recent trends for utilization of pomegranate plant parts and their extracts. However, it is important to note that positive aspects of pomegranate can be further explored and fortified, magnifying the economic importance of this crop through nanotechnology applications may well elevate it to industrial crop.

## ORIGIN, HISTORY AND DISTRIBUTION

Pomegranate, *Punica granatum* L. is native to Persia and perhaps some surrounding areas of Afganistan and Baluchistan, but with time it has diffused and got adapted to a wide range of climatic conditions. Moreover, it is believed to be originated in Central Asia and adapted the Mediterranean regions of Central Asia, Africa and Europe (Verma *et al.*, 2010). It is believed to be one of the first five fruit crops (date palm, fig, olive, grape and pomegranate), which was domesticated by mankind in 2000 B.C., but spread to the Mediterranean countries was at very early date (Hays, 1957; Chandra *et al.*, 2010)

According to Melgarejo and Martínez(1992) and reviewed by Chandra *et al.* (2010), pomegranate is also mentioned in the Bible and the Koran and is often associated to fertility. The first description of pomegranate trees was given by Theophrastus, to whom Linnaeus recognized as father of botany for about 300 years before the birth of Christ (Linnaeus, 1753). It travelled to Central and South India from Iran around 1st century AD and was reported growing in Indonesia in 1461. In time it spread into Asia (Turkmenistan, Afghanistan, India, China, etc.), North Africa and Mediterranean Europe. The domestication process took place independently in various regions and not only in the Mediterranean region (Zukovskij, 1950; Melgarejo and Martínez, 1992).

The most important growing regions in the world are Egypt, Spain, Turkey, Morocco, Tunisia, Georgia, China, Afghanistan, Pakistan, Bangladesh, Iran, Iraq, India, Saudi Arabia Turkmenistan and Tajikistan. It is also found place in Israel on the coastal plains and Jordan valley in the Western world. Globally India has the largest share in area and production of pomegranate, while with respect to productivity, Spain ranks first (18.5 t/ha) followed by the USA (18.3 t/ha), while Iran is the greatest exporter (60,000 t) followed by India (35,176 t), reviewed by Chandra *et al*., 2010). Presently, due to rapid increase in area expansion and production since one decade, exact figures on area and production of pomegranate in world is not available, however it is estimated that more than 1.5 million tonnes of pomegranate fruits are produced annually at global level (Holland and Bar-Ya'akov, 2008; reviewed in Silva *et al*., 2013.

In India, pomegranate is cultivated on 1.93 lakh hectares with annual production of 21.98 lakh metric tonnes. The fruit occupied 3.01 per cent of area and 2.40 per cent fruit prodcution with 11.4 mt/ha productivity.

## TAXONOMICAL AND BOTANICAL DESCRIPTION

The genus *Punica* with 2n =16 or 18 chromosomes, is having only two species, *Punica granatum* L. and *P.protopunica* Balf. The latter is presumed to be ancestor of cultivated pomegranate and is endemic to Yemen (Socotra Islands) (Levin, 2006). Patil and Karale (1985), reported *Chlorocarpa* and *Prophyrocarpa* as two subspecies under *granatum* species, which are endemic to the Transcaucasus and central Asiatic regions.Besides this, an ornamental type, Japanese Dwarf pomegranate (*Punica granatum* var. nana) is also popular in Asian countries particularly Iran, is highly suitable for pot plant production since it grows to a maximum height of 1.5 to 2m. A wild type pomegranate, 'Daru' produces small sized fruits, hard seeds with sour taste and is highly suitable for preparation of 'anardana' (Beach, 1980).

Pomegranate is a shrub (5-10 m high) that naturally tends to develop multiple trunks and has a bushy appearance. The stem is smooth with dark grey bark, having polygonal young branches with thorns at tip off young branches. Leaves are an oblanceolate shape with an obtuse apex and an acuminate baseopposite, short-petioled, simple, entire, exstipulate, 2–8 cm long, bright green, glabrous and glandular (reviewed in Silva *et al*., 2013). Flowers are terminal or axillary, solitary, pairs or in clusters, short peduncle or sessile, Sepals, 5-8 with orange-red to deep colour, petals are obovate, orange-red or pink, stamens are long and more than 300 per flower having orange red filaments and yellow anthers attached to the prominent calyx, carpels are usually 8 in number, superimposed in two whorls forming syncarpic ovary (Levin, 2006; Babu, 2010). There are three types of flowers *viz.* hermaphrodite, male and intermediate flowers were observed in same pomegranate tree. Male flowers are smaller with companulate (Bell-shaped) calyx, whereas hermaphrodite flowers are urceolate (Vase shaped) with well developed ovary. The intermediate flowers are of tubular shape with degenerating ovary. The fruit develops from the ovary and is a fleshy berry. The nearly round fruit is crowned by the prominent calyx. The apex of this crown is almost closed to widely opened, depending on the variety and on the stage of ripening. The fruit is connected to the tree with a short stalk.The mature pomegranate fruit is large, usually 3 inches in diameter, and sometimes as large as 4 to 5 inches. Fruit generally mature in 5 to 8 months and often change from round to a slightly squared-out shape.

## CLIMATIC AND SOIL ADAPTABILITY

The species is primarily mild temperate to subtropical in habitat and adapted to regions with cool winter and hot summer. It can grow from the plains to an elevation of about 2000 m. It is drought tolerant and favors semiarid climate, but can be injured severely by temperature below -10° C, although higher temperature is beneficial at the ripening period which produces sweeter fruits.

Pomegranate thrives well on varying soil conditions (calcareous, alkaline, deep loam, etc). In Himalayan regions, it is even grown on rock strewn gravel soil. However, it grows best on deep, rather heavy loam and alluvial soils which are ideal for its cultivation. Soils rich in organic carbon proved highly beneficial. It can tolerate soils which are limy and slightly alkaline. It can also be grown in medium or light black soils.

## RECOMMENDED AND POPULAR CULTIVARS

Most of varieties are evergreen while those of temperate habitat are deciduous. Most of the cultivars grown today are the result of human selection from naturally occurring variation. There are more than four hundred cultivars of pomegranate differing in the habit of the tree, leaf type, fruit form, size, colour, aril characters, keeping quality etc., in India until last decade, Ganesh variety was by far the most popular one. This is a seedling selection from a hard seeded 'Alandi'. It produces large size (400-450g), fruits with sweet (16-17 °B) arils containing soft seeds. But the arils are pink or light pink in colour. Hence, in the last decade two varieties of pomegranate *viz*., Ruby and Arakta or Mrudala with red aril colour were developed using Ganesh variety as base.The varieties of commercial importance are listed below.

**G-137**: Clonal selection from 'Ganesh' with yellowish pink, big (289.9 g), juicy (88.7%), less acidic and sweeter fruits than 'Ganesh'.

**Ruby**: Released from IIHR, Bengaluru. It is selection in the progeny of {[(Ganesh x Kabul) x Yercaud] X [Ganesh x Gul Shah Red] – $F_2$}. It is early maturing with thin skin, red and non sticky arils, sweet juice, soft seeds, very sweet and low in tannin content.

**Mridula**: A selection in F2 of cross 'Ganesh' X 'Gul Shah Red', it is evergreen bush with dark red fruits weighing about 250 gm, sweet in taste with 16.32 % TSS,0.47 % acidity and seeds are softer than Ganesh.

**RCR-1**: Seedling selection from 'Alandi'. Fruits are yellowish to red in colour, with intense red blush, large size and yielding 267 fruits (58.6 Kg) / tree.

**Amlidana:** It is progeny of cross between 'Ganesh X 'Nana'. It is superior to sour variety 'Daru'. Its fruit provide more acidic (16%) anardana (acidulant product). Higher yielder, short statured and amenable for high density orcharding.

**Bhagwa**: In recent years almost all the new orchards are with Bhagwa variety due to bigger fruit size, sweet, bold and attractive arils, glossy, very attractive saffron coloured thick skin, with better keeping quality so making it suitable for distant markets. This variety was found less susceptible to fruit spots and thrips as compared to other varieties of pomegranate. Bhagwa variety of pomegranate is heavy yielder (30 to 35 kg fruits/tree) and possesses desirable fruit characters and matures in 180-190 days.

**Ganesh:** It is a popular variety of Maharashtra. It possesses pink colored arils, soft seeds and is sweet with agreeable taste and medium to large sized fruits. It is a high yielding variety and good cropper. Ganesh is a seedling selection by Dr. G.S Cheema at Pune.

**Kandhari:** The fruits are large in size with deep red rind.The seeds are hard.The juice has TSS 12 per cent, acidity 0.61 per cent and well blended. The variety is a regular bearer. It is grown in Himachal Pradesh.

In addition to this, there are several other seedling selections which are grown on a limited scale across the country like Yercaud, CO-1 (Tamil Nadu), Dholka (Gujarat), Jaloore Seedless, Jodhpur Red (Rajasthan), Muscat (Maharashtra) and Panji (Goa). In Himachal Pradesh a sour pomegranate type-Daru, comes abundantly in wild. Several temperate introductions like Gulsha Rose Pink, Kali Shirin, Kazaki Anar, Lupania, Shirin Anar, Sunni Bedana etc., do not have commercial significance in tropics.

## PROPAGATION TECHNOLOGY

Vegetative propagation by cuttings (softwood and hardwood), air layering and micropropagation can be used to grow new pomegranate trees which produce fruit identical to the parental tree. Commercially, pomegranates are propagated using both softwood or hardwood cuttings, but hardwood cuttings are most commonly used. Softwood cuttings are taken from wood late in the season and require mist and greenhouse conditions for rooting to occur. In contrast, hardwood cuttings (25-30 cm) treated with 2000 pm IBA are taken from one year old wood or suckers, are also used for clonal propagation. Air layering parent plants and transplanting suckers is also an effective method of propagation (Morton, 1987).

Micropropagation could be a beneficial approach to mass produce cultivars with ideal characteristics, including insect and disease resistance. Research has demonstrated the most efficient way to micropropagate is by enhancing axillary bud branching (Singh *et al.*, 2012). However, further studies are needed to refine micropropagation methods.

## PLANNING AND PLANTING

Land is prepared by ploughing, harrowing, leveling and removing weeds. Pomegranate is propagated vegetatively by cuttings, tissue culture, air layering or gootee. Air layering is usually done during the rainy season and also in November-December. Planting is usually done in spring (February-March) and July-August. High density planting with spacing gives 2-2.5 times more yield than that obtained when the normal planting distance of 5 X 5 m. is adopted. Farmers have adopted a spacing of 2.5 X 4.5 m. Closer spacing increases disease and pest incidence. Pits of 60 X 60 X 60 cm. size are dug about a month prior to planting and kept open under the sun for a fortnight. Each pit is filled with top soil mixed with FYM (20 kg) vermi-compost (2 kg), neem cake (1 kg), pongamia cake (1 kg), furadon (20g) and super phosphate (500g). After filling the pit, watering is done to allow soil to settle down. Cuttings/air layers are then planted and staked. Irrigation is provided immediately after planting by drip irrigation.

## SOIL CULTURAL PRACTICES TECHNOLOGY

### Nutritional Need

Pomegranate cultivated in marginal and poor soils requires replenishment of nutrients removed by the plant. The quantity of fertilizers and farmyard manure to be applied depends on the nutrient status of soil and plant. The common fertilizer dose in kg per plant / year is as follows in Table 1:

**Table 1:** Common fertilizer schedule in pomegranate

| Tree Age (Years) | FYM (Kg) | CAN (Kg) | SP (Kg) | MOP (Kg) |
|---|---|---|---|---|
| 1 | 10 | 0.5 | 0.60 | 0.25 |
| 2 | 15 | 1.0 | 0.70 | 0.50 |
| 3 | 20 | 1.5 | 1.00 | 0.75 |
| 4 | 20 | 2.0 | 1.50 | 1.00 |
| >5 | 20 | 2.5 | 1.5 | 1.00 |

Full dose of FYM, superphosphate and half dose of muriate of potash and calcium ammonium nitrate has to be applied during winter (December-January) and remaining dose of CAN and MOP may be applied in two splits-once at time of fruit set and remaining a month after that in a band about 15 cm away from the tree.

Micronutrients are also essential for pomegranate. About 50 g each micronutrient in the form of $ZnSO_4$, $FeSO_4$, $MnSO_4$ and $CuSO_4$ are recommended to be added in each year to avoid deficiency. Copious irrigation after fertilization is beneficial. Leaf nutrient standards in pomegranate (Ttable 2) are given below for efficient nutrient management.

**Table 2:** Leaf nutrient standards in pomegranate

| Sl.No. | Nutrient | Range |
|---|---|---|
| 1. | Nitrogen (N) | 1.26-2.03 % |
| 2. | Phosphorus (P) | 0.16-0.25 % |
| 3. | Potash (K) | 0.74-2.03 % |
| 4. | Iron (Fe) | 75.4-135.5 ppm |
| 5 | Maganese (Mn) | 21.3-80.4 ppm |
| 6. | Copper (Cu) | 17.8-55.4 ppm |
| 7. | Zinc (Zn) | 17.1-28.5 ppm |

## Fertigation and foliar sprays of nutrients

Where the farmers are having venturi system, instead of soil application of fertilizers they can go for fertigation with following schedule (Table 3).

**Table 3:** Fertigation schedule in pomegranate

| Name of the Fertilizer | Quantity/ha | Days of application from pruning- alternate days | | | | |
|---|---|---|---|---|---|---|
| Mono Ammonium Phosphate (MAP)- 12:61:0 | 1.5 kg | 10 to 20 | | | | |
| Calcium Nitrate | 2.0 kg | 21 to 30 | | | | |
| Complex 19:19:19 | 500 g | 31 to 40 | | | | |
| Potassium Dihydrogen Ortho Phosphate -0:52:34 | 1.0 kg | 41 to 50 | 61 to70 | 91 to 100 | 111 to 120 | 131 to 140 |
| Potassium Nitrate- KNO3-13:0:45 | 1.0 kg | 51 to 60 | 81 to 90 | | | |
| Potassium sulphate- 0:0:50 | 1.0 kg | 71 to 80 | 101 to 110 | 121 to 130 | 141 to 150 | |
| Zinc sulphate (2g)+Borax (1g)+Calcium Nitrate (2g) | | 50th day | | | | |
| Zinc sulphate (2g)+Borax (1g)+Calcium Nitrate (2g) | | 80th day | | | | |
| Magnesium sulphate (2g)+ Ferrous sulphate (2g)+ Manganese sulphate (2g)+ Boric Acid (1g) | | Before flowering | During flowering | After fruitset | | |

## Water Need

Xerophyte nature of pomegranate makes it to grow in the varied climatic conditions ranging from temperate regions to semi-arid agro ecosystem in India. Although it is a drought tolerant crop, supplemental irrigation is necessary for realizing higher fruit yield with better quality. Pomegranates have a higher salt tolerance than most fruit crops. However, the water quality should be less than 1000 mg/kg (ppm) total soluble salts for best results, but plants will tolerate more than 2000 mg/kg (ppm) total soluble salts (Burt, 2007). Regular irrigation must be applied to achieve better establishment of plants, from onset of spring growth until the onset of monsoon. Irrigation may be given for one hour on alternate days through two drippers per tree each of 4 l/hr flow rate capacity (Nath *et al.*, 2007). However, most critical stages for irrigation are flowering and fruit ripening, as moisture stress may lead to flower and fruit drop and fruit cracking at mature stage.The month, season and age-wise water requirement of pomegranate is listed herewith (Table 4).

**Table 4:** Water requirement (l/day) of pomegranate

| Month | Age of tree (Years) | | | | |
|---|---|---|---|---|---|
| | 1 | 2 | 3 | 4 | 5 |
| January | 2.70 | 7.59 | 20.93 | 29.90 | 38.87 |
| February | 2.83 | 10.39 | 28.66 | 40.95 | 53.23 |
| March | 2.96 | 11.10 | 31.08 | 44.40 | 57.72 |
| April | 3.24 | 12.15 | 34.02 | 48.60 | 63.18 |
| May | 3.20 | 12.75 | 35.70 | 51.00 | 66.30 |
| June | 1.99 | 7.69 | 21.95 | 31.35 | 40.75 |
| July | 1.54 | 5.94 | 16.94 | 24.20 | 31.46 |
| August | 1.33 | 5.13 | 14.63 | 20.90 | 27.14 |
| September | 1.33 | 5.13 | 14.63 | 20.90 | 27.17 |
| October | 1.57 | 6.07 | 18.90 | 27.00 | 35.10 |
| November | 1.80 | 6.75 | 18.90 | 27.00 | 35.10 |
| December | 1.68 | 6.30 | 17.64 | 25.20 | 32.76 |

*Source:* Bangar and Kadam (2002), reviewed by Meshram *et al.*, 2010

In pomegranate, irrigation is also an important factor for crop regulation and the total water applied therefore depends on the desired bahar.

## Weed Control

Several studies indicated that weeds are shelter for many pests and diseases and also competes for water and nutrients. Therefore, mowing or spraying of herbicides (diuron or glyphosate) to manage weeds is recommended in older plantings to avoid root damage. Planting or a cover crop mixture of legumes and grasses in the intera row is advisable to manage weeds and improve soil structure (Burt, 2007). Weeds in the inter row area can be controlled using a directed glyphosate spray provided that suckers have been removed prior to spraying since it is systemic herbicide. The basin of trees should be cleaned by removal of weeds and grasses.

### Inter Culture

During initial years of planting intercrops such as *Solanaceous* crops, cucurbits, legumes and beans, cabbage, cauliflower, peas etc can be grown successfully. It will not only generate additional income but also improves soil physical properties. The intercropping may be discontinued, when plants starts commercial yielding so as to avoid completion between main crop and intercrops for water and nutrients (Pareek *et al.*, 2000).

It has also been observed that mulching with organic materials (banana trash, coconut husk, paddy husk, etc) or with black polythene sheet and spraying with anti-transpirants like Kaolin (8-10%), Phenyl mercuric acetate (10-5m) and liquid paraffin (1%) were found beneficial for enhanced productivity by reducing evaporation losses.

## PLANT CULTURAL PRACTICES TECHNOLOGY

### Training

Pomegranate plants can be trained on a single-stem or in multi-stem system. The single stem training has its own disadvantages. The plants have a tendency to produce ground suckers, making the plant bushy. As such it is rather difficult to train the plant to a single stem. The crop is highly susceptible to stem-borer and shoot-hole borer. Moreover, this system is hazardous. Thus single-stem training is uneconomical for commercial cultivation. Therefore multi-stem training is more prevalent in the country. Allowing too many stems also comes in the way of intercultural operation. The varying of stem number of 3–4 does not affect the yield significantly in early years of bearing and a multi-stem training with 4–5 stems/hill is beneficial.

### Pruning

Pomegranate plants do not require pruning except removal of ground suckers, water shoots cross branches, dead and diseased twigs and giving a shape to the tree. Pomegranate fruits are borne terminally on short spurs, arising from matured shoots, which have the capacity to bear fruits for 3–4 years. With advance in age they decline. A little thinning and pruning of old spurs to encourage growth of new ones is required. Sometimes non-bearing twigs and water shoots along with dried and diseased twigs have to be removed. Pruning can be achieved by spraying of Chloremquat (1000 ppm) (Lihosin 2ml/Ltr.) or TIBA (500 ppm) or Maleic hydrazide (500 ppm) through reduced growth and leaf fall.

## Pollination and Fruit Thinning

The blossom biology of pomegranate differs with agro-climatic conditions, where it is grown. It flowers throughout the year under tropical conditions however, in low temperature areas, it behaves as deciduous and flowers once in a year (Parmar and Kaushal, 1982). The flowering in pomegranate occurs on current season growth about 30 days after bud burst, mostly on spurs or short branches (Holland *et al.*, 2009).Pomegranate tree produces both hermaphroditic (bisexual) flowers and functionally male flowers on the same plant(andromonoecy). The hermaphroditic flowers have well-formed female (stigma, style, ovary) and male (filaments and anthers) parts and have been referred to as "fertile". The male ûowers produce well-developed male parts with reduced female parts or may have degenerated female parts (Holland *et al.*, 2009).

According to Mars (2000), there are three waves of flowering. Thus, in subtropical conditions of Northern states, flowering occur from last week of March to second week of May (Singh *et al.*, 1978). However, in Southern hemisphere, flowering was observed during June, October and March (Nalawadi, *et al.*,1973). Furthermore, three distinct flowering seasons were observed in subtropical and western India *viz,AmbeBahar* (flowering in January-February), *Mrigbahar* (flowering in June-July) and *HasthBahar* (Flowering in September-October) (Nalawadi *et al.*, 1973). The economic viability of Ambebahar is higher due to higher flowering intensity as compared to other flowering seasons (Prasanna Kumar, 1998). Both self and cross pollination by insects, mainly bees has been reported in pomegranate, however according to Singh and Rana (1993), it is often cross pollinated. Further, several studies on bagging and emasculation indicated that pomegranate flowers can self pollinate and produces normal fruits, however the degree of fruit set varies with cultivars (Mars, 2000; Levin, 2006).

## Crop Regulation

The pomegranate plants flower and provide fruits throughout the year in Central and Southern India. However, it needs to be thrown into rest period so as to enable prolific harvest at a given time. Looking at patterns of precipitation, flowering can be induced during June–July (*mrig bahar*), September–October (*hasth bahar*) and January–February (*ambe bahar*). In areas having assured rainfall where precipitation is normally received in June and continues up to September, flowering in June is advantageous; where monsoon normally starts in August with erratic pattern, flowering during August is beneficial; the areas having assured irrigation potential during April–May, flowering during January can be taken; and where monsoon starts early and withdraws by September induction of flowering in October is possible. In Rahuri, flowering during January–February is better in quality followed by October flowering. Considering comparable yields, prices and irrigation needs it is recommended that October cropping could be substituted for January flowering.

## Process of Bahar Treatment

- Withhold the irrigation two months prior to the bahar
- In light sandy and shallow soils, withhold water for 4–5weeks
- Due to water stress, leaves show wilting and fall on the ground
- The trees are medium pruned 40-45 days after withholding irrigation.
- Give ethrel spray at 2 to 2.5 ml/l mixed with 5 g/l of DAP
- At this stage cover the roots with a mixture of soil and FYM and irrigate immediately.
- Apply the recommended doses of fertilizers immediately after pruning
- Consequently, new growth, profuse flowering and fruiting is observed
- Resume the normal irrigation.
- The fruits are ready for harvest 5 months after flowering.

## Crop and Grade Regulation

A grown-up, well-managed tree at three years age gives 80–100 fruits annually, and increases by ten per cent annually of which 8–10 % are of 'A' grade; 20–25% are of 'B' grade and the remaining are of 'C' and 'D' grades, and cracked fruits.

### Improve average grade by crop regulation

- After the fruit set, do not allow fruits to develop in clusters and keep only solitary fruits.
- Allow flower set on inner/thicker shoots to develop in to fruits, remove those which are developed terminally on weaker shoots
- After getting set, remove all the flowers coming thereafter.

## Operation to be done during Pre Pruning and Pruning

- Spray 1% Bordeaux mixture 2 days before defoliation.
- Prune the twigs carefully
- Sterilize the secateurs with sodium hypochlorite (2 to 3 ml/l)
- Defoliate with a mixture of ethrel (2 to 2.5 ml/l) + DAP 5g/l.
- Remove weeds and suckers
- Collect & burn fallen leaves /debris from the orchard
- Harrowing in interspaces is advocated.
- Apply full dose of well rotten FYM and P, 1/3rd N&K fertilizers + Micronutrients ($ZnSO_4$, $FeSO_4$, $MnSO_4$ each 25g and 10g Borax (Boron) /tree)+ Neem Cake 1-1.5 kg/tree + Vermicompost 2 kg/tree+ Phorate10G @25g/tree or Carbofuran 3G at 40g/tree in shallow trench or ring (15-20 cm wide of 8-10cm depth) at 45-60cm away from the stem, cover the trenches properly with soil and give light irrigation immediately after fertilizer application.
- After 45 days apply 1/3rd dose of Nitrogen + 1/3rd dose of potash
- After 90 days apply 1/3rd dose of Nitrogen + 1/3rd dose of potash

## Physiological Disorders

### Fruit cracking

Fruit cracking, a serious chronic production problem in pomegranate industry, is more intense under dry conditions of the arid zone. The fully-grown, mature cracked fruits though sweet, lose their keeping quality and become unfit for marketing. The cracked fruits show reduction in their fruit weight, grain weight and volume of juice. It is mainly associated with fluctuation in soil moisture, diurnal differences in day and night temperatures and nutrient imbalances. Several authors reported that it might be due to lack of moisture content in soil at the time of fruit developmental stage and due to deficiency of calcium, boron and potash (Josan *et al.*, 1979; Phadnis, 1974). However, varietal character plays an important role in fruit cracking (3.57% in Kazaki to 76.67% in Suni Bedana). It has also been reported that fully developed pomegranates crack due to moisture imbalance, as they are very sensitive to variation in soil moisture and also to day and night atmospheric moisture deficit. Prolonged drought causes hardening of peel. If this is followed by heavy irrigation or rains, the pulp grows and the peel cracks. The percentage of cracked fruits is also related to season. Fruit cracking was maximum in *Mrig bahar* crop (30-35%) followed by *Ambe bah*ar (25-30%) (Randhawa *et al.*,1958; Hiwale, 2005).

There are some cultivars/strains–PS 75 K 3, Appuli, Shirvan, Burachni, Apsherconskil, Krasnyl, Sur-Anar, Kyrmyz-Kabukh and Francis–which are tolerant/resistant to cracking. Cracking can be managed through maintaining soil moisture and not allowing wide variation in soil moisture depletion, cultivating tolerant types, applying copious and regular irrigation during fruiting season using $GA_3$ (15ppm) and applying boron (0.2%) reduces cracking of fruits and improve fruit color.

### Internal breakdown

This is also known as blackening of arils and characterized by disintegration of arils in matured fruits. The first reference on occurrence of this serious malady was given by Ryall and Pentzer in 1974. They observed that some of arils in matured fruits become underdeveloped and posed abnormal flavor. This disorder cannot be identified externally, whereas the arils become soft, light creamy-brown to dark blackish-brown and unfit for consumption. It is increasing rapidly in the pomegranate-growing pockets in western Maharashtra.

The incidence of internal breakdown occurs 90 days after anthesis. Its intensity increases if the fruits are left on the tree for 140 days onwards. It is evident in evergreen and deciduous cultivars. The incidence is more in *Ambe bahar*. It increases with increase in weight of fruits from 150–200g (26.60%) to more than 350g (60%). No insect or organism is associated with this malady. The TSS, acidity, ascorbic acid, reducing sugars, calcium, phosphorus and enzyme catalase are reduced, whereas non-reducing sugars, starch, tannins, nitrogen, potassium, magnesium, boron and enzyme polyphenol oxidase and peroxidase increase in the affected arils compared with the healthy ones. The exact causes are not known and remedial measures are difficult to advocate. Therefore pomegranates should be harvested at 120–135 days after fruit set.

## HARVESTING AND PRODUCTION OF FRUITS

Pomegranate being non-climacteric fruit should be harvested when Total Soluble Solids (%TSS) and size meets the market requirements. Harvesting of immature or over-mature fruits affects quality. The fruits become ready for picking after 120–130 days of fruit set. The calyx at the distal end of the fruit gets closed on maturity. Ripe fruits give a distinct sound of grains cracking inside when slightly pressed from outside. At maturity fruit gives a characteristic metallic sound when pressed gently and attains colour specific to the variety (yellowish-red) and get suppressed on sides. Fruit colour is not sure guide to maturity. A grown-up, well-managed tree gives 60–80 fruits annually, with a life span of 25–30 years.

## POST- HARVEST FRUIT TECHNOLOGY

Pomegranate fruits can be graded depending up on the size (weight) into different categories as stated below Table 5:

**Table 5:** Grading of fruits pomegranate

| Size code | Weight in grams (minimum) | Diameter in mm (minimum) |
|---|---|---|
| A | 400 | 90 |
| B | 350 | 80 |
| C | 300 | 70 |
| D | 250 | 60 |
| E | 200 | 50 |

Being non climacteric fruit, pomegranate has low respiratory pattern. The fruits can be held in cool storage (5-7°C) before the extraction of arils. The fruits will suffer from chilling injury if they are stored at temperature below 5°C (Wasker *et al*., 1999). However, storing the fruits at 2°C with intermittent warming was found optimum for minimizing chilling injury and maintaining the fruit quality up to 13 weeks (Artes *et al*., 1996). Further, according to Artes *et al*. (1996), storage of fruits under controlled atmospheric storage at 10 % $O_2$ and 5 % $CO_2$ resulted in improved juice color with better aril quality.

## INSECT- PESTS AND MANAGEMENT

**Pomegranate butterfly, *Deudorix isocrates* (Fab.):** Pomegranate butterfly lay eggs on the flower and young fruits. The larvae on hatching enters inside young fruits and start feeding on aril. The conspicuous symptoms of damage are offensive smell and excreta of caterpillars oozing out of the entry holes, with the excreta found stuck. The affected fruits eventually fall down.

- Remove and destroy all the affected fruits (fruits with exit holes).
- Spray Decamethrin @ 1ml/litre at the time when more than 50% fruits have set. Repeat after two weeks with Fenvalerate @ 1ml/litre in non-rainy season. Quinalphos @ 2.5ml/litre is also effective. The number of sprays depends on severity of infestation. For suppression of menace of pomegranate fruit borer, β-cyfluthrin 0.025% and cypermethrin 0.01% were found equally effective.
- Remove flowering weeds especially of compositae family.

**Shot hole borer, *Xyleborus* sp. (Scolytidae: Coleoptera):** This is becoming a major pest nowadays on pomegranate in Karnataka. Early diagnosis with symptoms is a must. Hence, regular visit to orchards by growers is suggested. Signs of lateral branch yellowing to quick drying of full tree, should be immediately brought to notice of specialists and treatments be undertaken as follows:

- Drench soil around main trunk with a mixture of chlorpyriphos 2.5 ml with altering recommended fungicides (carbendazim 1g/litre or propiconazole 2ml/litre) at monthly intervals.
- If pest is severe, repeat the above drenching after a month.
- If infestation is low, drench with Azadirachtin (0.15%) 3ml/litre around main trunk 2-3 litres of mixture/tree with either of the above fungicides.
- Avoid water logging and keep soil raked and aerated.
- Infested trees should be uprooted and burnt, especially the root zone.
- Pits of uprooted trees should be treated with chlorpyrifos 2.5 ml/litre, by thoroughly drenching.
- Drench soil with chlorpyriphos 2.5ml/litre around all un-infested trees prophylactically once in six months, followed by a spray on trees with Quinalphos 2.5ml/litre, followed by Azadirachtin 1500 ppm 3 ml/litre. Avoid leaving infested trees in field after uprooting.
- In case of nematode occurrence, need based application of either phorate 25g/ plant or carbofuran 40g/plant to the basins have to be supplemented to bring down nematode population.

**Stem boring beetles, *Coelosterna spinator* (Fab.), *Zeuzera coffeae:*** The grubs bore into the trunk, primary and secondary branches. The bored stem show yellowing followed by drying, leading to dieback.

Detect early infestation by periodically looking out for drying branches.

- For the conrol of coelosterna apply chlorantraniprole 0.4 GR @ 15g per plant plant at the root zone.
- Cut out the drying portion of the branch and swab copper-oxychloride 50% WP on cut end.
- Spray all surrounding trees with Quinalphos 2.5ml/litre or chlorpyrifos 2.5 ml/litre.
- It is good to give a prophylactic swab during May/ June on the main trunk with following mixture: Copper-oxychloride (50% WP) 10g + (sticker 1 ml + Neem oil 1 ml) (all per litre of water).

**Thrips, *Rhipiphorothirps cruentatus* Hood, *Scirtothrips dorsalis* Hood:** Thrips rasp tender fruits; causing scab on them and thereby reducing market and export value. Thrips infestation is often seen on leaves and also on young fruits causing characteristic scab on fruits

- Spraying dimethoate 2ml/litre or fipronil 1 ml/litre or imidacloprid 0.3ml/litre or thiamethoxam 0.3g/litre prior to flowering is important. If serious, a spray of acephate 1.5 g/litre should be repeated after fruit set. The subsequent sprays for borer limit thrips build-up.
- Keeping the basins clean also reduces damage due to thrips.
- A follow-up spray of Azadirachtin @ 3ml/litre is useful.
- The numbers of sprays depend on the pest severity.

**Pomegranate aphid, *Aphis punicae* Shinji:** Sap sucking by aphids lead to shriveling of shoots. If serious, honey-dew accumulates on leaves and sooty mold develops affecting photosynthesis.

- Spray dimethoate 2ml/litre or imidacloprid 0.3ml/litre or acetamiprid 0.3g/litre as new shoots emerge.
- If predators like syrphids and coccinellids are found, delay spraying and in some cases, natural enemies sufficiently suppress the aphids.

**Mealy bugs:** Many mealybugs like ***Icerya purchase* Maskell, *Planococcus* sp., *Pinnaspis* sp., *Maconellicoccus hirsutus* (Green)** etc have been reported on pomegranate. Prune the affected parts of the plant and destroy.

- Spray chlorpyriphos 2.5 ml/litre +neem oil or pongamia oil 10 ml/litre after pruning infested parts.
- If on a small scale, release *Cryptolaemus montrouzieri* Mulsant near the site of infestation.

**Fruit sucking moth, *Othreis* sp. :** Adult moths of both sexes cause damage to fully ripe fruits by piercing and sucking the juice usually during night times. Damaged fruits are completely unmarketable and must be removed at packing to avoid contamination of sound product.

**Management:** The adult moths preferred to feed on ripe guava or banana fruits to pomegranate. These bait fruits can be tied to wooden stakes singly using thread. Commence baiting during flag end of harvest as maximum damage is been caused to mature fruits. The fruit baits can be erected in the borders where moth damage is severe @ 50 bait fruits per hectare. This baiting and trapping is found to be an effective method for controlling the fruit sucking moth damage. In areas where fruit sucking moth poses serious problem using regular use of fruit baits or erection of fly-proof netting around the trees or orchard gives the best control.

## DISEASES AND MANAGEMENT

**Leaf and Fruit spot** (*Pseudocercospora puniceae* (Henn.) Deighton): This is a very serious disease of pomegranate plants affecting leaf and fruits. Pathogen causes irregular, scattered, yellowish spots on leaves having a halo around at the early stage, subsequently the spots increase in size and form bigger patches and become blackish. The lesions are covered with dull white crust of fungal growth. Severe infection of the disease leads to defoliation and hampers plant growth. On flower buds brownish black irregular small spots appear which grow in size along with the fruit. Spots become deeper and more prominent. At greenish stage of fruit, reddish irregular dots/spots develop on the fruits. The spots get covered with deposit of ash colored spores under favorable humid conditions. Mature fruits show numerous black, irregular, slightly corky patches of the infected hard necrotic tissue. Severely infected fruits show cracking at times. Disease is severe in high humidity areas and during rainy season. Hexaconazole (Contaf 0.1%) or carbendazim (Bavistin 0.1%) or thiophanate methyl (Topsin M or Roko 0.1%) sprayings control the disease. First spray at flower bud stage and subsequently 8-9 sprays at 7-10 days interval depending upon the weather conditions. Among non-systemic fungicide Chlorothalonil (Kavach 0.2%) or Mancozeb (Indofil Dithane M 45 0.2%) also check the disease.

**Anthracnose (*Colletrotrichum gloeosporioides* (Penz.) Penz.&Sacc):** The disease manifest in various form of symptoms causing leaf blight, fruit spot, wither tip, die back and fruit rotting as described below:

(i) **Leaf blight:** The disease starts as minute, dull violet black or black spots surrounded by yellow necrotic areas. The spots enlarge; coalesce to form depressed large spots of aniline black in color. Severely infected leaves show necrotic areas extending from leaf margins to cover whole leaf blade. Leaves show curling at some points and severely infected leaves, become dry and fall off. Sometimes shot hole stage is also observed.

(ii) **Fruit spots**: The pomegranate fruits are highly susceptible to attack by this pathogen from early stages of development (flower bud stage - (Reddish) and fruit enlargement stage before attaining greenish color and subsequently at color breaking stage. Smaller brownish spots, which become depressed subsequently, increase in size to form the large spots. During greenish stage of fruit development infection remains quiescent and exhibit symptoms at color breaking stage. Mature fruits show numerous brownish black depressed spots, which coalesced to form bigger patch. The entire infected portion changes color to yellowish brown and fruits start rotting which becomes serious during storage. Severe infection of fruits results in flower bud and fruit drop and in mummified fruit and at times cause rupture of the fruit exocarp also.

(iii) **Wither tip and die back:** Younger, developing shoot tips are killed due to attack of pathogen and some developing branches show drying from the tips backward with necrotic areas extending downwards, such branches become dry and devoid of foliage subsequently and give die back appearance. These symptoms are more pronounced in older trees and neglected orchards. For the disease management, prune the dried twigs and branches and burn them. After pruning a general spray of copper oxychloride (Blitox 0.3%) and pasting of cut ends with copper fungicide should be practiced. Subsequently spraying with hexaconazole (Contaf 0.1%) or carbendazim (Bavistin 0.1%) or thiophanate methyl (Thiophanate methyl or Roko 0.1%) should be given. First spray at flower bud stage and subsequently 5 – 6 sprays at the interval of 7-10 days interval has to be provided. chlorothalonil (Kavach 0.2%) or mancozeb (Indofil Dithane M 0.2%) in non-systemic category also control the disease.

**Alternaria leaf spot and fruit rot: (*Alternaria alternata* (Fr.) Keissl.):** Minute pale brown to reddish brown circular to irregular spots develop on the ventral surface of the leaves. The spots, enlarge coalesce to form bigger spots, dark brown in colour having concentric rings. Within a few days of infection by pathogen, leaf brightening become more prominent and such affected pale yellow leaves dry and fall down.

On fruits, the pathogen cause external as well as internal rotting. Initially minute brown coloured spots circular to irregular in outline confined to rind appear on the fruits. As the disease advances the spots turn reddish brown and later on dark brown to black in colour. Fruits lose their natural lustre and severe infection of the pathogen leads to internal rotting of fruits discolouration of central core and the seeds For the disease management, spraying with protectant fungicides namely mancozeb (Dithane M45 0.2%) or zineb (Dithane Z 78 0.2%) or chlorothalonil (Kavach 0.2%) or ziride (Cuman L 0.4%) or with systemic fungicide - iprodione (Rovral 0.2%) should be taken up as the symptoms appear.

**Leaf, flower and fruit spots: (*Phytophthora nicotianae var nicotianae*):** Brownish lesions develop on the leaf surface which extends in area causing blightening of leaves. Severely infected leaves fall off. Brownish black lesions develop on the flowers and fruits. The affected flowers shed off prematurely. In fruits brownish to black rotten areas extend from calyx end to the periphery. The fruits remain on the plants but become useless as infections spread inwards from the rind causing rotting. Disease is serious during high rainfall period. Spraying with Metaxyl + Mancozeb (Ridomil MZ 0.2%) or Al Phosphonate (Alliette 0.2%) controls the disease.

**Wilt (*Ceratocystis fimbriata* Ellis & Halts/*Fusarium oxysporum*):** In some pomegranate orchards, the few branches start dying back, the foliage turn sickly, dropping down turn yellow and shed off. After some time, the affected branches become dry and dead. In severe form the whole mature tree dies all of a sudden even when bearing fruits. For the management of the disease, orchards should be kept clean and managed scientifically with proper plant care approach. Branches showing die back symptoms and dried ones should be pruned and destroyed. Soil drenching with benomyl (Benlate 0.1%) or carbendazim (Carbendazim 0.1%) or propiconazole (Tilt 0.1%) should be practiced in plants showing symptoms and nearby apparently healthy plants also.

**Bacterial blight or Nodal blight (*Xanthomonas axonopodis* pv. *punicae*):** The symptoms of the disease are noticed on fruits, leaves, branches and stems. On fruits circular brown to black lesions develop with L or Y or star shaped crack. In advanced stage entire fruit was covered by the lesions, which lead to cracking of the fruits. On leaves water soaked pin head size lesions with yellow halo appear. In advanced stage, the spot enlarge covering entire lamina. Defoliation of infected leaves are noticed in severely affected orchards. On the node brown to black water soaked elliptical lesions near the auxiliary bud covering entire node is seen resulting into nodal blight . In the advanced stage of nodal infection, affected areas become flattened and depressed with raised edges. Cracking of nodes will lead to death of branches from the infected portion. For the management of the disease;

- Clean cultivation and orchard sanitation has been advised to prevent the disease.
- Strict quarantine measures have to be followed and infected cuttings should not be supplied or imported.
- Remove the affected fruits, leaves and twigs and destroy by burning them far away from the main field.
- Spray antibiotics, viz. Streptocycline or Bacterinol-100 @ 500ppm and Copper oxychloride (0.3%) two to three sprays at ten day interval followed by one application of Bordeaux mixture (1%).
- Prune severely affected plants up to soil level 12.5cm above ground level. Cover the pruned part with Bordeaux paste.

## MARKETING OF FRUITS AND EXPORT POTENTIAL

The fruits of Bhagwa variety produced in the country are famous for their quality. However, it is equally significant that merely increased production would not bring revolution in agricultural economy unless these productions are associated with the efficient marketing system. Therefore, marketing of fruits assumes a special significance in the pomegranate cultivation, when the economy of small and marginal farmers. No doubt, pomegranate is a profitable venture but with the rapid increase in acreage and production several issues in marketing have emerged. Existing trade of pomegranate fruits is characterized by high transportation, packing costs and malpractices in market and lack of storage facilities etc.

At global level, total trade of pomegranate is 1 to 1.2 million tons per year. Spain is the biggest exporter contributing 70 to 75 % of World's trade of pomegranate while Iran is the 2nd largest exporter shares 15 to 20 %. In spite of the largest producer in the World, India lies at 3rd position and exports 5 -10 % only (NRCP, 2015).The greatest month of commercial pomegranate production in Spain is from September to December and the demand for Indian pomegranates decreases once the pomegranates from Spain and Iran starts flooding the markets. But in India pomegranate fruits are produced throughout the year, therefore it has a comparative advantage to supply pomegranate fruits even during off-season to European countries. Netherlands, U.K., Saudi Arabia, Bangladesh, Belgium, Germany, Bahrain, Oman, Sri Lanka, Nepal, Cyprus, Qatar are some of the major countries where pomegranate fruits from India are exported. Bhagwa is one of the best varieties suitable for export purpose and it is gaining popularity among consumers in foreign markets.

## FUTURE THRUSTS

The versatile adaptability, hardy nature, low maintenance cost, steady but high yields, better keeping quality, fine table and therapeutic values and possibilities to throw the plant into rest period when irrigation potential is generally low, indicate the avenues for increasing the area under pomegranate in India.Development of new varieties/ hybrids with resistance to biotic and abiotic stresses using Marker Assisted Selection, Marker Assisted Back Cross, Marker Assisted Recurrent Selection, gene pyramiding, Genome wide selection for interested traits for crop improvement; Gene silencing technique to address the problem of bacterial blight, fruit borer, etc.; Development of complete linkage map for chromosomes of pomegranate; Understanding the population structure and evolutionary relationship of accessions;Use of genomics (structural, functional & comparative) for basic, applied & strategic research, although challenging holds the key to success if researchers come together and work in concerted mode.

## LITERATURE CONSULTED


Al-Zoreky, N.S. 2009. Antimicrobial activity of pomegranate (*Punica granatum* L.) fruit peels.*Int. J. Food Microbiol.*, 134: 244–248.

Artes, F., Gains, M.J. and Martinez, J.A. 1996. Controlled atmospheric storage of pomegranate. J. Zeitschriff. Fuer-Lensmittel-untersuchung-und-Forschung, 203(1): 33.

Babu, K.D. 2010. Floral biology of pomegranate (*Punica granatum* L.). In: Chandra,R. (Ed.), Pomegranate. *Fruit Veg. Cereal Sci. Biotechnol.*, 4(2): 45–50.

Bangar, A.R. and Kadam, J.R. 2002. Fertilizer and Water requirement of pomegranate. Bulletin of Mahatma Phule KrishiVidyapeeth, India, 415: 17-24.

Beach, M. 1980. Dwarf and other pomegranate, *Light and garden*, 17(1): 13-15.

Burt, J. 2015. Growing Pomegranates in Western Australia; WA Department of Food & Agriculture, Oct. 2007, http://www.agric.wa.gov.au/PC_92669.html?s=1001.

Chandra, R., Jadhav, V.T. and Sharma, J. 2010. Global scenario of pomegranate (*Punica granatum* L.) culture with special reference to India. In: Chandra, R. (Ed.),Pomegranate. *Fruit Veg. Cereal Sci. Biotechnol.*, 4(2): 7–18.

Dell'Agli, M., Galli, G.V., Corbett, Y., Taramelli, D., Lucantoni, L., Habluetzel, A.,Maschi, O.,Caruso, D., Giavarini, F., Romeo, S., Bhattacharya, D. and Bosisio, E. 2009.Antiplasmodial activity of *Punica granatum* L. fruit rind. *J. Ethnopharmacol.*, 125: 279–285.

Ender, P., Vural, G., Nevzat, A. 2002. Organic acids and phenolic compounds in pomegranates (*Punica granatum* L.) grown in Turkey. *J. Food Compos. Anal.*, 15: 567–575.

Grove, P. and Grove, C. 2008. Curry, Spice and All Things Nice – The What, Where and When. Grove Publications, Surrey, UK http://www.menumagazine.co.uk/book/azpomegranate.html

Hayes, W.B. 1957. Fruit Growing in India, 3rd Edn. Vanguard Press, Allahabad, India.

Hiwale, S.S. 2005. Effect of Fruit thinning on growth, yield and quality of pomegranate cv. Ganesh. Paper presented in *First Hort. Congress* at Pusa, New Delhi, 6-9 Nov. 2005.

Holland, D. and Bar-Ya'akov, I. 2008. The pomegranate: new interest in an ancient fruit. *Chron. Horti.*, 48: 12–15.

Holland, D., Hatib, K.and Bar-Ya'akov, I. 2009. Pomegranate: botany, horticulture, breeding. In: Janick, J. (Ed.), *Horticultural Reviews*, John Wiley and Sons,New Jersey, pp. 35: 127–191.

Ismail, T., Sestili, P., Akhtar, S., 2012. Pomegranate peel and fruit extracts: a review of potential anti-inflammatory and anti-infective effects. *J. Ethnopharmacol.*, 143: 397–405.

Josan, J.S., Jawanda, J.S. and Uppal, D.K. 1979. Studies on the floral biology of pomegranate. I. Sprouting of vegetative buds, ûower bud development, flowering habit, time and duration of flowering & floral morphology. *Punjab Hort. J.*, 19: 59–65.

Julie, J.M.T. 2008. Theraupatic applications of pomegranate (*Punica granatum* L.): a review. *Alter. Med. Rev.* 13: 123–144.

Kashiwada, Y., Nonaka, G.I., Nishioka, I., Chang, J.J., Lee, K.H. 1992. Antitumor agents,129. Tannins and related compounds as selective cytotoxic agents. *J. Nat. Prod.*, 55: 1033–1043.Khan, S.A., 2009. The role of pomegranate (*Punica granatum* L.) in colon cancer. *Pak. J.Pharmaceut. Sci.*, 22: 346–348.

Khan, S.A. 2009. The role of pomegranate (*Punica granatum*) in colon cancor. *Pak. J Pharm. Sci* 22: 346-348.

Levin, G.M. 2006. Pomegranate Roads: *A Soviet Botanist's Exile from Eden*, 1[st]Edn. Floreant Press, Forestville, California, pp. 15–183.

Linnaeus, C. 1753. Species Plantarum, Stockholm, Sweden, 1: 472.

Mars, M. 2000. Pomegranate plant material: Genetic resources and breeding, a review. *Opt.Mediterr. Ser. A Sem. Mediterr.*,42:55–62.

McFarlin, B.K., Strohacker, K.A., and Kueht, M.L. 2009. Pomegranate seed oil consumption during a period of high-fat feeding reduces weight gain and reduces type 2diabetes risk in CD-1 mice. *Br. J. Nutr.*, 102: 54–59.

Melgarejo Moreno, P. and Martínez Valero, R. 1992. *El Granado*. Ediciones Mundi Prensa, Madrid.

Meshram, D.T., Gorantiwar, S.D., Silva JAT, Jadhav V. T. and Chandra R. 2010. Water Management in Pomegranate. *Fruit Veg. Cereal Sci. Biotechnol.*, 4(2): 106-112.

Mori-Okamoto, J., Otawara-Hamamoto, Y., Yamato, H., Yoshimura, H. 2004.Pomegranate extract improves a depressive state and bone properties in menopausal syndrome model ovariectomized mice. *J. Ethnopharmacol.*, 92: 93–101.

Morton, J.F. 1987. Fruits of warm climate. Miani, F1.p 269-271.

Nalawadi, U.G., Farooqi, A.A.,Dasappa, M.A., Reddy, N.,Gubbaiah, G.S.,Sulikeri, and. Nalini, A.S. 1973. Studies on the floral biology of pomegranate (*Punica granatum* L.). *Mysore J. Agr. Sci.*,**7**:213–225.

Nath, V., Singh, R. S. And Das, B. 2007. Pomegranate. *In: Fruit production technology* (Eds. Yadav, P.K.). International Book Distributing Co., Lucknow, UP, pp. 307-322.

Newman, R.A., Lansky, E.P. and Block, M.L. 2007. Pomegranate: The Most Medicinal Fruit, 1[st]Edn. Basic Health Publication, Laguna Beach, CA, pp. 1–120.

NRCP, 2015. *Annual Report 2015-16.* National Research Centre on Pomegranate, Solapur, India, pp. 1-23

Pareek, O.P., Vashishtha, B.B., Nath, V. and Singh, R.S. 2000. *Shuska Keshtramein-Anar Utpadanki Unnat Taknik*, Extension bulletin-2, NRCAH, Bikaner, Pp-16.

Parmar, C. and Kaushal, M.K. 1982. *Punica granatum*L. In: Wild Fruits of Sub-Himalayan Region. Kalyani Publishers, New Delhi, India, pp. 74–77.

Patil, A.V. and Karale, A.R. 1985. Pomegranate, In: *Fruits of Tropical and Subtropical* (Bose, T.K.-Ed), Naya Prokash, Calcutta, pp. 537-548.

Phadnis, N. A.1974. Pomegranate for desert and juice. *Indian Hort*. 19(3): 9-13.

Prasanna Kumar, B. 1998. Pomegranate. In: Chattopadhyay TK (Ed) *A Text Book on Pomology*(Vol3), Kalyani publications, Ludhiana, India, pp. 161-188.

Pruthi, J.S. and Saxena, A.K.(1984. Studies in anardana (dried pomegranate seeds). *J. Food Sci. Technol.*, **21**:296-299.

Randhawa, G.S., Singh, J.P. and Malik, R.S. 1958. Fruit cracking in some trees with special reference to lemon. (*Citrus limon*). *Indian J. Hort.*, 15: 6-9

Ryall, A.L. and Pentzer, W.T. 1974. Handling, transportation and storage of fruits and vegetables. In: Fruits and Tree Nuts (Vol 2). Avi Publishing Westport, Connecticut.

Silva, J.A.T., Rana, T.S.,Narzary, D., Verma, N., Meshram, D.T. and Ranade, S.A. 2013. Pomegranate biology and biotechnology: A review. *Scientia Horticulturae,* 160: 85–107.

Singh, B.P. and Rana, R.S. 1993. Promising fruit introductions. In: *Advances in Horticulture* (Chadha, K.L. and Pareek, O.P.-Eds), Malhotra Pub, New Delhi, 1: 43-66.

Singh, N., Singh, S., and Meshram, D. 2012. Pomegranate: *In Vitro* Propagation and Biohardening. Lambert Academic Publishing, Saarbrucken, Germany, pp. 160.

Singh, R. P., Kar, P.L. and Dhuria H.S. 1978. Studies on behavior of flowering and sex expression in some pomegranate cultivars. *Plant Sciences*, 10: 29-31.

Verma, N., Mohanty, A., and Lal, A. 2010. Pomegranate genetic resources and germplasm conservation: a review. In: Chandra, R. (Ed.), Pomegranate. *Fruit Veg. Cereal Sci. Biotechnol.*,4(2): 120–125.

Wang, R.F., Liu, R.N., Ding, Y., Xiang, L. and Du, L.J. 2010. Pomegranate: Constituents, bioactivities and pharmcokinetics. In: Chatra, R. (Ed), Pomegranate, Fruit /Veg. Cereal Sci. Bioitechnol., 4, Special Issue pp: 77-87.

Waskar, D.P., Khedkar, R.M. and Garnade, V.K. 1999. Effect of post harvest treatments on shelf life and quality of pomegranate in evaporative cool chamber and ambient conditions. *J. Food. Sci. Technol.*, 36(2): 114.

Wetzstein, H.Y., Ravid, N., Wilkins, E. and Martinelli, A.P. 2011. A morphological and historological characterization of bisexual and male flower types in pomegranate. *J. Am. Soc. Horti. Sci.*, 136: 83–92.

Zukovskij, P.M. 1950. Cultivated Plants and their Wild Relatives. State Publishing House, Soviet Science, Moscow, pp. 60–61.

# 4

# The Ber (Indian Jujube)

*J. S. Bal*

## INTRODUCTION

The ber or Indian jujube (*Ziziphus mauritiana* Lamk.) in the most common fruit crop grown in arid and irrigated zones of Indian sub-continent. It is one of ancient fruits of India which is cultivated practically all over the country for its fresh consumption. Ber is a tropical as well as subtropical fruit and can be successfully cultivated even in the most marginal eco-system of tropics and subtropics. Ber is a hardy fruit tree and is best known for growing under adverse soil and climatic conditions. It can thrive well even in drought, water-logged and salinity conditions. It is common notion among the people that ber is a poor man's fruit. It is one of the most nutritious fruit compared to other ones. The grafted varieties of ber fetch even higher price in the market than several other fruits. Hence, ber stands no longer the poor man's fruit. Now, ber growing has received a great attention and preferred as a commercial fruit crop in North Indian states like Punjab, Haryana, Rajasthan and Uttar Pradesh on account of its potential for higher yield and good economic return to the growers. Moreover, the tree has less water need in the summer months during the period of dormancy, thus considered an ideal fruit for cultivation in the arid and semi-arid regions of North India. Apart from ber, no other fruit trees grow wild so extensively in this region, thus it is becoming popular among the growers.

## NUTRITIVE AND CULTURAL SIGNIFICANCE

**Nutritive Significance**: The ber is palatable and delicious fruit and has high nutritive value as compared to many other fruits. Thus, it is rated as one of the most nutritious fruit. The fruit contains total soluble solids 13 to 20 per cent and acidity 0.20 to 0.60 per cent in different varieties at ripe stage. Ber is a rich source of vitamin C next to aonla and guava fruits. The ripe fruits of ber cv.

Umran contains 110-120 mg ascorbic acid per 100 g of fruit. The fruit contain vitamin A (β-carotene) 81 μg per 100 of fruit. The ripe fruit contains sucrose 5.6 per cent, glucose 1.5 per cent and fructose 2.1 per cent. The fruit pulp of cv. Umran contains total sugars 10.6 per cent, starch 1 per cent and total carbohydrates 13.6 per cent. The fruit is a good source of minerals and protein. At peak maturity fruit contains 0.030 per cent calcium, 0.036 per cent phosphorus, 1.14 per cent iron and 1 per cent protein. The fruit of ber cv. Umran contains amino acids like asparagine, aspartic acid, arginine, glutamic acid, glycine, serine, threonine, α-alanine, valine, methionine, leucine and isoleucine.Ber fruits are higher in calorific value.

**Uses:** The ber fruit is mostly eaten fresh but can be utilised for the preparation of several delicious products. The fruit of the choice grafted cultivars is very tasty, juicy in some varieties, sweet and rich in flavour. The fruit is easily digestible and can act as a mild laxative. The fruit is dried in the sun for off season use. It is a common practice in north India as the fruit is easily available locally. The excellent ber Chhuharas can be prepared and the fruits of Umran variety are highly suitable for sun drying and dehydration. Fully mature unripe fruits of ber can be used for making murabba, candy, chutney and pickle. The juicy ber varieties like Sanaur Selections can be converted into pulp and serve as base material for squash, nectar and RTS beverage. The fully mature ber fruits can be canned in sugar syrup. Jelly can be prepared from the ripe fruit which retain vitamin C in good amount for long period. Ber candy can be best consumed during off season. The fruits of Umran cultivar can be used for the preparation of excellent candy. Ber Chhuharas can be prepared from different grafted varieties when harvested at early ripe stage. The fruits are dried in a mechanical dryer at 65±2$^{0}$C for 31 hours. Organoleptically fruits of Sanaur varieties were acceptable on 6 months storage as ber Chhuharas.

The ber tree is chiefly utilized for its edible fruits. The tree is one of the principle host plant for rearing lac insects to prepare fine quality of silk. The tree is also used for carpentry and building wood, fodder for goats, hedges, fences and fuel. The different plant parts and fruits are used for medicinal purposes. The bark is considered to be a good remedy for diarrhoea. The roots are used as decoction in fever and its powder is applied to ulcers and old wounds. Ber fruit has been used in traditional medicines as an emollient expectorant, coolent, anodyne and tonic. During pregnancy period, the fruit is given to the women to receive abdominal pains. The fruit is useful for the patients of open hurt surgery for speedy recovery and quick healing. Its fruit is known for lowering blood pressure and cholesterol level. Certain triterpenoid acids isolated from the ber fruits are also considered useful in curing HIV virus and fighting against cancer.

**Cultural Significance:** The ber is mentioned in the earliest Sanskrit literature thus has a long historic background. In different Indian languages ber is referred also with many other names. In Valmiki's Ramayana use of ber is mentioned in Pind dan (a ritual after death). The ber is intimately connected with the folklore of the people in Northern India. In Punjabi culture, many typical folk songs citied ber tree and fruit have heart touching feelings. Rural people are well connected with these songs and relish them in social ceremonies. The abundant presence of ber trees in wild state in India proves its ancient domestication.

The ber tree is considered the most sacred one in Sikh religion. The Sikh Gurus preferred to plant ber trees in Holy religious places (Gurdwaras). The old historic ber trees in the premises of Golden Temple, Amritsar are the sign of rich heritage. The oldest ber tree of India now about 450 years old is Ber Baba Budha Sahib which is located near a holy pond (Sarowar) in the Parkarma (periphery). This ber tree is associated with highly respected Saint Baba Budha Ji, hence it is called Ber Baba Budha Sahib. The other oldest ber plant located in the premises is Dukh Bhanjani Ber.Under the periphery of this tree, the leprotic husband of Bibi Rajni took both in the holy Sarowar and became healthy. Therefore, people from all casts and creeds take bath in holy Sarowar under this tree and get relieved from ill-health problems. Lachhi ber is another old ber plant located just near the main gate of Golden Temple known as Darshani Deori. The tree bear round shaped fruits hence called Lachhi ber. Guru Nanak Dev Ji, the first Sikh Guru also planted ber plant in the Gurudwara at Sultanpur Lodhi, Kapurthala (Punjab) which is now known as Ber Sahib. Guru Nanak Dev Ji also planted another ber plant at Achal Sahib Gurdwara near Batala (Punjab). Sikh Gurus felt that the ber tree is very useful and considered it a mother tree because it gives shade, shelter and sweet fruits during summer months which are highly beneficial for human health. The health of all these plants is considerably improved after giving consistent care by the Scientists of Punjab Agricultural University, Ludhiana. A healthy environment needs to be provided to all sacred ber trees to preserve the rich heritage.

## ORIGIN, HISTORY AND DISTRIBUTION

The ber is said to be indigenous to the area stretching from India to South-Western China and Malaysia. It has been closely associated with the cultural ethos of India since the ancient times. The antiquity of ber in India is evident from mentioning by several references in the ancient Sanskrit literature. The ber has special significance in the folklore of the Punjabi community of North India particularly in the Punjab. The commercial cultivation of ber is said to have started when a Muslim grower won prize by presenting delicious ber fruits from a budded plant to Raja Raghoji Bhonsle-II. It is believed that the ber tree from Garhwal (U.P.) has been introduced more generally into other parts of India. Then it spread gradually from India to South,South-East Asia, Zanizibar and the East-African coast and then further into the African continent. The process of spreading ber have started only a few hundred years ago.It is found in many tropical countries of both hemispheres often in naturalized state.

Most of the *Ziziphus* species are distributed in tropical and subtropical regions. Ber is distributed between 30$^0$S and 30$^0$N latitude mainly in India, Pakistan, Bangladesh, Nepal, Bhutan, Sri Lanka, Central and Southern Africa and North region of Australia. Ber plantation on small or large scale exist in China, Afghanistan, Russia, Iran, Iraq, United Arab Emirates (UAE), Syria, Italy, France, Spain, Burma and USA.China ranks first among ber growing countries in the world. The total growing areas under ber (jujube) in China is 20 lakh ha with total production of 60 lakh tonnes. Recently commercial cultivation of Chinese jujube (*Ziziphus jujuba* Mill.) is emerging in Australia, USA, Italy and Israel.

India ranks second among ber growing countries in the world after China with an area approximately 47000 hectares. The annual production is estimated 4.31 lakh tonnes, thus average productivity is 9.2 metric tonnes per hectare. In India, the important provinces for ber cultivation are Maharashtra, Gujarat, Uttar Pradesh, Tamil Nadu, West Bengal, Haryana, Rajasthan, Madhya Pradesh, Punjab, Karnataka, Bihar, Chhattisgarh, Andhra Pradesh and Jammu & Kashmir. In Gujarat, ber orchards are found around Banaskantha, Sabarmati, Bhavnagar, Surendranagar, Patan, Ahmedabad, Bharuch, Vododra, Sabarkantha and Mehsana. In Haryana, Hisar, Rohtak, Jind, Panipat, Mohindergarh and Gurgaon districts are famous for ber cultivation. In Rajasthan, ber orcahrds are spread around Bharatpur, Jaipur and Jodhpur. In Punjab, Sangrur, Bathinda, Ferozepur, Ludhiana, Patiala and Muktsar districts are the most famous for ber cultivation. In Uttar Pradesh, ber orchards are found around Varanasi, Aligarh, Faizabad and Agra. The districts of Murshidabad, Malda, Bankura and Birbhum are famous for ber growing in West Bengal. In Tamil Nadu its cultivation is done in Tirunelveli, Ramanathapuram, Dharamapuri and Salem districts. In Karnataka, Bijapur, Bellary, Gulbarga, Belgaum, Raichur and Bidar are quite popular for ber cultivation.

## TAXONOMICAL AND BOTANICAL DESCRIPTION

The ber belongs to the genus *Ziziphus* of the family *Rhamnaceae*, order *Rhamnales* and class *Dicotyledonae*. The family Rhamnaceae has about 50 genera and more than 600 species. The genus *Ziziphus* has about 170 species throughout the world. About 40 species of plants are found in tropical and subtropical regions of Northern hemisphere. Some species of the family are cultivated either for edible fruits or for ornamental purposes. The status of important species is given below for general understanding.

| S.No. | Scientific Name | Common Name | Generals status |
|---|---|---|---|
| 1. | *Ziziphus mauritiana* Lamk. | Ber or Indian Jujube | Major cultivated species |
| 2. | *Ziziphus jujuba* Mill. | Chinese Jujube | Major cultivated species |
| 3. | *Ziziphus spina-christi* (L.) Desf. | Christ's Thorn Jujube | Minor cultivated species |
| 4. | *Ziziphus lotus* (L.) Lamk. | Lotus Jujube | Minor cultivated species |
| 5. | *Ziziphus sativa* Gaertn. | The common Jujube | Wild growing species |
| 6. | *Ziziphus xylopyrus* (Retz.) Willd. | The goth ber | Wild growing species |
| 7. | *Ziziphus oenoplia* (L.) Mill. | The makoh ber | Wild growing species |
| 8. | *Ziziphus mistol* Griseb | The Mistol | Wild growing species |
| 9. | *Ziziphus joazeiro* Mart. | Jua | Wild growing species |
| 10. | *Ziziphus rugosa* Lamk. | The suran ber | Wild growing species |
| 11. | *Ziziphus xylocarpa* Willd. | The kat ber | Wild growing species |
| 12. | *Ziziphus funiculosa* Buch.-Ham ex Wall. | Bons-boguri | Wild growing species |
| 13. | *Ziziphus glabrata* Heyne ex Roth | Karkala | Wild growing species |
| 14. | *Ziziphus oxyphylla* Edgew. | Kokan ber | Wild growing species |
| 15. | *Ziziphus vulgaris* Lamk. | Chinese date palm | Wild growing species |
| 16. | *Ziziphus nummularia* (Burm. f.) Wight & Arn. (Syn. *Ziziphus rotundifolia*) | The Mallah ber | Wild growing species |
| 17. | *Ziziphus abyssinica* Hochst. Ex ARich. | Catch thorn | Wild growing species |
| 18. | *Ziziphus mucronata* Willd. | Buffalo thorn | Wild growing species |

**The ber or Indian Jujube *(Z. mauritiana)*** is commonly grown in India and widely distributed in warm countries. The tree is summer deciduous in nature.It is a thorny shrub or small tree upto 10-12 metres in height. Its leaves are simple, alternate, lengthwise 8cm and breadthwise 5cm, densely tomentose underside and has stipular thorns.Inflorescence is axillary cyme and sub-capitate in the leaf axil. Flowers are greenish cream in colour, small in size (0.8cm), bisexual, actinomorph, pentamerous and acrid-scented. Petals 5 arising between and smaller than the calyx lobes. Calyx 5 lobed, shortly tubular and lobes valvate. Stamen 5, opposite to the petals. Style thrum, anthors 2-locular opening lengthwise and stigma bilobed. Ovary is superior to half inferior and 2-locular. Fruit drupe, ellipsoid to subglobose, medium to large in size (2-5 cm long, 2-3.5 cm broad) and greenish yellow to golden yellow in colour. Epicarp and mesocarp is the edible portion.

**The Chinese Jujube *(Z. jujuba)*** grows in China, Southern Europe and South and East Asia. The tree is winter deciduous in nature. Tree is 9-10 metres in height and spiny. The branchlets are often slender and frequently give appearance of pinnate leaves. Leaves are glabrous at the underside, serrate, 2.5-5.0 cm long oblong, ovate to lanceolate. Fruits are dark red or brown in colour.

**Jhar ber or Mallah ber** ***(Z. nummularia)*** is a small shrub and found growing in Wastelands of Rajasthan, North-Western India and Andhra Pradesh. Fruits has very little pulp, poor in quality and has peculiar acrid taste. The species is not commercially important, however, its seedlings are used as a rootstock for propagating ber varieties in different regions of India.

**Chirst's Thron Jujube** ***(Z. spina-christi)*** grows as a commercial crop in Iran and Iraq.

## CLIMATIC AND SOIL ADAPTABILITY

### Climatic Adaptability

The ber tree grows well under varying climatic conditions. It is a hardy fruit tree and can grow successfully even under adverse climatic conditions where most other fruit trees fail to grow well. The cultivations of ber can be done in both subtropical and tropical climate. The ber tree can be grown in areas from sea level up to an elevation of 1000 meters. It does not bear well beyond this limit. Ber tree relish hot and dry climate for successful growing but the tree need adequate watering during the fruiting season. The trees can withstand high summer temperature upto 42$^0$C and can tolerate temperature up to 45$^0$C. The fruit-set is adversely affected when temperature goes above 35$^0$C. Temperature below freezing is injurious to the young twigs and developing fruits and result in substantial crop loss. The ber orchards are found growing in areas having a minimum temperature of 4$^0$ to 12$^0$C and occasionally reaching as low as -2$^0$C for short period. The ber trees can very well withstand hot and dry weather during summer months of May-June as the trees enters into dormancy by shedding its leaves.With the advent of rainy season new growth starts in July and continues till mid November when it is inhabited with the onset of cold weather.

Ber orchards grow and yield well in the regions receiving annual rainfall over 400mm. Excessive atmospheric humidity is considered a limiting factor for satisfactory fruiting. The trees produced better quality fruits in terms of fruit size and TSS under high atmospheric vapour pressure deficit conditions. Moreover, pests and diseases incidence also reported lower in dry regions.

### Soil Adaptability

Ber cultivation can be done on wide range of soils. Deep sandy loam soils with neutral or slightly alkaline reaction and good drainage are more desirable for good and healthy tree growth and better yield. Ber can be grown on marginal lands or even those which are considered unsuitable for growing other crops. The tree develops a deep tap root system and as such adapts itself to a variety of soils. Its tree has ability to thrive well under adverse conditions of salinity drought and water logging.

The ber trees are tolerant to salinity level upto 6 $dSm^{-1}$. It can flourish even in soils with pH up to 9.2. The ber fruits are considered major source of nutrient removal from soil and tree removes significant amount of nutrients during single cropping season. Therefore, judicious manuring and fertilization to the soil are necessary to replace these losses and to correct deficiencies.

## RECOMMENDED AND POPULAR CULTIVARS

The choice of suitable varieties is of paramount importance for successful ber cultivation. Maximum number of ber cultivar have been reported to be available in China. More than 100 ber cultivars are grown in India which have been developed by selection in different regions. Some of the cultivars are reputed for the specific traits they possess and few of them have been classified as early, mid and late on the basis of their maturity. Some cultivars are preferred for their early or late ripening quality, while some are liked for the amount and quality of fruit pulp. Certain cultivars are considered best when consumed at hard maturity stage and still others are prized for candying and drying.

The cultivars grown in ber growing states are as follows:

| | |
|---|---|
| Gujarat | Umran (Chameli), Randeri, Katha, Mehroon, Kaithli, Banarsi Karaka, Gola |
| Maharashtra | Umran, Meharun, Chhuhara, Badami, Darakhi, Gola, Shamber |
| Rajasthan | Gola, Goma Kirti, Mundia, Seb, Jogia, Umran, Banarsi Karaka, Kathli, Maharwali, Tikadi, Thar Sevika, Thar Bhubhraj |
| Haryana | Gola, Umran, Kaithli, Banarsi Karaka, Sandhura Narnaul, Sanaur-2, Seo |
| Punjab | Umran, Sanaur-2, Wallaiti, Kaithli, Zg-2, Gola, Dandan |
| Madhya Pradesh | Tiloki, Narma, Gol, Badam, Bachcha |
| Andhra Pradesh | Banarsi, Kaithli, Dudhia |
| West Bengal | Narikeli, Banarsi, Baruipur, Ghughudanga |
| Bihar | Banarsi, Nagpuri, Thornless |
| Karnataka | Umran, Banarsi, Gola, Sanauri |

### Salient Features of Commercial Ber Cultivars

**Umran:** This variety has been found to be best and is especially recommended for commercial cultivation on a large scale in Punjab, Haryana, Rajasthan, Gujarat and Maharashtra. The variety is native to Rajasthan and introduced from Alwar state. The fruits of Umran are tough and firm and can stand transportation better for long distances without spoilage. The fruit of Umran is large in size oval in shape and has round apex. It has an attractive golden yellow colour which turn into chocolate brown at maturity. The fruit is sweet with 14-19% TSS, 0.22% acidity. It has pleasant flavour and excellent dessert quality. Umran is a prolific cropping cultivar and yield 150-200 kg of fruits per tree. It is a late ripening cultivar and ripens from second fortnight of March to mid-April. It has good keeping quality and is suitable for making candy. This variety is susceptible to powdery mildew disease.

**Sanaur-2**: As the name denotes this is a selection from Sanaur a small town near Patiala, which is known for ber growing. The tree is spreading and semi vigorous. The fruit is extremely attractive in appearance with fine quality, thus recommended for commercial cultivation. The fruit is large and oblong with a roundish apex. The fruits attain a light yellow colour at ripening. The pulp is sweet with a typical flavour having 18-19 per cent TSS. Being a prolific bearing cultivar, it yields about 150 kg fruits per tree. It is mid- season cultivar and ripens during second fortnight of March. It is fairly resistant to powdery mildew disease. The cultivar is recommended for growing in Kandi area.

**Wallaiti:** This cultivar was selected from Sanaur area of Patiala district in Punjab. On account of its exotic appearance, it is called 'Wallaiti. The fruit is medium in size, oval in shape, skin smooth and attain light golden yellow to golden yellow colour at maturity. The pulp is soft, sweet with 13.8-15 per cent TSS. It is an early cultivar, ripens during first fortnight of March. The average yield 114 kg per tree. It is moderately susceptible to powdery mildew disease. The cultivar is excellent for growing in kitchen gardening.

**Gola:** It is an early variety which starts ripening in second fortnight of January and continues up to end of February. The variety is popularly grown in Haryana, Rajasthan, Gujarat and Punjab. The fruits are round in shape, medium in size and attain golden yellow colour at maturity. The pulp is sweet and soft having TSS 16-20 per cent. The fruit quality is excellent but cannot stand for long transportation. The average yield is 80 kg fruits per tree.

**Kaithli:** This cultivar is a selection from Kaithal in Haryana. The tree is erect and vigorous growing in nature. Its fruit finds favour on account of its sweet taste, soft pulp with 14-16% TSS and nice flavour. The fruit is medium in size, oval in shape and has a tapering apex. The bearing capacity is average and yield 120 kg fruits per tree. The fruits of this cultivar does not stand long distant transportation. The cultivar is susceptible to powdery mildew disease. It is recommended for growing for local markets.

**Banarsi Karaka:** The tree of this cultivar are tall and vigorous. It is very popular mid- season cultivar of Banaras area of Uttar Pradesh. The fruits are globose, oblong to long in shape with tapering to pointed stylar end. The average fruit weight is 40-50 g. It develops greenish yellow to yellow colour at maturity.

**Dandan:** This cultivar is grown commonly in Patiala district of Punjab and in Hansi and Jind areas of Haryana. The fruit has a pointed apex like teeth. The fruit is medium in size, oval in shape with round base, smooth surface and has an attractive yellow colour at maturity. The variety is suitable for diabetic patients as its TSS content ranged from 9 to 11.2 per cent. It has 8.8 per cent sugar, 0.24 per cent acidity and 109 mg/100g pulp vitamin C. It is mid-season cultivar ripens during second fortnight of March. The average yield is 94 kg fruits per tree. It is moderately susceptible to powdery mildew disease.

**Goma Kirti:** The variety is a clonal selection from 'Umran' which ripen about 3-4 weeks earlier than Umran. The fruit apex is broadly pointed and fruit base is grooved. Fruit surface is rigid and mature fruits are greenish yellow in colour. The pulp is creamy white with medium in texture. The fruit size is medium (length 3.60 cm and breadth 2.90cm) with average weight of 17.6g. It is an early maturing variety which is resistant to various pests and diseases by virtue of its earliness.

Chhuhara, ZG-2, Thar Sevika, Thar Bhubhraj, Darakhi, Sanaur-5 and Sandhura Narnaul are other promising varieties of ber.

### Hybridization in Ber

Hybridization work in ber was undertaken in different Agricultural Universities in India to develop a suitable powdery mildew resistant hybrid. The existence of polyploidy and entomophily as well as a range of self and cross incompatibilities in *Ziziphus* species has resulted in wide hybridization. Some genotypes are widely accepted on account of their productivity and quality traits but such genotypes lack certain important traits without which desired result are not achieved.

At Rahuri, selection of Darakhi-1 and Darakhi-2 resistant to powdery mildew disease was made. Kakrola Gola and CAZRI Gola selections were made from cultivar Gola in Rajasthan. Goma Kirti a clone developed from 'Umran' mature earlier than this cultivar. Thar Sevika and Thar Bhubhraj are the varieties developed from Seb x Katha. At Ludhiana, F1 hybrid between Chinese and Sanaur-2 develop light red tinge at the tip which is quite appealing in appearance.

## PROPAGATION TECHNOLOGY AND ROOTSTOCKS

The propagation in ber by budding is the most successful method. The shield or T-budding method is employed because it is easier to perform. Budding is done during March-April or June to September when there is proper flow of sap in the stock. The bud take efficiency in ber is high (85%) in the month of June. The higher budding success is also noted in the second fortnight of August and during September. Ring budding is another method which can also be used for multiplication of ber plants. The budding through this method should be done during June-July when the new growth starts. The mother plant from which the scion bud sticks are obtained for budding should be healthy, vigorous and free from insect-pests and diseases and should be true-to-type. Two to three months old shoots with plump buds should be selected.

Protocol for clonal propagation of ber has been developed by various research workers. Different explants such as nodal segments, seedling and root suckers can be used for clonal propagation. Application of biotechnology has ample scope in solving problems like powdery mildew and black leaf spot diseases and attack of fruit fly.

## Selection of Rootstock

Rootstock evaluation has been attempted and done at PAU Ludhiana. The commercial cultivar 'Urman' was raised on twelve rootstocks belonging to the genus *Ziziphus*. It was concluded that ber plants should be budded on *Z.mauritiana* (Elongated Dehradun) for greater plant vigour and higher yield at 7.5x7.5m planting. The highest yield (146 kg/tree) was recorded on this rootstock. *Z. Mauritiana* (Coimbatore) produced semi-vigorous plants of 'Umran' scion which are suitable for closer spacing of 6x6m. Umran tree budded on this rootstock showed a spread of 6m as compared to vigorous rootstocks like *Katha* Punjab and Elongated Dehradun. Thus about 50% more plants per unit area can be accommodated with over 20% increase in yield of equally good quality fruits.

***Katha* ber *(Z. mauritiana)*** is generally used for raising rootstock seedlings in North India. The seeds of *Katha* ber are easily available from the wild growing tree and possess the qualities of good rootstock. Jhar ber or Mallah ber *(Z. nummularia)* can also be used for raising of rootstock. It is a slow growing and seedlings become buddable after longer period then the seedlings of *Katha* ber and Elongated Dehradun. The cultivar 'Umran' raised on *Z. nummularia* from Punjab and Rajasthan strains can be successfully grown under alkaline soil conditions. Umran cultivar raised on *Z. joazeiro* and *Z. xylopyrus* rootstocks produce greater diameter and resulted in inverted bottleneck union and on *Z. Jujuba* develop typical proliferation of cambium at the bud union.

At IARI New Delhi, the rootstock Jhar ber showed incompatibility with inverted bottleneck symptoms budded with 'Gola' cultivar. *Z. mauritiana* Ecotype 291 and *Z. mauritiana* Ecotype-Assam Gauhati with cvs. Banarsi Karaka, Ponda and Gola were compatible with perfect union. *Z. rotundifolia* and *Z. mauritiana* were found suitable rootstocks for ber cv. Gola under rainfed condition of arid region of India. In Rajasthan, ber tree on Barodi (*Z. mauritiana* var. *rotundifolia*) produced vigorous growth and higher fruit yield. Plants on *Z. nummularia* resulted an inverted bottleneck graft union.

**Raising of Rootstock:** The germination of ber seeds is quite a difficult process due to stony nature of endocarp which contains the seed. Seed stones of *Katha* ber *(Z. mauritiana)* or Elongated Dehradun collected from dropped fruits contains 50-70 per cent non-viable seeds. Therefore, seeds should be dipped into a salt solution of 17% concentration for 24 hours before sowing for better germination. The ber seeds can be sown by cracking the hard shell. They germinated rapidly after about 8-10 days. The seed should always be collected from healthy and vigorous growing ber tree to get best rootstock material. The ber stones are sown during March-April after fresh extraction from the fruits in a well-prepared nursery field at a distance of 15cm within rows and 30 cm between rows. Germination starts after about 3-4 weeks and seedlings attain buddable size of lead pencil by August, while the rest are ready for budding by next April.

The ber seeds can also be sown in polythene bags (30x22 cm size) in April for better transportation of budded plants. The ber plants can also be raised in the polythene tubes (25x10cm) of 300-gauge thickness. The seeds are sown in the polythene tubes filled with well-prepared compost in the first fortnight of April. After that, the tubes should be protected from direct exposure of sun. Three months old seedlings are budded in July and plants become ready for transplanting in August-September or next spring season.

## PLANT AND FRUIT PHYSIOLOGY

The ber tree is summer deciduous in nature under the subtropical climatic conditions of India. The tree develops sylleptic branching pattern. Polymorphism is a characteristic feature in the proleptic branches which are produced after annual pruning. They produce vigorous shoots with sylleptic branches, normal shoots without branches and spur type branches. It sheds its leaves during the hot summer months of May-June and enter into dormancy. The tree put forth new growth with the advent of rainy season in July and comes into flowering by the end of September. The flowering period lasts for about two and a half months from September to November. The fruit is ready for harvest in the months of March-April. In Rajasthan, flowering starts in July and continue upto September. In South India under tropical climate flowering continue throughout the extended summer i.e. May to September. Flower buds appear both on current season growth and on one year old mature shoots. The number of flowers per cluster varies from 16 to 28. Bud development period is divided into eight stages and takes 21-22 days for passing through various stages. The pollen viability ranged from 87 to 91 per cent and pollen germinability from 36 to 48 per cent in different ber cultivars. Peak receptivity in stigma is found on the day of anthesis. Ber did not set any fruit by self-pollination, thereby showing self-incompatibility.

The fruit setting in ber starts in second week of October and it continues up to mid November. The growth of ber fruit is most active during the first six weeks and the last twelve weeks. During the middle six weeks period, the growth is very slow. The increase in fruit length is faster than the increase in diameter during the first phase, whereas in the third phase the fruit increase in diameter than in length. The fruits of Umran cultivar reached in ripe stage in 190 days and of Sanaur-2 in 180 days after first set. The fruit growth showed three distinct phases and followed a pattern of 'Double Sigmoid' curve. The fruits of ber cv. Umran contained sucrose, glucose and fructose in approximate ratio of 3:1:1. The chlorophyll content of ber peel reduced to the minimum as the fruit reached maturity. The ber has a somatic chromosome number of 2n=48. This might indicate polyploidy and may be tetraploid.The Chinese jujube has chromosome number 2n=24 and has polyploidy variants with 2n=48 and 2n=96.

## PLANNING AND PLANTING

**Planting:** The ber plants are planted generally twice in a year first during spring season (February-March) and again during rainy season (August-September). However, the latter season of planting gives a better success. The plants are dug out from the nursery with good-sized earth balls so that their root system is not much disturbed. The ber plants can also be transplanted bare-rooted (without earth ball) from mid January to mid February in North Indian conditions with equal success. Such plants are difoliated just before lifting them from the nursery. These plants are also headed back at a height of 60-75 cm from ground level.

Before planting, the orchard land should be properly laid out, cleaned and leveled with Laser land leveler to save the irrigation water and restrict the soil borne disease. One meter deep and one meter wide round pits should be dug for each plant. Refill the pits with a mixture of top soil, about 20kg well rotten farmyard manure and 1 kg superphosphate. The each pit acid 15 ml of chlorpyriphos 20EC mixed in about 2kg soil against white ants. The refilled pits should be watered a few days before planting the trees.

**Planting density:** Planting density in ber trees varies from place to place. It depends upon agro-climatic conditions, growing habit of cultivar, rootstock used, irrigation facilities and pruning practices followed. However, the most important aspect of successful ber growing is to provide optimum spacing at planting. It is desirable that optimum spacing should be given between rows and plants so the ber trees may grow and bear crop properly. The grafted ber trees are spreading in habit and grows into a big trees. The tree require proper spacing for its healthy growth and fruiting. To obtain good income, ber plants should be planted 7.5 m apart in square system thus accommodates 180 trees in a ha. The ber plantation under rainfed arid conditions can be made at 6x6m distance. The income of ber crop can be enhanced by employing the semi-vigorous rootstock than the usual plantation. Therefore, the ber plants on *Ziziphus mauritiana* Coimbatore rootstock can be planted at a distance of 6x6m in square system with 275 plants per hectare.

However it will be difficult to suggest exact spacing for ber trees which will suits to every locality or soil conditions. The spacing following ber in different region in given below in table.

| Region | Spacing (m) | No. of plants (per ha) | Remarks |
|---|---|---|---|
| North Indian | 7.5x7.5 | 180 | Planting on vigorous rootstock |
| | 6.0x6.0 | 275 | Planting on semi-vigorous rootstock |
| Arid North-West India | 6-7x6-7 | 275-200 | |
| Semi-arid regions | 7-8x7-8 | 200-156 | |
| | 7.2-9.6x7.2-9.6 | 190-105 | |
| | 11-12x11-12 | 82-70 | |

## SOIL CULTURAL PRACTICES TECHNOLOGY

### Water Need

The transplanting of ber is usually done during monsoon season (July to September) under semi-arid agro-cliamatic conditions of North-West India. The adequate soil moisture availability is the key factor in initial establishment of ber plants. The little depletion of soil moisture may cause high mortality and poor growth of plants. The frequent irrigation after 4-5 days for first two months after transplanting of ber is necessary. The water supply is the single most important factor determining the productivity and quality of ber fruits in dry irrigated areas.

Application of 8-11 irrigations on the basis of climatological approach at CPE (Cumulative pan evaporation) 75mm during the entire cropping season are essential for young ber plants for initial establishment and better production. The better absorption and utilization of source-sink relationship leads to maximum yield from ber plants irrigated at CPE 75mm. The development of ber fruit takes place from October to February, therefore irrigation is essential during this period at interval of 3 to 4 weeks depending upon the weather. The fruit becomes large, quality improved and fruit shedding is minimized by irrigating the trees during fruit development period. However, if no irrigation is applied during this period, trees will continue to bear. The yield is reduced because of heavy fruit drop and results into smaller sized poor quality fruits. Irrigation should be stopped in second fortnight of March, as fruits on the branches lying on the ground get damaged and their ripening is delayed.

The harvesting of ber fruit is over by middle of April and trees become dormant in May-June by shedding their leaves. At this time, they need little or no irrigation. With the application of irrigation during the dormant period, the trees would continue to put forth a growth haphazardly which is unwanted. There is sufficient rain during July to September and the ber trees produce sufficient fresh growth. The trees come into flowering during second fortnight of September and continue to flower in October. Light irrigation is recommended during flowering period. Heavy irrigation needs to be avoided at flowering which leads to flower drop and subsequently fruit yield is adversely affected.

## Nutritional Need

Proper nutrition of ber tree is necessary to get good crop over the years. Application of manures and fertilizer annually is essential to replenish the nutrient removal by the tree through fruit harvest and pruning in addition to losses from the soil. It is reported that a full grown ber tree of Umran cultivar removed total of 142-191g N; 57-87g P and 467-684g K during one growing season.

Ber tree bears on the fresh wood which sprout during June to September, thus manuring during July helps the plant to put out the maximum growth. The fruit becomes large and attractive and get decent price in the market. Farmyard manure at 20kg and nitrogen at 100g per tree should be applied to one year old ber plant. Similar amount of farmyard manure and nitrogen should be increased every year up to the age of five years. The quantity of farmyard manure and nitrogen should be stabilized at 100kg and 500g, respectively after the age of five years. Farmyard manure to be applied in May-June. Half of the nitrogen may be applied during rainy season (July-August) and the other half at the time of fruit-set (October-November). The fertilizer should be evenly spread in the tree basins up to periphery. Light hoeing with spade or khurpa should be done after adding the fertilizers to the basins. While doing hoeing mix it thoroughly with the soil.

In Haryana state, urea and superphosphate each @1250g/plant to full grown Umran tree has been recommended. Urea is given in two split doses, half in

July and second half during November. Superphosphate is applied in June-July to enhance the yield. Under Punjab conditions, the higher yield have been obtained by application of 400g nitrogen, 100g phosphorus and 200g potassium. To full grown trees of cv. Gola, application of 1kg nitrogen, 1kg phosphorus at 150g potassium has been suggested.

**Leaf Sampling:** To determine nutrient status in ber leaf, leaf sampling techniques should be adopted. Leaf samples in ber are collected during November to January. Five to seven months old two leaves from middle of shoot are collected. Four to eight leaves are picked from each direction at working height (1-2m) removing one leaf per shoot. Cover 10-12 per cent trees from the block for properly diagnosing the nutrient status. The leaf samples should be collected in polythene bags and put the bags in ice-box. Send the samples immediately to leaf analysis laboratory.

## Weed Control

Excessive weed growth is a serious problem in ber cultivation. Tropical and sub-tropical climate is very conducive for the rapid multiplication of weed in the ber orchards. Manual weeding is very expensive and time consuming while the mechanical inter culture is not much convenient in the ber orchards. Hence, the use of herbicides seems to be the only efficient, convenient and more economical method of controlling weeds in ber orchard.

In ber orchards, cultivation of soil helps to kill the weeds like baru, kahi, motha etc. which propagate negatively through rhizomes/bulbs. Without cultivation it is indeed very difficult to eradicate such obnoxious weeds. Shallow cultivation should be carried out in ber orchards to avoid any damage to fibrous roots which are mostly in the upper soil. After pruning in May-June, one summer ploughing is helpful in exposing the harmful insect- pests and fungal organisms. Similarly, ploughing at the time of application of manure (July-August) and in September-October proves useful in checking the growth of weeds.

Research on weed management in ber orchard has revealed that pre-emergence application of Hexuron 80WP (diuron) at 3kg per hectare can be made during the first fortnight of August after making the field free from growing weeds and stubles. Glycel 41SL (Glyphosate) ate 3*l* per hectare or gramoxone 24 SC (paraquat) at 3*l* hectare as post emergence are very effective when the weeds are growing actively preferably before weeds flower and attain the height of 15-20cm. Herbicides should be dissolved in 500 litres of water per hectare to give complete coverage of weeds/field.

## Inter Culture

Inter cropping in ber orchards is a useful practice during the initial pre-bearing period. The ber tree begins to bear after two three years of its planting in the orchard. To develop proper frame work of the tree, no fruit should be taken at least for the first two three years. Ber trees cover the entire inter row spaces after about five years of planting. Therefore, inter cropping can be successfully practiced in the vacant land in young orchard during the first three four years.

In ber orchards, only leguminous crops of short stature like gram, moong and mash can be grown in the initial years to earn some income from the vacant space. These crops also enrich the soil by fixing atmospheric nitrogen. It is advisable that the exhaustive and tall growing crops should not be raised in the orchard as they deplete the soil of its nutrients to a greater extent and compete for light with the trees. The ber trees and intercrops should be given manures and fertilizers, irrigation and plant protection measures independently as per their requirement.

## Mulching in Ber Orchards

The ber tree is xerophytic in nature. It is very hardy plant due to its deep tap root system and needs comparatively less care and irrigation after establishment in the orchards. But, ber plants in arid region face a water deficiency problem and suffer severely due to weed competition during rainy season. To overcome this problem, the use of different kinds of mulches like paddy straws, dry grass, other organic residue and polythene prove useful to prevent loss of water evaporation, prevention of soil erosion and control of weed growth. Mulching has been found to be an effective method in maintaining the growth and increasing the yield of ber trees. Moreover, different mulching practices are known to regulate soil hydrothermal regime and ultimately reduce the weed growth.

Black polythene is the ideal choice of mulch for ber orchards.Paddy straw is also found to be the best natural mulch for successful cultivation of ber crop. Black polythene, black polythene + glyphosate @1*l*/ha and black polythene + gramoxone @1*l*/ha reduced the weed population by cent per cent in ber orchard. Soil moisture content in the orchard was observed maximum under black polythene mulch. Similarly, this mulch material proved useful in fostering soil NPK availability and leaf nutrient levels in ber.

## PLANT CULTURAL PRACTICES TECHNOLOGY

Canopy management is the most important fruit plant cultural practice. Both training and pruning of ber trees are the specialized horticultural operations.

## Training of Plants

The ber plant is precocious in bearing. It start bearing within 2-3- years after planting and bear commercial crop in the fourth year. Training of the ber trees should be done in such a way that sufficient air and light must gets penetrated inside the foliage to facilitate proper coloration and development of superior quality fruits. The ber trees are best trained as per modified leader system during the initial 2-3 years for having strong framework. The young grafted plants after planting in the field should be given support with a bamboo stick to avoid breakage of the bud union and to support the main stem. Most of the commercial ber cultivars are spreading in nature, therefore, staking is absolutely essential during the first two years to train the trees properly. While training the trees, the branches up to 75cm should be removed. The laterals (4-5) most favourably located around the main stem should be selected for having proper framework. The main stem is then headed back just close to outgrowing lateral to develop modified leader system.

## Pruning of Plants

Pruning of ber trees is an important horticultural practice but it is often neglected by the growers. Ber is a summer deciduous in nature. The main objective of canopy management in ber tree is to control the tree size and to regulate the crop. This consists of removal of non-productive parts, diseased, dried inter crossing and broken branches so as to allow the light to enter the interior of the tree. Thus, regular pruning is essential to maintain vigour and productivity of the tree. By giving regular pruning over the year, the trees remain young up to thirty years. Pruning saves the fruit from being affected by the strong winds and powdery mildew disease. Pruning is a useful practice in removing the lac infested twigs as lac insect devitalize the trees.

Ber fruit is borne in the axils of leaves on the young growing shoots of the current years. Therefore, annual pruning is essential to induce a good and healthy tree growth which will provide a maximum fruit bearing area on the tree and improve fruit size and quality. Unpruned trees occupy large space in the orchard as the canopy get unnecessarily enlarged and ultimately become economically unproductive. Thinning out of the branches is also desired to avoid too much crowding and to admit sufficient sunlight and facilitate proper aeration.

**Nodal Pruning:** The pruning in ber delayed the occurrence of full bloom in trees. The fruit size and weight in cultivar Umran is achieved maximum in trees pruned by retaining 6 buds and yielded superior quality fruits. Pruning of Sanur-2 cultivar by heading back at 8 buds length of previous year's growth gives higher yield of better quality fruits.

**Amount of Pruning:** The pruning in ber appreciably improve the fruit quality. The ber trees of Umran, Sanaur-2 and ZG-2 pruned with 4 pruning levels by retaining past season's, shoots 20, 40, 60 cm and no pruning resulted slightly increased fruit-set and fruit retention with light pruning (60 cm removal). Heavy pruned trees produce significantly heavier and bigger sized fruits. In general, light annual pruning i.e. heading back of 25per cent previous season growth (branchlets, shoots etc) is desirable to obtain heavy yield, good fruit size and better quality. The lower branches should be pruned suitably to prevent them from spreading on the ground. The diseased, broken and inter crossed branches should also be thinned out. Severe pruning after every 4-5 years is recommended.

**Time of Pruning:** The ber tree enter into dormancy by shedding their leaves during summer months after harvesting the crop. The best time for taking up pruning in the second fortnight of May in North India. The early pruning advances bud sprouting and early harvest and improves fruit yield and quality. Little deviation in time of pruning from the dormancy phase results into lower yield and poor fruit quality. In Sanaur-2 pruning should be done during the third week of April when the trees are dormant. The cultivar Umran pruned on 9 May and retaining 6 buds yielded superior quality fruits. In South India, the regions having mild winter particularly in Tamil Nadu, pruning can be carried out any time from January to April. Thus, it is possible to regulate fruit maturity so that bearing occurs at the desired time. In Maharashtra, the best time for pruning is before the end of April. In Andhra Pradesh, pruning during first fortnight of April results in early flowering and early harvesting. Pruning done during second fortnight of April produce higher yield of good quality fruits.

The fruit setting from later opening flowers resulted in smaller fruits of poor quality. Selective pruning of late flowering (21 October to 5 November) branches slightly improve fruit size, weight and quality rating in ber cv. Umran. TSS and vitamin C in different ber varieties also little improved after selective pruning.

## Rejuvenation of Old Trees

Ber trees needs to be rejuvenated after the age of about 25 years. This can be done by heading back the main limbs to about 30cm during second fortnight of May. Four to five main limbs are maintained on each tree. During the rainy season, number of shoots will emerge from the stub. The first thinning of shoots should be done during fourth week of August by retaining 12-18 shoots on the basis of vigour of the plant. The second thinning of shoots should be done during second fortnight of September before flowering and on an average 8-12 shoots should be retained in each tree depending upon the bearing behaviour of individual cultivar. The rejuvenated tree bear small crop during first year and about 80-100kg fruits per tree during second year. The tree starts giving commercial crop in third fruiting season with higher yield (100-200kg fruit per tree) of excellent fruit quality.

## Top Working

Top working of wild ber trees is a quite simple procedure and can be practised by the growers easily. One can notice inferior ber trees growing in the villages on neglected lands and along the borders of the fields which yield small sized and poor quality fruits. Such trees can be top worked by budding with improved grafted varieties. The wild ber trees are headed back to a height of one metre during their dormancy period in the month of May. After heading back, apply Bordeaux paste to the cut ends of thick branches. New growth will emerge below the cut ends during rainy season (July). Two strongest shoots should be retained and others are removed. The selected shoots of pencil thickness (2-3 cm dia) should be T-budded in July-August with buds from choice grafted varieties. Soon after the buds begins to sprout, portion of shoots above the budding joint should be cut off smoothly. Top worked trees developed enough growth within a year. They start bearing normal crop of good quality fruits in the subsequent years.

## Fruit Quality Improvement

Umran a commercial cultivar of ber in North India ripen rate i.e. end March to first fortnight of April. As all the ber fruits on the tree do not ripen simultaneously, thus four to five picking have to be made. The ripening time of ber in North India coincides with the harvesting time of wheat crop which leads to higher labour costs for its picking. Ethephon has been found effective in accelerating the ripening and improving quality of different fruits.The techniques to induce early and uniform ripening permit staggering the harvest which avoid the bothersome operation of picking the fruits in 4-5 lots. Pre-harvest spray of ethephon accelerated the maturity of Umran ber and reduced the ripening period. Ethephon at 400ppm (250 ml in 300 litres of water) at colour break stage i.e. first week of March, advances ripening of Umran ber by two weeks. It produced attractive, uniform, better quality and deep golden yellow with chocolate coloured fruits. Ethylene produced by ethephon either cause synthesis of the pigment or caused degradation of chlorophyll, thereby cause increased colour development in fruits.

Ethephon application gives more profit per unit area due to advance marketing. The use of 400 ppm ethephon enable early picking of late ripening cultivar Umran allow the growers to benefit from relatively higher prices of the early market and to diffuse seasonal gluts. Labour costs of 4-5 picking is saved due to early picking and on watch and ward. Harvesting period can be stretched from two weeks to one month and ease the strain of marketing at peak ripening time.

## Use of Plant Growth Regulators

To obtain better income from ber orchard comparable with that of other horticultural crops, it is necessary to formulate a new cropping pattern with the use of growth regulators. The harvest period of ber can be extended from one to about two months and to make available quality fruits through improved post-harvest means for considerable length of time.

Spray application of NAA 10-25 ppm at full bloom stage appreciably increased the fruit size and weight and fruit quality of Umran ber. NAA at 10ppm applied at fruit set and subsequently at slow growth phase improved the fruit size and weight significantly in ber. The fruit quality in terms of TSS content was also improved with NAA application. In another study, higher fruit size and weight was reported with $GA_3$ 30-40ppm and NAA 10ppm sprayed first in $2^{nd}$ fortnight of October and then superimposed in second fortnight of November in ber cv. Umran. Consumer acceptance with respect to palatability rating was observed maximum in fruits treated with $GA_3$ 30ppm and NAA 50 ppm. Both $GA_3$ and NAA @30 ppm resulted better in checking physiological fruit drop.

Preharvest spray with 0.03% boron and 0.05% zinc along with 50ppm NAA improved fruit quality in Pewandi ber. Similarly, pre-harvest spray of ethephon at 300ppm to Umran ber induced uniform ripening. Such fruits harvested at optimum maturity could be stored upto 30-40 days in cold storage (0.3.3$^0$C temp. 85-90%RH).

The fruit yield in ber cv. Umran is recorded higher from the trees sprayed after fruit setting with 30ppm $GA_3$. The fruit size and weight is reported better with 20ppm $GA_3$ and 30ppm NAA. Likewise previous study, 30ppm $GA_3$ or 30ppm NAA is suggested to prevent physiological fruit drop effectively in Umran ber.

## SPECIAL PROBLEMS

### 1. Fruit Drop

The fruit drop in ber is a common problem and bulk of immature fruits dropped during initial stage of growth and development due to various factors like hormonal imbalance, abortion of embryo, incidence of diseases and inclement weather. Nearly 50% fruits of ber in Umran cultivar remained very small which adversely affects the yield and profit to the growers. It is reported that only 20-30% fruits attained marketable size of 25-30g per fruit. The fruit drop in ber may be caused due to the following various reasons.

i) Physiological fruit drop
ii) Pathological fruit drop
iii) Drop due to pest attack
iv) Drop due to adverse climate

**i) Physiological fruit drop:** After fruit setting a heavy drop takes place in November-December which might be attributed due to two main reasons. Firstly, it may occur largely due to lack of fertilization or degeneration of ovules. The drop is mainly due to shriveling of fruits which occurred upto December. Out of the total dropped fruits about 30% have aborted embryo. Secondly, fruit drop may occur due to premature formation of a abscission layer in the stalk of mature fruit. The possible reason of occurrence of abscission resulted due to natural exhaustion of growth regulators present in the fruit or when they cease to flow from their production centre because of some disorders in the tree or climatic conditions. Higher fruit drop about 60% of the total fruit set takes place from November to January. Mean temperature declined and relative humidity increased in this period. Later on from early February to mid-March per cent drop decreased considerably. The synthesis of promoters decreased during the slow growth phase due to formation of inhibitors which cause heavy fruit drop. Auxins inhibits abscission of fruits by preventing physiological break down of calcium pectate of middle lamella. Foliar spray of NAA at 30ppm in the second fortnight of October and again in the second fortnight of November effectively check physiological fruit drop in Umran ber.

**ii) Pathological fruit drop:** This drop takes place due to attack of powdery mildew fungus. The fruits are covered with whitish powdery mass of causal fungus. The fruits become cankered, corked, disfigured and misshapened. Such fruits drop heavily before reaching maturity. To check the disease and subsequently pathological fruit drop, the affected plants should be sprayed with 0.05% Karathane or Bayleton in September (at flowering), mid October, mid-November and mid- December.

**iii) Drop due to pest attack:** In ber orchards attack of white ant is very common. It attack the root system, main stem and main limbs. With the result, this block smooth flow of sap in xylem and phloem. This leads to reduction in supply of available carbohydrates to the fruits and results into fruit drop. Apply chlorpyriphos to the tree basin @1ml in one litre of water once in April and again after rainy season in September. Apply thick solution of chlorpyriphos (100ml in 1 litre water) with brush in the form of paint to the trunk and main limbs after scrapping the affected portion from white ants.

**iv) Drop due to adverse climate:** December-January are the coldest months in north India. Severe cold and cold waves blow during this period leads to heavy leaf fall and fruit drop. Sparse foliage on the trees results into poor photosynthesis activity in the plant. The leaves fail to provide proper food substance to the fruits. The fruit stops in growth, shriveled and falls. To cope with severe cold conditions, apply light irrigation to the orchards when there is chance of occurrence of frost. Collect and burnt the heaps of dried leaves slowly to provide thick smoke screen over the field. In this way atmospheric temperature will be raised and can save to some extent the heavy drop occurs during cold months of December-January.

## 2. Fruit Cracking

The fruit cracking is a preharvest physiological disorder which cause considerable reduction in fruit yield and quality in certain cultivars. The possible cause of fruit cracking in ber might result due to various reasons.

i) High rise in temperature in March
ii) Endogenous level of growth hormones
iii) Irregular irrigation interval
iv) Varietal character

The soft texture ber varieties like Sanaur-2, Gola, ZG-2 and Kaithli are prone to splitting due to their skin and high total soluble solids at maturity. Higher level of gibberellic acid and abscisic acid in the skin, stone may leads to fruit cracking. The uptake of water and solutes is mostly governed by the presence of certain chemicals. The enhanced water uptake causes accumulation of solutes within the fruit with consequent rupturing of fruit skin.

Applying irrigation to ber trees at proper interval helps in regulating water uptake and solutes. Foliar spray of NAA 20-30 ppm at pit hardening stage first in second fortnight of October and superimposed during second fortnight of November increase the hormone (promoter) level and thus reduce the cracking of fruit skin in soft ber varieties.

## HARVESTING AND PRODUCTION OF FRUITS

The ber tree grows fast and within 2-3 years of planting, it starts giving fruits. The fruit of ber requires about 22-26 weeks to mature after fruit setting. The peak period of harvesting ber in North India is mid-march to mid-April but some early cultivar may ripen by end February. In Gujarat, the harvesting season of ber is December to March and in Rajasthan it is January to March. The fruits are ready for harvest in October to November in south India.

Ber is a late peak type climacteric fruit. The fruit should always be picked when it has acquired normal size and characteristic colour of the variety i.e. deep golden yellow in Umran. Normally 4-5 pickings have to be made in the fruiting season to harvest the total fruits on the tree. The fruits should not be allowed to become over ripe on the trees, as they deteriorate in taste and quality and thus fetch lower price in the market.

**Maturity indices:** Several physiological and biochemical changes in ber at maturity have been suggested as indices for picking of fruits. Such indices include changes in specific gravity, fruit colour development, increase in TSS, decrease in acidity, increase in ascorbic acid and carotenoids content, loss of chlorophyll and decrease in total phenolic content.

Development of fruit colour on the tree is a good indication for picking. The commercial cultivar Umran attain deep golden yellow colour with chocolate tinge at harvest maturity. The specific gravity of ber fruits at maturity should be less than one. TSS, acidity and ascorbic acid content should range between 14-19%, 0.20 to 0.22% and 100-120mg per 100g pulp respectively in Umran cultivar. The fruits at maturity contains sucrose, glucose and fructose in the ratio of 3:1:1.

During growth and development period of ber fruit, biochemical changes takes place continuously. The content of the fruits increases throughout the growth period and reaches at maximum value at maturity. Contrary, acid content of the fruit registered high during the initial stage of fruit growth in green coloured immature fruits and with incipient ripening B-carotene content increased which gives the fruit golden yellow colour. Ascorbic acid and sugar content continued to increased till the fruit reached market.

**Production of Fruits:** Ber tree are regular and heavy bearer and the mature trees of different grafted varieties can bear one to two quintiles of fruit per tree. Peak harvesting of ber in different cultivars spread over a period of one and half month from second fortnight of February to mid-April in north India. The fruits are picked in several pickings. The annual production of ber fruit is approximately 4.31 lakh mt in India. Productivity of fruits is 9.2 mt per hectare which is quite high compared to many other fruits. Umran is a prolific bear cultivar yielding 150-200 kg fruits per tree. Sanaur-2, ZG-2 are other high yielding varieties and bear yields about 150 kg fruit per tree.

## POST- HARVEST FRUIT TECHNOLOGY

### Ripening of Fruits

Studies conducted at PAU has shown that ethephon 400ppm advanced maturity in Umran ber by two weeks and induces uniform ripening. Ber fruits harvested at early maturity on 15 March ripen unevenly. Post-harvest problems of ber fruit include variation in ripening time, uneven ripening of immature fruits after harvest and poor retention of mature fruits on the tree. Ethephon at 2000-3000 ppm and hot water dip ($52^0C\pm1^0$)regardless the stage of maturity enhance ripening percentage and develops the best fruit colour after ripening. Hot water treatment controlled the development of pathogens thus enhancing the shelf life and maintained fruit quality.

## Grading and Packaging of Fruits

The grading of ber fruits is carried out manually and graded on the basis of size and colour development. To sell the fruit at premium in the market grading of fruits should be practised regularly. Sorting should be done to remove culled, undersized, under/over riped, misshapened, cankered and bird-damaged fruits. The fruits are graded into four grades viz. A grade-large sized fruits, B grade-medium sized fruits; C grade – small sized fruits and D grade- under/over riped, deformed, misshapened and cankered fruits. Distribution of fruits is found highest in B grade (33%), which is followed by A grade (27%) fruits. Distribution is found 21 and 19% in C and D grade fruits, respectively. In Umran ber the fruits of A and B grades accounted for 60% of the total crop have deep golden yellow colour and are more acceptable to the consumers and thus, considered the best for efficient marketing.

The fruits can be packed in corrugated fibre board (CFB) boxes, wooden crates, poly nets, baskets and gunny bags of convenient sizes on the basis of grades. Packing of ber fruits in CFB cartons of 8kg capacity (40x25x20 cm size) with 6 holes of 1cm diameter on two sides and spreading paper shavings as cushion material retained good quality for 9-12 days. CFB boxes of 4 and 8 kg capacity are preferred for packing of ber fruits. The fruits packed in poly nets of 1kg capacity consumer pack is good choice for local sale of fruits. Packing fruits in bamboo baskets putting dry grass as a cushion could be useful for local/short distance market. For long distance transportation of ber fruits, gunny bags of 16-40 kg sizes are preferred.

## Storage of Fruits

The normal storage of the ber under ambient conditions is very short. The shelf life of ber fruits needs to be extended so as to stagger their marketing over longer periods. The ber fruits remain in good condition in perforated polythene bags for one week at ambient temperature. The shelf life of fruits can be enhanced upto 10 days in Umran and 12 days in Sanaur-2 at room temperature in perforated polythene bags after treating with 6% wax emulsion.

Umran ber could be stored for about 3 weeks in the home refrigerator (0-4$^0$C) in perforated polythene bags. The ber fruits of Umran cultivar harvested at colour break stage can be stored at 7.5±1$^0$C and 90-95% RH for two weeks with acceptable colour and quality. The shelf life of Umran ber fruits can be extended up to 20 days with least browning percentage after treating with $GA_3$ 25ppm, packing in CFB boxes and storing in commercial cold store (temp 0-3.3$^0$C and 85-90% RH). Treatment of ber cv. Seb with 1-methlcyclopropene (100-400nl.L$^{-1}$) considerably prevented the decrease in fruit firmness and vitamin C content thus delayed fruit ripening and extended shelf life of ber fruits. The shelf life of fruits can be enhanced up to 30 days in Umran and 40 days in Sanaur-2 by storing them in perforated polythene bags of 100 gauge thickness in the commercial cold store (temp 0.3.3$^0$C and 85-90% RH) after treating with 6% wax emulsion. Ber fruits cv Umran retained with pedicel and treated with ascorbic acid (2%) could be stored in cold chamber (7$^0$C) for three weeks in good condition. This combination can be useful in the supply chain management of ber fruits to extend storage life and marketability.

The colour development of ber fruits packed in HDPE bags and CFB boxes proceed slowly and retain its original colour upto 20 days in cold store. The fruits in paper bags develops browning after 30 days in cold store. The overall rating of fruits in CFB boxes and HDPE bags remain excellent upto 20 days of cold storage with better consumer acceptance. Pre-harvest spray of ethephon 300-400 ppm and morphactin 10-25ppm registered lower pectin methylesterase (PME) activity in paper bag after 40 days of cold storage in Umran ber.

### Processing of Fruits

The ber fruits harvested fresh can be processed into a number of desirable products for off-season utilization. The ber products like candy and preserve, canning in sugar syrup, fruit pulp and squash can be prepared. The fruit can be dried and dehydrated in excellent condition. The fruits of Sanaur -2 harvested at early stage can be dried best in a mechanical dryer at 60±2$^0$C for 31 hours for making Chhuharas. Organoleptically Chhuharas were acceptable up to 6 months storage.

## INSECT-PESTS AND MANAGEMENT

**Fruit-fly *(Carpomyia vesuviana* Costa*)*:** Fruits fly is one of the serious pests of ber and causes great damage to ber fruits. Fruit fly deposits eggs in the epicarp of developing fruits. After hatching, the larvae enter into pulp and start feeding. The infested fruits become deformed, turn brown, rot and drop off. Fully grown maggots emerge out of the fruit by making a hole and pupate in the soil.

To control the pest, pick and destroy the infested fruits. To escape egg laying on fruits, harvest at green and firm stage and do not allow the fruit to ripe on the tree. Rack the soil around the tree during summer to expose the pupae to heat and natural enemies. Spray 500 ml of Rogor 30 EC (Dimethoate) in 300 litres of water during February-March. Stop sprayings at least 15 days before fruit harvest.

**Leaf-eating caterpillar** ***(Euproctis*** **spp.)and ber beetle** ***(Adoretus pallens*** **Arrow):**These pests should be watched carefully during rainy season for their attack on the trees. The young caterpillars initially remain gregarious and scrap leaves and tender fruits. Later, instars disperse and devour leaves, fruits and tender shoots. The affected fruits become unmarketable.

For their control, proper sanitation should be maintained in the orchard. At early and smaller scale, pluck the leaves with egg masses and young caterpillars and destroy.

**Lac insect** ***(Kerria lacca):***It causes serious damage by sucking the sap from the twigs which usually dry up. Their presence on ber trees is considered as harmful because they devitalize the trees and affect adversely the yield of fruits.

The ber orchards should be watched regularly to control the pests timely. Remove and destroy the infested dry part and scrap off the infested twigs before treatment. Spray the trees with 250 ml of Rogor 30 EC (Dimethoate) in 250 litres of water in April and again in September.

## DISEASES AND MANAGEMENT

**Powdery mildew:** The disease is caused by ***Oidium erysiphoides f.sp. Ziziphi*** .It appears from September to December and has become a big menace to ber orchards in north India. Young developing leaves and fruits are covered with whitish powdery mass of the causal fungus. The disease causes premature defoliation and heavy fruit-drop. Affected fruits remain small and become cankered and disfigured. Sometimes the attack is so severe that the entire crop is lost either through drop or rendered unmarketable, thus causing heavy economic losses to the growers. The disease can wipe out the entire crop if not checked in time. The disease can be controlled by 3-4 sprays of 0.05 per cent Karathane 40 EC (50 ml in 100 litres of water) or 0.05% Bayleton 25 WP (50 g in 100 litres of water) or 0.25 per cent wettable sulphur (250 g in 100 litres of water). First spray must be given at flowering (September), second spray after fruit-set in mid October, third in mid-November and fourth to mid-December. If needed another spray can be given in January.

**Leaf spots**: Phoma Leaf Spot is caused by ***Phoma macrostoma*** Mont, and 'Black Mould of Leaf is caused by ***Isariopsis indica.*** Both are very common in ber growing regions and are caused by different fungi. The Phoma leaf spot appears on the upper surface, the black mould make its appearance only on the lower surface of leaves. In Phoma leaf spot symptoms appear when the leaves have fully expanded but in the Black mould, the symptoms can appear even on young leaves. Phoma leaf spot appears with grey centre, yellow margin and dark fungal growth on the mid-rib, main vein, petioles and the leaves. Black mould spot appears as small circular, small finger-like projections like softy tufts.

Leaf spots of ber can controlled by spraying Bordeaux mixture 2 : 2: 250 or with 0.3% copper oxychloride 50% (300 g in 100 litres of water) on leaves with the appearance of disease in August or when the leaves have expanded. After 14 days of the first spray give another spray with 0.2 per cent Dithane M-45 WP 75% (200 g per 100 litres of water) on leaves. First and second spray should be repeated alternatively at 14 days interval till the fruits are fit for marketing. Sprayings are stopped a week before harvesting the fruits.

**Black fruit spot**: The disease is caused by ***Altemaria altenata***. Disease produces small, irregular, sunken, black spots on fruits. Sometimes, ***Phoma macrostoma*** is also associated with the disease at later stages. The infected fruits become disfigured and may drop off before harvest. Black fruit spot start appearing in January. It become very severe during February.

Spray with Mancozeb 75Wp @0.25% (2.5g per litre of water) during end January and mid-February.

## MARKETING OF FRUITS AND EXPORT POTENTIAL

### Marketing of Fruits

The marketing of ber fruit is not properly organized in India. The process of ber marketing involves many constraints like cost of transportation, packing and loading and unloading of the produce. Ber is a perishable crop and resulted substantial spoilage to fruits during transportation and marketing. The harvesting season of ber coincides with harvest of wheat and other *Rabi* crops. The producer has to deal with various marketing hurdles and functionaries like commission agent and wholesaler. This all is a cumbersome process and involve some uncertainties in smooth marketing of the produce. The growers generally auction their crop to the fruit contractors. They harvest the fruits according to their convenience and sell the produce in the local or distant market. But, now few progressive ber growers have started doing marketing of fruits at their own level.

Different grafted varieties normally bear 100-200kg fruit per tree depending upon the cultural practices followed. It has been observed that large and medium sized fruits (A and B grades) comprises 60 per cent of the total yield of a tree. These fruits should preferably be packed in CFB boxes of 4kg capacity for fetching higher premium in the market. For marketing of remaining fruits of C grade (small sized) and culled type fruits can be packed in baskets or gunny bags. Presently, the ber fruits of A and B grades fetch higher price in the wholesale market ranging from Rs. 10 to 15 per kg. Small sized (C grade) fruits of different varieties generally earn Rs. 5-8 per kg in the market. The return from ber orchards can exceed the profit from the cultivation of several other fruits. The growers can net on income of Rs. 1 to Rs. 2 lakh per hectare or even more by growing selected cultivars and following latest recommended practices.

The ber plants are normally planted at 7.5x7.5m apart in square system. However, plantation can be made closer at 6x6m on semi vigorous rootstock Coimbatore. Thus, nearly 50% more plants can be accommodated per unit area and 20% higher yield can be obtained. It is estimated that with enhance yield of 50 quintal fruits per hectare with semi-vigorous rootstock, can give net profit of about Rs. 50000/-.

### Export Potential

The ber fruits are generally sold in local markets. Demand of ber fruits for export is growing in recent years. During the year 2014-15, 2015-16 and 2016-17, 1629.9 mt, 752.49 mt and 1074.23 mt fresh ber fruits with value of Rs 3.66 crores, Rs 3.63 crores and Rs 5.38 crores, respectively were exported to different countries.

## FUTURE STRATEGIES

### Short Term

- Managing fruit drop with chemical or other means.
- Technology device for improving seed germination.
- Biological weed control in orchards.
- Extending shelf life at low temp for stretching market period.
- Management of insect-pests more efficiently and effectively.

### Long Term

- Hybridization studies for developing cultivars tolerant to biotic stresses and for processing.
- Varietal evaluation and introduction from different sources.
- Evaluation of dwarf and high yielding rootstock.
- High density planting on trellis for easy harvesting.
- Canopy management for effective implementation of different orchard cultural practices.

## LITERATURE CONSULTED

Abbas, M.F.1997. Jujube. In: Post harvest Physiology and Storage of Tropical and Subtropical Fruits (ed. S.K. Mitra):405-415. Cab International Wallingford, U.K.

Aggarwal, P., Kaur, B., and Bal, J.S. 1997. Studies on dehydration of different ber cultivars for making ber chhuharas. *J. Food Sci.Tech.*, 34:534-536.

Azam-Ali, S., Bonkoungou, E., Bowe, C., dekock, G., Godara, A. and Williams, J.T. 2006. Ber. International Centre for Underutilized Crops, Southampton, U.K.

Bajwa, G.S., Brar, S.S., Minhas, P.P.S., and Bal, J.S. 1992. Chemical weed control in ber (*Zizyphus mauritiana* Lamk.) orchard. *Indian Journal of Weed Science*, 24:11-16.

Bal, J.S. 1981. A note on sugars and amino acids in ripe ber. *Prog. Hort.*, 13 (3-4):41-42.

Bal, J.S. 1982. Ber the poor man's fruit. *Science Reporter*, 16(2):126-127.

Bal, J.S. 1992. Identification of ber *(Zizyphus mauritiana* Lamk.*)* cultivars through vegetative and fruit characters. *Acta Horticulturae*., 317:245-254

Bal, J.S. 2001. Evaluation of ber varieties. *Indian J. Plant. Genet. Resources*, 14:159-160.

Bal, J.S. 2010. Advanced technologies in ber. Paper presented at: National Seminar on Impact of Climate Change on Fruit Crops (PAU Ludhiana).

Bal, J.S. 2013. The utilization of rootstocks in Indian jujube. *Acta Horticulturae*, 993:47-50

Bal, J.S. 2014. Fruit Growing, 3rd Edn., Kalyani Publishers, New Delhi.

Bal, J.S. 1992. Ber for health and profit. *The Tribune,* 112(84):6.

Bal, J.S. 2016. Development and production of Indian jujube (ber) in India. *Acta Horticulturae,* 1116:15-22.

Bal, J.S. and Boora, R.S., 2016. Studies on diversity in ber varieties under North-West sub-tropical condition of India. *Acta Horticulturae,* 1116:55-60.

Bal, J.S. and Gill, K.S. 2016. Canopy measurement, nodal pruning and planting density in Indian jujube (ber) *Acta Horticulturae,* 1116:99-104.

Bal, J.S. and Randhawa, J.S. 2007. Ber. Punjab Agricultural University, Ludhiana.

Bal, J.S. and Singh, P. 1978. Development physiology of ber *(Z. mauritiana* Lamk.*)* var. Umran. Part-III. Minor chemical changes with reference to total phenolic, ascorbic acid (Vit. C) and minerals. *Indian Food Packer*, 32(3):66-69.

Bal, J.S. and Singh, P., 1978. Development physiology of ber *(Z. mauritiana* Lamk.*)* var. Umran. Part-I. Physical changes. *Indian Food Packer*, 32(3):59-61.

Bal, J.S., and Dhaliwal, J.S. 2003. A note on effect of selective pruning of late flowers on size and quality of fruit in ber cultivars. *Haryana Journal of Horticultural Sciences*, 32:232-233.

Bal, J.S., and Singh, S. 2012. Mulching in Ber Orchards. Saarbrucken, Germany: Lambert Academic Publishing, GmbH & Co. KG. pp.110.

Bal, J.S., and Uppal, D.K. 1992. Ber Varieties. Associated Publishing Company, New Delhi. pp.90.

Bal, J.S., Bajwa, G.S., and Singh, S.N. 1996. Role of ethephon in inducing early ripening and marketing of ber *(Zizyphus mauritiana* Lamk.*)*. *Indian J. Hortic*., 53: 38-41.

Bal, J.S., Gill, K.S., and Singh, J. 2013. The vegetative and fruit characteristics of hybrid of Indian jujube. *Acta Horiticulturae*, 993:43-46

Bal, J.S., Gill, S.S., Sandhu, A.S. 2016. Raising of rootstock and propagation techniques-Ber. In: Raising Fruit Nursery, 2[nd] Ed. Pp: 108-109. Kalyani Publishers, New Delhi.

Bal, J.S., Jawanda, J.S. and Singh, S.N. 1979. Development physiology of ber *(Z. mauritiana* Lamk.*)* var. Umran. Part IV. Changes in amino acids and sugars at different stages of fruit ripening. *Indian Food Packer*, 33(4):33-35.

Bal, J.S., Randhawa, J.S., and Boora, R.S. 2010. Studies on post-harvest life of ber fruits treated with growth regulators and stored in different packages at cool temperature. *Proc. National Seminar on Impact of Climate Change on Fruit Crops* (PAU, Ludhiana). pp:333-338.

Bal, J.S., Randhawa, J.S., and Singh, H. 2002a. Standardization of packaging for extending shelf life of ber fruits in cold storage. *Proc. National Workshop Post- harvest Management Horticultural Produce* (NPC, Chandigarh).Pp:367-371.

Bal, J.S., Randhawa, J.S., and Singh, H. 2002b. Grade specification and colour variation studies in ber fruits. *Proc. National Workshop Post- harvest Management Horticultural Produce* (NPC, Chandigarh).Pp:161-167

Bal, J.S., Randhawa, J.S., and Singh, J. 2004. Studies on the rejuvenation of old ber trees of different varieties. *Journal of Research* (Punjab Agricultural University), 41:210-213.

Bal, J.S., Singh, M.P., and Sandhu, A.S. 1997. Evaluation of rootstocks for ber *(Zizyphus mauritiana* Lamk.*)* cv. Umran. *Journal of Research* (Punjab Agricultural University), 34:60-63.

Bal. J.S. 1981. Some aspects of development physiology of ber *(Z. mauritiana* Lamk.*)*. *Prog. Hort.*, 12(4):5-12.

Gill, K.S., and Bal, J.S. 2008. Effect of the nodal pruning on the fruit quality of ber cv. Umran. *Haryana Journal of Horticultural Sciences*, 37: 223-224.

Kaundal, G.S., Grewal, S.S., and Bal, J.S. 1984. Influence of meteorological factors on the bud-take efficiency of ber *(Zizyphus mauritiana* Lamk.*)* cv. Umran. *Journal of Research* (Punjab Agricultural University), 21:372-374.

Kaur, S and Bal, J.S. 2016. Influence of pedicel retention and ascorbic acid on post-harvest life of ber in cold chamber. *Acta Horticulturae,* 1116:125-130.

Pareek, S and Yahia, E.M. 2013. Post harvest biology and technology of ber fruit. *Horticultural Reviews*, 41:201-240.

Sidhu, A.S., Bal, J.S., and Singh, H. 2010. Studies on irrigation scheduling in young ber plants. *Proc. National Seminar on Impact of Climate Change on Fruit Crops* (PAU, Ludhiana).Pp;196-202.

Singh, H., Bal, J.S., and Singh, G. 2004. Standardization of pruning technique in ber. *Indian J. Hortic.*, 61:259-260.

# 5

# Litchi

*Alemwati Pongener, S.D. Pandey, and Vishal Nath*

## INTRODUCTION

Litchi(*Litchi chinensis*) is an important sub-tropical evergreen fruit crop. The tree bears heart-shaped, conical, or spherical drupaceous fruits with a thin, indehiscent pericarp that surrounds the sweet, translucent, succulent, and juicy aril. Although litchi originated in the Asia-Pacific region, orchards are found in more than 80 countries. Having introduced into India by the end of the 17th century, litchi is an important fruit crop in India and is a source of livelihood to millions especially in the Gangetic plains along the foothills of the Himalayas where the crop is grown popularly. Bihar is the major litchi-producing state in India accounting for about 37% of both national acreage and production. Muzaffarpur is famous for its Shahi litchi and is popularly referred to as the 'Litchi City' or 'Litchi Hub'.In India it is grown in an area of about 85000 ha with total production of 528000 MT. Bihar is followed by West Bengal, Jharkhand and Assam in terms of production, while the highest productivity of litchi (16 MT/ha) is reported from Punjab.

## NUTRITIVE AND CULTURAL SIGNIFICANCE

Litchi fruits are juicy and delicious, and are a summer delight. Litchi is an excellent source of vitamin C providing about 71.5 mg/100g (USDA, 2013), thereby the recommended dietary allowances of 40 mg/day for Indians can be easily met by consuming a few litchi fruits. The consumption of a single litchi fruit would meet 2-4% of the daily dietary requirements for P, K, Mg, Fe, Zn, and Mn, and about 22% of Cu. Litchi is high in carbohydrates and is therefore a good source of energy. There is already a large body of evidence associating intake of dietary antioxidants, such as vitamin C, to reduced risk of chronic diseases. Litchi also contains good amounts of potassium, copper, dietary fiber, vitamins, antioxidants, and polyphenols. Research suggests that oligonol, a low molecular weight polyphenol from litchi fruit, possesses anti-influenza effects,and weight reduction and beneficial effects in preventing obesity-induced metabolic syndrome.Litchi is believed to relief coughing and decoctions of the root, bark and flowers are gargled to alleviate ailments of the throat.

## ORIGIN, HISTORY AND DISTRIBUTION

Litchi has its origin in southern China and northern Vietnam where it is believed to have been cultivated for thousands of years. Ancient documents and living specimens indicate to the existence of litchi culture by 200 BC in Hainan, Guangdong and Guangxi regions of China. Its cultivation spread from China to the rest of the world and commercial orchards are present in sub-tropical regions. China is the largest producer of litchi in the world followed by India, the two countries accounting for almost 91% of world production. Other major countries where the fruit is grown include Taiwan, Thailand, Vietnam, South Africa, Madagascar, Indonesia, Israel, and Australia. Commercial cultivation of litchi was limited to China and Vietnam before the $17^{th}$ century, but has since spread slowly to different parts of the world over the past 400 years. Litchi was introduced to Thailand from China about 300 years ago. Chinese migrants and merchants have also been instrumental in the distribution of litchi. Important routes in this spread include Australia, Mauritius, India, Madagascar, Florida, Hawaii, South Africa etc. The spread of commercial production has accelerated over the last few decades because of increasing interest in exotic fruits in Europe and up-markets.

Litchi trees are believed to have introduced into India in 1798 from Thailand (Singh and Babita, 2002). Since then, the fruit has spread throughout the country especially in the sub-tropical belt along the foothills of the Himalayas. Much of the litchi-growing regions are concentrated along the Indo-Gangetic plains, and organised orcharding has shown substantial growth in the last 50 years. Bihar continues to dominate in terms of area and production of litchi in India. Other states include West Bengal, Uttar Pradesh, Tripura, Uttarakhand, Jharkhand, Punjab, Himachal Pradesh, Assam, Nagaland, Mizoram etc. In recent years, litchi cultivation is picking up in hilly tracts of Southern India in the states of Kerala, Tamil Nadu, and Karnataka. Litchi is harvested in this region in the month of December, an off-season time when no litchi is available anywhere in the rest of India. Thus, growers fetch high income from their produce.

## TAXONOMICAL AND BOTANICAL DESCRIPTION

| | | |
|---|---|---|
| Kingdom | : | Plantae |
| Sub Kingdom | : | Tracheobionta |
| Division | : | Magnoliophyta |
| Class | : | Magnoliopsida |
| Subclass | : | Rosidae |
| Order | : | Sapindales |
| Family | : | Sapindaceae |
| Sub family | : | Sapindoideae |
| Genus | : | Litchi |
| Species | : | *Litchi chinensis* |

There are reportedly three subspecies of *Litchi chinensis* namely *L. chinensis* subspecies *chinensis*, *L. chinensis* subspecies *philippinensis* and *L. chinensis* subspecies *javensis*. However, the latter two are not commercially grown, and popular cultivars today belong to subspecies *chinensis*. The scope of this chapter, therefore, only pertains to *L. chinensis* subspecies *chinensis*.

The litchi is a medium to large evergreen tree that usually grows to about 10-12 m. The crown is round, dense, compact, and symmetrical. Normally, trees have a thick, straight, and short trunk and dark brown-grey bark. Litchi tree puts forth several flushes of growth throughout the year. Vegetative flush comprises several leaves and internodes, while floral flush is a terminal inflorescence compound dichasium. A new flush begins only after the leaves of preceding flush have matured or been removed.

Leaves are pinnately compound with 4-7 leaflets. Matured leaflets are glossy dark green on upper surface and grey green on the under surface, and are elliptical to lance-shaped. The inflorescence is determinate containing several panicles on current-season shoots. Panicles are mixed with lower buds producing leaves only, middle buds producing floral buds and topmost buds producing only floral branches with occasionally small leaves that do not persist. Panicles are 10-40 cm long and produce hundreds of small white, green or yellow flowers.

Flowers are 3-6 mm wide and possess a cup-shaped calyx with 4-5 serrated sepals, but have no petals. Each flower has 5-10 stamens. There are three types of flowers that open in succession on the same panicle. Type-I flowers lack ovules and are functionally male (referred to as $M_1$). Type-II flowers are structurally hermaphrodite but functionally female as the stamens do not dehisce. Type-III flowers are functionally male ($M_2$) although they have rudimentary pistil.

Fruit takes 80-112 days to mature depending on cultivar and weather. The fruit are drupes and may be round, ovoid, or heart-shaped varying upto 5 cm long and 4 cm wide. The edible part of the fruit is an aril formed from cells in the outer seed coat. The aril is white to off-white, translucent, juicy and sweet.

## CLIMATIC AND SOIL ADAPTABILITY

Litchi is an environmentally sensitive crop, and has highly specific climatic requirements. It has adapted most successfully in warm tropics with hot, humid summers and dry, cool winters, corresponding to tropics and sub-tropics between 13° to 32° N and 6° to 29°S latitudes. It flourishes best in a moist atmosphere, having abundant rainfall, and free from frost. Flower initiation in litchi is best below 20°C, while optimum temperature for leaf and fruit growth is about 30°C. Litchi requires a frost-free environment. Winter and autumn temperature between 8°C and 13°C for about 200 hours coupled with moisture deficit induce profuse flowering. Areas with winter temperature above 25°C are not suitable for litchi cultivation. Litchi can tolerate high temperature and grows satisfactorily between 20-35°C. Temperatures below 2°C damage new leaves and those below -2°C can kill trees.

Light is an important component for optimal growth of litchi trees. Flowering and fruiting are reduced once adjacent trees start to touch each other. Therefore, pruning becomes essential, at this stage, to allow sufficient light inside the canopy. Overcast conditions during flowering reduce activity of pollinators and decrease fruit set.

Traditionally, litchi is grown in areas where rainfall is adequate to maintain good plant and fruit growth. Under Indian conditions, annual rainfall between 1250 and 1500 mm is considered optimal. However, plant water needs can be managed if adequate irrigation facilities are available. Litchi can be grown on a range of soils. Deep, well drained, non-saline calcareous soils with proper texture and fertility and high organic content are ideal for growing litchi. Litchi is fairly tolerant to soil pH and are found growing successfully in both acidic (<5.5 pH) and saline conditions (up to 8.5 pH). However, litchi ideally prefers acidic soils between 5.5 and 7.0 pH.

## RECOMMENDED AND POPULAR CULTIVARS

There are reportedly more than 200 varieties of litchi in China. However, being an introduced fruit crop, the genetic base in India is narrow. Nearly 40 varieties have been reported from different parts of the country, but only a few are grown commercially. These include Shahi, China, Bedana, Rose Scented, Seedless, and Bombai. Litchi cultivars in India can be classified based on the region in which they are popular (Table 1) or based on the time of fruit maturity (Table 2).

**Table 1:** Popular cultivars of litchi in different states of India

| State | Cultivar |
|---|---|
| Bihar | Shahi, China, Deshi, Purbi, Rose Scented, Kasba, Mandraji, Late Bedana, Early Bedana, Trikolia, Swarna Roopa |
| Uttar Pradesh & Uttarakhand | Early Large Red, Early Bedana, Late Large Red, Late Bedana, Muzaffarpur (Shahi), Rose Scented, Culcuttia, Extra Early Green, Gulabi, Pickling, Khatti, Dehradun, Piazi |
| West Bengal | Bombai, Ellaichi, Early Ellaichi, Late China, Deshi, Purbi, Kasba, Muzaffarpur |
| Punjab & Haryana | Saharanpur, Dehradun, Muzaffarpur, Late Seedless, Early Seedless, Rose Scented, Early Large Red, Late Large Red, Culcuttia, Khatti, Gulabi |

**Table 2:** Classification of litchi cultivars based on period of fruit maturity

| Period | Cultivars |
|---|---|
| Early | Shahi, Rose Scented, Deshi, Muzaffarpur, Dehradun, Ajhauli, Green, Dehra Rose, Trikolia |
| Mid | China, Purbi, Culcuttia, Bombai, Bedana, Swarna Roopa, Kasba, Sabour Bedana, Sabour Mathu |
| Late | Late Bedana, Kaselia, Longia |

Shahi and China are leading commercial cultivars in Bihar. Recently, three new varieties of litchi have been released by ICAR-NRC on Litchi. Some characteristic features of these varieties are listed in Table 3.

**Table 3:** New varieties of litchi released by ICAR-NRC on Litchi

| Variety | Plant stature | Maturity | Fruit | Other features |
|---|---|---|---|---|
| Gandaki Yogita | Dwarf, suitable for HDP | Late maturing | Pulp recovery > 75%<br>TSS 18-20°B | Free from fruit cracking and fruit borer, suitable for Indo-Gangetic plains |
| Gandaki Lalima | Robust | Mid late maturing | Fruit weight 28-32 g<br>Pulp recovery >60%<br>TSS 18-19°B | High yielder; Free from fruit cracking and fruit borer, suitable for Indo-Gangetic plains |
| Gandaki Sampada | Spreading | Mid late maturing | Fruit weight 35-42 g<br>Pulp recovery>83%<br>TSS 18-20°B | High yielder; Free from fruit cracking and fruit borer, suitable for Indo-Gangetic plains |

**Plate 1:** *Gandaki Yogita, Gandaki Lalima*, and *Gandaki Sampada* : New litchi varieties released by ICAR-NRC on Litchi, Muzaffarpur, Bihar (See colour version on page 314 )

## PROPAGATION TECHNOLOGY

Litchi can be propagated through seeds or through asexual means. Sexual propagation, although easy, has several disadvantages mainly long juvenile phase and the resulting plants are no longer true-to-type. Seed propagation is only used for raised plants under crop breeding and hybridization programmes. Asexual methods of propagation in litchi include air-layering, grafting and cutting with varying degree of success. In all litchi-growing regions, air-layering is the most widely accepted method for successful propagation of litchi. Nursery men prefer this method as it is easy to perform and gives the highest success rate among all methods.

### Air-layering

Air-layering is done when leaves of previous growth flush have attained proper maturity. Straight branches of 10-25 mm diameter and 45-60 cm length from tip are selected. If sufficient moisture is available, air-layering can be done any time of the year; however, best results are obtained when air-layering operation is done during the spring or rainy season. On selected branches, a ring of bark 2.5 cm in width is removed and the thin cambium layer beneath the bark is scraped away. The ringed area is then covered by a layer of moist peat moss of about 2.5 cm thick and 10 cm long, and then sealed by polyethylene bags. Auxins (IBA @ 2-10 g/L) are sometimes used to promote rooting, but are not essential. Under congenial environmental conditions of 25-30°C, several roots are visible through the polyethylene that change colour from white to creamy brown within 2-4 months after layering. The rooted air-layered branch is then cut off below the rooted portion and given special care to avoid damage to the root system. Use of root trainers and rooting media promotes secondary and tertiary root, thus improving survival and establishment. Leaves are halved to reduce transpirational loss and placed in polybags for initial care and establishment under shade in the nursery. Such plants are allowed to put forth 2-3 growth flushes before they are finally transplanted to the permanent field.

Survival of air-layers after detachment is a challenge in litchi nurseries. In order to overcome high mortality of air-layers in nurseries, preparation and use of suitable potting medium becomes essential. Mycorrhizal association and colonization of roots of litchi has been well documented, and several species of mycorrhiza, especially *Glomus mosseae* and *G. intaradices* have been isolated in abundance under different litchi orchard ecosystem. Mycorrhiza in consortia with *Azotobacter* and *Trichoderma* improves survival of air-layers. It is recommended to add 1 kg mycorrhiza + 500 g *Azotobacter chroococcum* + 500 g *Trichoderma viride* to prepare 500 kg of potting mixture. Consequently, this helps in survival and establishment of air-layers, and improves growth and development during initial years of plant establishment in the orchard.

## PLANT AND FRUIT PHYSIOLOGY

Litchi trees grow in recurrent terminal flushes. They grow from the end of branches in periodic spurts separated by short intervals of dormancy. Young trees can flush for as many as 5-6 times a year, or even more. Vegetative flushing is also the norm in adult bearing trees where a bloom flush leads to flowering in winter. A period of cold is therefore a prerequisite for flowering to be initiated in litchi, otherwise litchi trees continue to remain vegetative and not produce flowers. Flower bud formation and differentiation usually occurs between November and February in the Northern hemisphere. Under Bihar conditions, the process of panicle emergence and flower opening is complete by the first fortnight of March in cv. Shahi. Fruit set takes place around the third week of March. Initially, fruit growth comprises of only pericarp and seed growth, while aril formation commences after about one month of fruit set around the last week of April. Thereafter, aril growth shows exponential growth and colour break stage occurs around first week of May as evidenced by the accumulation of anthocyanins in the pericarp tissue. Fruit growth follows a sigmoid pattern throughout. By the end of third week of May, the fruit attains harvest maturity and are ready for picking (Fig. 1).

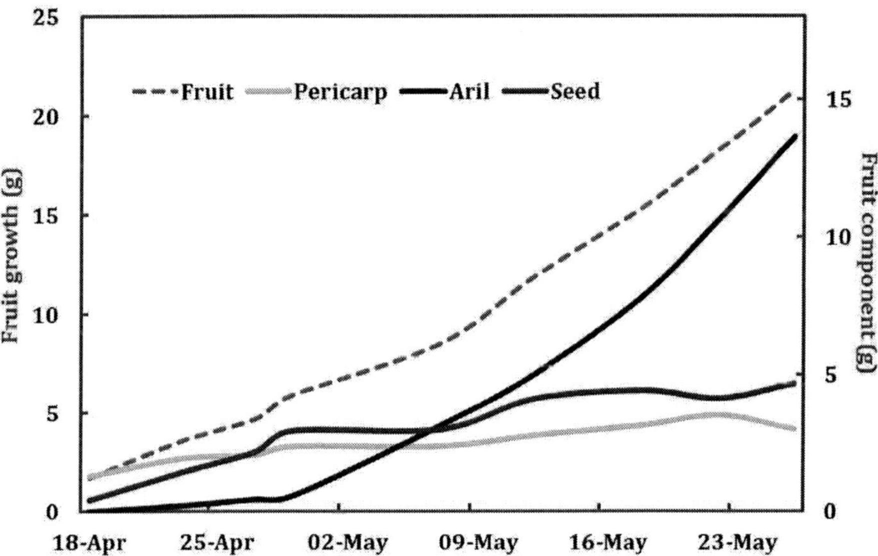


**Fig. 1:** Depiction of fruit growth including fruit components in litchi cv. Shahi

## PLANNING AND PLANTING

Careful initial planning is essential for success in any fruit culture as it becomes very difficult to rectify orchard layout at later stages. Poor land preparation and inappropriate lay out of orchard causes recurring problems during subsequent years. Selected field should be deeply ploughed and harrowed to break compact clods and remove roots of perennial weeds. The field should then be properly leveled to ensure ease of irrigation and drainage of excess water. Layout can be done as per the slope of the field, and popular systems include square, rectangular, quincunx, hexagonal and contour system. Pits of size of 60 cm × 60 cm × 60 cm for light soils should be dug atleast two month before planting (bigger pit size can be made for heavier soils). Dug out soil and open pit are left exposed for 2-4 weeks to ensure solar disinfection.

Well-decomposed farmyard manure, cakes and vermicompost, *Trichoderma*-rich manures and bio-fertilizers along with soil from litchi rhizosphere are usually mixed with soil for pit filling during planting. As a common dose, 30-40 kg FYM, 2 kg neem/karanj cake, 250 g of SSP or bone meal is sufficient for one pit. Fertilizers mixed with top soil should be placed at the bottom of the pit, followed by mixture with lower soil. The upper level of pit is kept 15 cm above the field level, which is then settled down through irrigation.

If irrigation facilities are available, litchi can be planted any time of the year accept during very hot, cold, or windy weather. Otherwise, the onset of monsoon is usually preferred for planting litchi, when there is sufficient moisture.Proper plant spacing is required for provide optimum light, water, and nutrients to the growing plants, as well as for ease of intercultural operations and mechanization. A planting distance of 10 m × 10 m is usually followed under conventional planting system. However, with increasing pressure on limited land resources, orchardists adopt closer spacing of even up to 5 m × 5 m. Such plantations require intensive management involving severe pruning of limbs to restrict canopy to available space. It is recommended to apply mulch around young trees to suppress weed growth and help in plant establishment. Additionally, newly planted litchi trees can be staked with bamboo or wood for proper anchorage, protection from winds and ruminants.

## SOIL CULTURAL PRACTICES TECHNOLOGY

### Water Need

Assured availability of water and appropriate irrigation system is a vital component of modern litchi orcharding. The method of irrigation can be decided based on water availability. Although flooding and furrow system of irrigation can be adopted where water is available in plenty, micro-irrigation system such as drip and sprinkler improves water use efficiency and conserves water. Irrigation can be scheduled as per the soil moisture content and sufficient moisture should be maintained in the root zone. As a rule of thumb, litchi orchards should be irrigated when 40-45% of available water in potential root zone has been depleted. During the initial two years of plant establishment plants can be irrigated weekly during summer and at fortnight interval during winter. Mild water stress in conjunction with cold stress is known to induce flowering in litchi. Therefore, it is advisable to withhold irrigation during the period preceding anticipated flowering (November – March). Once fruit set is over, the orchards are irrigated at weekly interval until harvest. Litchi fruit cracking or splitting is a result of improper soil water management. Regular irrigation maintains optimum moisture levels and also raises humidity in the orchard, thereby reducing fruit cracking.

## Nutritional Need

Nutrients play vital roles in growth and development of litchi plants. Considerable amount of soil nutrients are depleted during fruit production. It has been estimated that 90-250 g of N, 30-50 g P, 240-260 g K, and 20-60 g Ca are used up to produce 100 kg of litchi fruit. Therefore, nutrient replenishment is important for sustained yield year after year. There is no single nutritional recommendation that fits all litchi orchards, because many factors,*viz*. health and stage of plant growth, inherent nutrient status of soil, topography, and climate in which the plant is grown, decide the quantum of nutrient application. Nutrition programme, therefore, must begin with soil and plant tissue analysis. Table-4 enlists the recommended soil nutrient concentrations for litchi (FAO).

**Table 4:** Recommended soil nutrient concentrations for litchi

| Parameter | Concentration | Parameter | Concentration |
|---|---|---|---|
| Ph | 5.5 to 6.0 | Copper | 1.0 to 3.0 mg $kg^{-1}$ |
| Organic carbon | 1.0 to 3 % | Zinc | 2 to 15 mg $kg^{-1}$ |
| Electrical conductivity | < 0.20 dS $m^{-1}$ | Mn | 10 to 50 mg $kg^{-1}$ |
| Chlorine | < 250 mg $kg^{-1}$ | Boron | 1.0 to 2.0 mg $kg^{-1}$ |
| Sodium | < 1.0 meq 100 $g^{-1}$ | Potassium | 0.5 to 1.0 meq $100g^{-1}$ |
| Nitrogen | 10 mg $kg^{-1}$ | Calcium | 3.0 to 5/0 meq $100g^{-1}$ |
| Phosphorus | 100 to 300 mg $kg^{-1}$ | Mg | 2.0 to 4.0 meq $100g^{-1}$ |

*Source*: www.fao.org/DOCREDP/005/AC681/E/

Soil and plant nutrient status should be ascertained before application of fertilizers. Objective of nutrient application is also decided by the stage of plant, therefore younger plants are treated differently from bearing and adult plants. The common dose of manure and fertilizer recommended for litchi under Bihar conditions is presented in Table-5

**Table 5:** Recommended manure and fertilizer application dose for Bihar state

| Year | Quantity/annum/tree | | | | |
|---|---|---|---|---|---|
| | FYM (kg) | Cake (kg) | N (g) | P (g) | K (g) |
| 1 | 10 | 1.0P | 50 | 25 | 50 |
| 2 | 15 | 1.5 | 100 | 50 | 100 |
| 3 | 20 | 2.0 | 150 | 75 | 150 |
| 4 | 25 | 2.50 | 200 | 100 | 200 |
| 5 | 30 | 3.0 | 250 | 125 | 250 |
| 6 | 35 | 3.50 | 300 | 150 | 300 |
| 7 | 40 | 4.00 | 350 | 175 | 350 |
| 8 | 45 | 4.50 | 400 | 200 | 400 |
| 9 | 50 | 5.00 | 500 | 225 | 450 |
| 10 | 55 | 5.00 | 600 | 250 | 500 |
| 11 or more | 60 | 5.00 | 600 | 250 | 600 |

Source: Chauhan, K.S. 2001.Litchi: Botany, Production and Utilization.

Correct placement of fertilizer is necessary for obtaining high application efficiency. Nutrients are applied during pit-filling for establishment and vigorous growth of young plants. In older plants the fertilizers should be applied in trenches (30 cm wide and 15-20 mm deep) dug around the tree in the drip area (about 2 m away from trunk). For best results, full amount of FYM, neem cake, phosphorus, potash, and 2/3[rd] nitrogen are applied immediately after fruit harvest in June-July, and the remaining 1/3[rd] of nitrogen during clove stage of fruit development. Micro-nutrients are generally applied through foliar sprays when deficiencies are detected. Usually Boron and Zinc are sprayed once a year.

Role of mycorrhiza and bio-pesticides in improving survival and establishment of air-layers has already been discussed. Colonization of litchi roots with mycorrhiza along with application of biopesticides act synergistically to improve plant health and growth. After plant establishment in the orchard, it is recommended to go for application of mycorrhiza consortia twice a year during July-August and in February to maintain plant health and vigour. For a bearing litchi plant 250 g mycorrhiza + 100 g *Azotobacter chroococcum* + 100 g *Trichoderma viride* per plant is applied into the feeder zone of top 10-15 cm plant basin for optimum plant growth and production.

## Orchard Floor Management

Since litchi is normally planted with wide spacing, the vacant space between plants and rows can be utilized to grow intercrops and cover crops. Raising intercrops during initial years of litchi plant growth not only generates orchard income but also controls weeds. Intercropping or cover cropping with suitable and synergistic crops improves soil fertility and supplements organic matter to the soil.Short duration annual crops such as vegetables are best suited for use as intercrops. Some crop combinations for interspaces of litchi plants include: Cabbage (Oct.-Feb.) – Cowpea (March-June) – Fenugreek (July-Sept.); Okra (Sept.-Dec.) – French bean (Jan.-March) – Tomato (April-July); Potato (Nov.-Feb.) – Amaranth (March-May) – Cowpea (June-Aug.) – Radish (Sep.-Oct.); Cauliflower (Aug.-Nov.) – Potato (Dec.-March) – Dolichos bean (April-July). Fast-growing and early-bearing fruits trees such as guava, papaya, citrus or phalsa can also be grown as intercrops/fillers during the pre-bearing years of litchi to provide additional income to growers.

Weeds compete for water and nutrients with litchi plants especially in the initial stages of plant establishment in the orchard. Therefore, timely control of weeds is very important for normal growth of litchi plants. If left unattended, weeds interfere in routine orchard operations like irrigation, spraying, pruning, harvesting, etc. Normally, hand weeding, mechanical weeding (mowing), and tillage are performed to remove weeds. Alternatively, establishment of weeds can be controlled by growing suitable intercrops. Another effective method of weed control is through use of mulch, organic or inorganic. Mulching not only provides a physical barrier for weed growth but also creates unfavourable condition for germination of weed seeds. Mulching also improves the physical properties and nutrient content as they decompose and gets incorporated into the soil. In addition to these methods, weed control can also be achieved through use of herbicides. Although safe to use, safety guidelines must be properly adhered to during herbicide application.

## PLANT CULTURAL PRACTICES TECHNOLOGY

### Canopy Architecture Management

Litchi is conventionally planted at wider distance to provide adequate space to the growing plant. However, if left unattended the litchi tree can grow a huge and dense canopy that is difficult to manage in terms of orchard operations. Canopy architecture management is a vital component of litchi production system. It involves development of strong tree frame in the initial plant growth stages, and maintaining optimum space for vegetative and reproductive growth in subsequent bearing stages. Poor canopy management results in overcrowding of branches, less air circulation, and poor light penetration inside the canopy. These increase the number of unfruitful branches or terminals. Proper canopy management helps in enhancing the harness of natural resources, eases orchard operations, and provides opportunity and scope for mechanization and adoption of modern approaches for higher fruit production and quality enhancement. Planning for canopy management in litchi includes understanding the tree‘s growth and fruiting behaviour, the tree architecture intended to be established, and striking a balance between vegetative growth and fruiting.

Scaffold/primary branches (4-5 in number)are selected about 60 cm from the ground in all directions in the second year of planting. Two secondary branches are selected on each of these primary limbs, and tertiary branches on the secondary branches. Care should be taken to ensure wide crotch angle so that a strong framework with spreading-type of canopy is obtained. Tree frame development and care is done up to the fourth or fifth year of planting.

Once litchi trees come into bearing, they are pruned annually immediately after harvest (June-July). Litchi is normally harvested by plucking the whole fruit bunch along with 20-30 cm shoot, which serves the purpose of pruning to some extent. Litchi bears fruit on new shoots; therefore, pruning off shoot terminals, dead wood, and internal branches maintains the canopy to manageable size and also prepares the tree for flowering and fruiting in the next season.

Litchi plants put forth several growth flushes throughout the year, and develop dense and overcrowded foliage within few years. This results in poor light penetration into the interior parts of the plant canopy, leading to unfruitful branches and high incidence of insect pests and diseases. Generally, the lower 2/3$^{rd}$ of plant canopy is productive, while the upper portion is comparatively less productive. Therefore, centre-opening is followed by many growers to allow sunlight, especially into the interior canopy.

Old, senile, and unproductive trees are common references made to litchi trees that are more than 50-60 years old. Such trees are not only unproductive but also pose challenges in undertaking normal management practices. Therefore, such trees are rejuvenated to bring such unproductive trees to economically viable condition. Rejuvenation or reiterative pruning is done by heading back the primary branches to about 1.5-2.0 m in the month of August-September. Care is taken to avoid splitting of bark and cut portion is treated with a suitable fungicide such as Bordeaux mixture immediately. Regular nutrient dose of 75-80 kg FYM, 2 kg neem/castor cake, 1 kg Urea, 1.50 kg SSP, and 500 g MOP should be applied in two split doses during August-September and February-March. Rejuvenated trees take 2-3 years to come to normal fruit bearing during which suitable orchard floor practices can be practised and intercrops can be grown to derive extra income from orchards.

Litchi plants are conventionally big in size, and pose huge difficulties in cultural operations such as annual pruning, harvesting, spray operations etc. Also, only the lower 2/3$^{rd}$ of plant canopy bears maximum fruitful terminals. High density planting and appropriate canopy architecture adoption has, therefore, become very pertinent in recent years. Plant density is increased significantly under HDP and orchard operations become easier compared to conventional planting. At the ICAR-NRCL, HDP and canopy architecture has been successfully

**Plate 2:** Canopy architecture design in litchi on Y-shaped trellies
(See colour version on page 315 )

demonstrated whereby litchi plants are maintained under Y-shaped trellies (Figure 2). Table 6 shows tree framing strategy to allocate optimum space to branches and development of fruiting terminals under different plant spacing in litchi.

## Fruit Quality Improvement

Production of good quality litchi fruits is affected by challenges such as fruit borer, fruit cracking, and sunburn injuries. Litchi fruit borer can cause heavy economic losses. Since litchi bears fruit on the outer canopy, fruits are highly prone to sunburn injury during the hot and dry summer. The resultant necrotic tissue becomes brittle and cracks easily under pressure from developing aril. Such cracked fruits then become port of entry and infection of fruits by decay and rot pathogens. Growers normally resort to treatment and spray of chemicals, growth regulators, and pesticides to manage these maladies, which increases production cost.

Bagging of litchi fruit bunch is a low cost, eco-friendly, sustainable, and economically viable technology that improves fruit quality. Non-woven polypropylene bags have been found superior in terms of conferring physical protection against entry of insect pests and reduce sunburn and fruit cracking by creating a shading effect and favourable in-package micro-climate. For best results bagging operation is done one month before harvest. Bagging also lowers production cost as growers can forgo chemical sprays.

**Table 6:** Tree framing and canopy design concept for litchi

| Plant spacing | Heading height (cm) | Canopy design concept | | | | |
|---|---|---|---|---|---|---|
| | | Primary branch (m) | Secondary branch (m) | Tertiary branch(m) | Fruiting terminals(m) | Gallery space(m) |
| 8×8 m | 70 | 1.00 | 0.80 | 0.70 | 1.00 | 0.50 |
| 6×6 m | 60 | 0.80 | 0.60 | 0.50 | 0.70 | 0.40 |
| 5×5 m | 40 | 0.70 | 0.50 | 0.40 | 0.60 | 0.30 |
| 4×4 m | 30 | 0.60 | 0.40 | 0.30 | 0.50 | 0.20 |
| 8×4 m | 30 | 1.00 | 0.80 | 0.70 | 1.00 | 0.50 |
| 6×3 m | 30 | 0.60 | 0.40 | 0.30 | 0.70 | 0.00 |
| | | 0.80 | 0.60 | 0.50 | 0.70 | 0.40 |
| Y-Shaped trellis (4x3m) | 70 | 0.50 | 0.40 | 0.30 | 0.30 | 0.00 |
| | | 0.60 | | 0.80 | | |
| | | | | Parallel training of I, II & III branches | | |

**Plate 3:** View of litchi orchard with bagged fruit bunches
(See colour version on page 315 )

### Use of Plant Growth Regulators

Plant growth regulators are used in litchi production system for different purposes. NAA @ 20-30 ppm and $GA_3$ @ 20-25 ppm is effective in reducing the incidence of flower and fruit drop, and fruit cracking. IBA @ 2-10 g/L is applied for promoting rooting in air-layers of litchi. Irregular bearing is a problem in certain cultivars of litchi especially cv. China. Application of paclobutrazol @ 2g a.i. per metre of canopy diameter is effective in inducing regular flowering and fruiting. Foliar application of ethrel @ 400 ppm or $KNO_3$ @ 1% at monthly interval from October-December induces panicle emergence, flowering and fruiting in litchi.

## SPECIAL PROBLEMS

Litchi cultivation is beset with a few problems, important among which are briefly discussed here.

### Flower and Fruit Drop

Poor yield is sometimes recorded in litchi despite profuse flowering and fruit set. This is due to heavy flower and fruit drop between flowering and fruit maturity. Maximum drop takes place during the first fortnight of flowering, and normally only 2-18% is carried to maturity. Flower and fruit drop can occur due to lack of pollination, failure of fertilization, abortion of embryo, or nutritional and hormonal imbalance. It is also induced and aggravated by external factors such as high temperature, low humidity, and desiccating westerly winds during flowering time.

Flower and fruit drop can be managed by increasing the population of honey bees to ensure adequate pollination and fertilization. Treatment with any of NAA @ 20-30, GA3 @ 20-25 ppm, or 2,4-D @ 10-20 ppm before flower opening also controls the malady. Foliar spray of boron @ 0.2% improves pollen germination, while two sprays of planofix @ 2 mL/5 L at 15 days interval from peanut size stage controls fruit drop.

## Fruit Cracking

Fruit cracking is a very common disorder in almost all litchi-growing regions. Losses due to fruit cracking can range in severity from 5-70%. The cracked portion becomes the port of entry of decay and spoilage pathogens, which makes such fruits unsuitable for consumption and marketing. According to Huang and Xu (1983) cracking of the skin commences before harvest due to the rapid expansion of aril which exerts pressure on the pre-grown pericarp, a phenomenon popularly referred to as '*ball skin and bladder effect*'. Although several reasons have been attributed to for inducing cracking in litchi, the foremost is water imbalance or mismanagement. Cracking is observed most when a prolonged period of soil moisture deficit,or drought,is followed by irrigation or sudden precipitation during aril growth stage of fruit development. Sudden availability of soil moisture results in quick growth of expanding aril, which exerts enough pressure for the pericarp to tolerate. Fruit cracking is favoured if temperature goes above 38°C and relative humidity goes below 60%.

It may be advisable to select sites and locations with less or no rainfall during harvest time. The most important management strategy against fruit cracking is to maintain adequate soil moisture throughout the fruit development period through regular irrigation. Mulching also stabilizes temperature and maintains soil moisture levels in the root zone. Calcium @ 2g/L, boric acid @ 2g/L, or GA3 @ 20 ppm can also be applied at initial stage of aril development to reduce incidence of fruit cracking.

## Sunburn

Sunburn, also known as lesion browning or pericarp necrosis, is a direct thermal injury due to exposure of fruits to sunlight. Sunburn is more common when orchard temperature is > 40°C and relative humidity goes below 50%. Losses due to sunburn may go as high as 20%. Growing windbreaks around litchi orchards offers protection from hot, desiccating winds. Irrigation, especially using sprinklers, increases humidity levels and brings down the orchard temperature, thereby, reducing sunburn. In recent times, bagging whole litchi bunch with non-woven polypropylene bags has given encouraging control of sunburn injury.

## Irregular Fruiting

Irregular bearing at young stage of the litchi plant is a constraint in litchi particularly in cv. China. This phenomenon happens in litchi due to failure to bloom owing of the continuous vegetative growth of the tree. Girdling is the complete removal of a strip of bark, and underlying cambial tissue, from around the entire circumference of a limb/branch of the plant. Girdling operation is performed in litchi after the plant has produced the second vegetative flush, which corresponds to last week of August or first week of September under Muzaffarpur conditions. The operation is done by removing 2-4 mm wide bark 3-4 mm deep around entire circumference of limb using a sharp pruning saw. Girdling ensures flowering and fruiting in the ensuing season. Alternatively, flowering and fruiting can be ensured through application as PGRs (Ethrel, $KNO_3$ and paclobutrazol) as already discussed.

## Pericarp Browning

Litchi fruit is popular for its attractive bright red colour and captivating flavour. However, the fruit is highly perishable and deteriorates rapidly if proper handling techniques are not employed. Browning of the skin after harvest (pericarp browning) and fruit decay are the most important limiting factors to successful handling and marketing of litchi.

The anthocyanin content in litchi pericarp ranges from 300-500 mg/kg FW. Anthocyanins are localized in the mid to upper mesocarp, specifically in the vacuoles (Underhill and Critchley, 1993). Owing to proton gradient across the tonoplast and resultant accumulation of organic acids, vacuoles have an acidic medium with a pH low enough ($pH < 3$) for stability of anthocyanins (Menzel, 2002). The stability of anthocyanins in hydrated litchi pericarp is largely attributed to the compartmentalization within intact cells. Anthocyanin-oxidative enzymes, such as polyphenol oxidase, are known to be localized within the chloroplasts or other plastids. This compartmentation prevents contact of enzymes with phenolic substrates such as anthocyanins in intact tissues, unless there is disruption at the cellular level. However, moisture loss or desiccation during storage initiates several processes that ultimately lead to pericarp browning in litchi. High correlation between water loss and pericarp browning due to skin desiccation has been established. Desiccation and senescence result in disruption of cellular compartmentalization, due to which the membrane permeability increases. The adverse effect on membrane stability causes an increase of pH inside the vacuoles. The rise in pH and disruption of compartments activate the oxidative enzymes and bring them in contact with the anthocyanins. The enzymes finally oxidises the substrates into a range of brown pigments.

Maintaining attractive red colour of fruit along with freedom from decay and insect-pests become the primary objectives of every stakeholder in the litchi supply chain.Any management strategy aimed at maintaining litchi fresh after harvest must take the following basic approaches into account:

- Prevent or reduce moisture loss or desiccation
- Store at low temperature and high humidity
- Maintain membrane integrity and reduce oxidative damage
- Maintain low pericarp pH
- Delay fruit senescence
- Reduce activity of oxidative enzymes (PPO,POD etc)
- Control or inhibit decay pathogens.

## HARVESTING AND PRODUCTION OF FRUITS

Litchi is a non-climacteric fruit, and does not exhibit appreciable improvement in fruit quality after harvest. Therefore, litchi should be harvested when they are fully mature and ready to eat. Delayed harvesting results in short shelf life and such fruits do not travel well. Also, fruits harvested too early may not develop attractive colour and flavour, thereby reducing market value and consumer appeal. Therefore, judging the correct stage of harvesting is very important to assure good quality fruits to consumers and optimum returns to growers.

A simple and easy method of judging harvest maturity is fruit colour; deep red skin colour being a determinant of fruit being harvest ready. However, colour formation is dependent on several factors and sometimes even a deep red fruit may be still immature. Litchi is also said to have attained harvest time when the soluble solids concentrate (SSC) of fruit pulp climbs to 18-21°B. A very reliable harvest index is the ratio of soluble solids concentrate (SSC) to acidity, which should be at least 40. Flattening of tubericles is another harvest indicator.

Since harvest time coincides with peak summer months, it is advisable to perform harvest operation during early morning hours. Litchi is harvested by breaking of cutting the whole panicle using secateurs/clippers. Picking platforms and ladders are used to pick fruits in the higher canopy. Harvested fruits are placed in plastic crates or baskets. Care should be taken to avoid mechanical injury during harvesting and subsequent handling.

India is the second largest producer of litchi in the world, next only to China. Harvest and availability of litchi from India orchards starts from mid-April in Tripura until mid-July in Punjab and Himachal Pradesh (Table 7). A small quantity of litchi grown in the hills of South India is available in December.

**Table 7:** Harvesting period of litchi in India

| State (s) | JAN | FEB | MAR | APR | MAY | JUN | JUL | AUG | SEP | OCT | NOV | DEC |
|---|---|---|---|---|---|---|---|---|---|---|---|---|
| Tripura | | | | | | | | | | | | |
| Assam | | | | | | | | | | | | |
| West Bengal | | | | | | | | | | | | |
| Bihar | | | | | | | | | | | | |
| Jharkhand | | | | | | | | | | | | |
| Uttarakhand | | | | | | | | | | | | |
| Punjab | | | | | | | | | | | | |
| HP | | | | | | | | | | | | |
| Kerala, TN, &Karnataka | | | | | | | | | | | | |

## POST-HARVEST FRUIT TECHNOLOGY

Litchi is harvested during hot summer time. It is a highly perishable fruit. Harvested fruits rapidly lose moisture and dehydrate, resulting in what is commonly known as pericarp browning. If left exposed and unattended, harvested produce exhibit complete change of colour from attractive red to dull brown within 48 hours. Such brown fruits lose market value due to severe impact on consumer's decision to purchase. Successful litchi enterprise therefore rests in adopting appropriate postharvest management practices.

Successful handling of litchi involves a series of sequential steps right from production factors till the produce reaches the consumer.

1. Harvesting at optimum maturity: Based on TSS/Acidity ratio, skin colour etc depending on cultivar.
2. Sorting and grading: Cracked, sunburnt, diseased, insect-infested, and other culls should be sorted from the lot. Fruits of uniform size and colour should be graded together.

3. Precooling: Immediate and fast precooling of harvested fruit is essential.
4. Packaging : Use CFB boxes with polyethylene liners.
5. Storage/Shipment: Maintain uninterrupted cool-chain

Proper temperature management is the single most important factor to reduce postharvest deterioration and maximize postharvest life of litchi. Immediate and fast precooling of harvested fruit removes the field heat and provides the basis for effective temperature management during subsequent storage or shipment. Thereafter, the fruit should be stored (or maintained) at a temperature of 2-5 °C and 90±5% relative humidity. Litchi fruit under such conditions has a shelf-life of 3-5 weeks. Low temperature storage reduces pathological decay of litchi but aggravate browning of skin.

Sulphur dioxide ($SO_2$) fumigation has been the most effective, and commercially followed, postharvest treatment for control of pericarp browning in litchi fruit. This is achieved by placing boxes of fruit in a gas tight sulphitation chamber where fumes of burning sulphur is introduced to the desired concentration. About 400-450 g of sulphur is sufficient for a tonne of fruits. Alternatively, $SO_2$ can also be introduced into the chamber through pressurized cylinders and treated @ 1.2% for 10 min. Fumigated litchi fruit has a bleached-yellow appearance. The red colour of skin can be recovered by dipping the fumigated fruit in dilute acid solution (0.1 N HCl for 5 min). Despite effective management of pericarp browning there is increasing objection to use of sulphur owing to sulphite residue in skin and aril. The United States FDA and EU stipulate that sulphite residue in edible portion of litchi should not exceed 10 ppm. Therefore, world-wide research is on to find safer alternatives to $SO_2$ fumigation in litchi.

Edible coatings are popularly used in several fruits and vegetables to increase shelf life. However, in case of litchi coating materials have not met with similar results. Chitosan @ 1-2% dissolved in 2% citric acid has been found to be an effective surface coating to maintain freshness of litchi.

Atmosphere modification is another technology being used in postharvest science. In litchi an atmosphere containing3-5% $O_2$ and 5% $CO_2$ is recommended. Modified atmosphere packaging (MAP) using polyethylene bags can reduce moisture loss and extend shelf life but care should be taken to ensure that excessive build-up of humidity inside the package is avoided otherwise rotting may aggravate.

Litchi fruits are also treated with solutions of dilute acids to maintain the red colour of skin. Acidic medium helps in maintaining low pH which is essential for stability of anthocyanin pigments.

The pleasant flavour of litchi makes it a choice fruit for processing into different value-added products. Litchi Nut, Dehydrated litchi pulp, Squash, Ready-to-Serve (RTS), Litchi Nectar *etc* are some popular processed products of litchi. The processing protocols for such product should be guided by the food safety standards of respective countries (FSSAI in India) or internationally accepted regulator such as the Codex Alimentarius.

## INSECT-PESTS AND MANAGEMENT

Litchi is infested by several insect pests. While several insects indirectly affect yield by damaging or feeding on vegetative parts such as leaves, shoots or branches, others severely impact litchi production through direct feeding on fruit. These include litchi fruit and shoot borer, litchi mite, bark eating caterpillar, litchi leaf roller, semi-looper, red and ash weevil, litchi bugs etc. Among these, three major pests are briefly discussed below.

**Litchi fruit and shoot borer(*Blastobasis* sp.):** Litchi fruit and shoot borer is the most important insect pest in litchi ecosystem, and causes widespread economic losses. Females lay eggs singly near the pedicel end of litchi fruit. On hatching the larvae bores into the fruit and starts feeding below the calyx. The larval excrement can be visible on peeling the skin of otherwise healthy looking fruit, making it unsuitable for consumption. Such direct damage to fruits can incur losses to the tune of 24-48%. Additionally, the pest causes indirect damage by mining into midrib of young leaves, or tunnel through young shoots. Maximum population on shoots is observed during September-October when the larva damage newly emerged shoot/flush.

Litchi fruit borer population can be kept at bay by adopting an integrated pest management schedule. Precautionary measures include adopting good orchard management like pruning of infested twigs in June, field sanitation, and removal of young fallen fruits. A prophylactic spray of neem-based formulation (4ml/l) when new flush emerges during September-October, and a preventive spray of neem oil (4ml/l) before flowering can be done. Integrated management schedule includes two sprays of systemic insecticide viz. thiacloprid or imidacloprid @ 0.5 – 0.7 ml/l during September at 15 days interval. This should be followed by spray of Novaluron 10 EC @ 1.5 ml/l when fruit attains clove size, cypermethrin 25 EC @ 0.5 ml/l 25-30 days after fruit set, and another spray of novaluron 10 EC @ 1.5 ml/l about 10 days before expected harvest.

***Bark-eating caterpillar (Indarbela tetraonis):*** Bark-eating caterpillar is another major pest of litchi in almost all litchi-growing regions in India. The larva bore into trunks or older branches especially in shaded areas of neglected orchards. The pest can be easily located by the presence of long-winding, thick, brownish, ribbon-like masses/galleries composed of frass or excreta adhered together. Under severe infestation, the entire branch or tree can die. Normal functioning of plant is impaired due to disruption in translocation of sap, and growth is arrested.

The caterpillars can be killed by inserting an iron spoke into the tunnels. Another common method followed by orchardists is to remove the insect gallery from tree trunk and injecting kerosene oil/petrol/ emulsion of DDVP (0.05%) into the hole before being plugged by clay/mud. As a preventive measure, the attacked trunk and branches can be sprayed with 0.05% DDVP.

***Litchi mite(*****Eriophyes sp.*****):*** Mite can be another major pest in litchi orchards. Both nymphs and adults damage leaves by sucking sap from leaves, shoots, and even inflorescence and developing fruits. The problem of mite infestation is most severe under improperly managed orchards. Symptoms of mite infestation appear as velvety growth on lower side of leaf which enlarge and turn to chocolate colour with deep lesions. The leaves become thickened, wrinkled, distorted, and puckered. Infestation can spread from neighbouring plants and results in huge losses due to poor flowering and fruiting.

Field sanitation must be maintained and infested shoots, leaves and plant parts should be removed. Also, general practice of pruning litchi plants after harvest should be done. This should be followed by two sprays of chlorfenapyr 10 EC (3ml/l) or propargite 57 EC (3 ml/l) at 15 days interval during July. If further infestation is seen, such twigs should be pruned off in October and a spray of chlorfenapyr 10 EC (3ml/l) or propargite 57 EC (3 ml/l) can be done.

## DISEASES AND MANAGEMENT

Fungal diseases are more common in litchi and common fungi include *Alternaria alternata*, *Colletotrichum gleoesporioides*, *Aspergillus* sp., *Fusarium solani*, and *Botryodiplodia* sp. Sudden wilting of young plants has been reported whereby initial symptoms appear as yellowing of foliage in plants that are less than five year old. Such leaves droop and wilt gradually leading to death of the plant. Application of biocontrol agents such as *Trichoderma herzianum* helps recover the plants. *Alternaria alternata* has recently been found to be a major pathogen causing disease in litchi at multiple phenophases – nursery, leaf and panicles in orchard trees, and during postharvest stages. Fungicidal spray of copper oxychloride (0.25%), difenconazole (0.05%), or carbendazim (0.1%) can be done if disease severity increases. Anthracnose on fruit appears as brown pinhead lesions which turn to dark brown to black sunken lesions. Fruit rots occur in postharvest stages which can be easily prevented by minimizing mechanical injuries to fruit during handling and following appropriate postharvest management measures. A pre-harvest spray of carbendazim (0.1%) 15-20 days before anticipated harvest reduces microbial load that carries to postharvest stages.

## MARKETING OF FRUITS AND EXPORT POTENTIAL

Litchi is a popularly traded fruit crop. While the short postharvest life has limited litchi marketing and trade within the place of production for a major portion of the produce, improvements in postharvest handling practices have allowed for sea shipment to distant markets. The major international markets include Hongkong, Singapore, Malaysia, Japan, U.K., Germany, France, USA, U.A.E., Saudi Arabia, Yemen, Lebanon, Dubai, and Canada. Countries that export litchi include China, Taiwan, Thailand, Australia, India, Madagascar, South Africa, and Israel. It is estimated that only about 5 per cent of the total litchi production is traded the world over, of which the major trade is in fresh form, while South Africa exports about 60% of production. This figure is likely to increase with development of suitable postharvest handling protocols, and growers' awareness of market intelligence. During the Indian litchi season (May-June), good quality litchi is not available elsewhere except in Thailand. This offers tremendous potential to exploit the international market for export.

## FUTURE STRATEGIES

Litchi is an introduced crop in India and it is therefore obvious that genetic base is narrow. The fruit is available only for a few weeks in India. Wide genetic variability is present in its place of origin. Efforts need to be directed for systematic crop improvement programmes to widen the country's genetic base to gear up for changing climatic conditions. Rising human population puts tremendous pressure on land resources, and farmers are under increasing strain to produce more from decreasing land resources. Modern canopy architecture needs to be designed and formulated to accommodate more litchi plants compared to conventional planting, and improve productivity. Emerging pest complex and diseases under changing climate need concerted efforts. Postharvest management needs to be improvised to extend shelf life and marketing of litchi. Under the food revolution, food processors will require tapping into the litchi industry to reduce postharvest losses through diversification into different value-added products.

## LITERATURE CONSULTED

Bose, T. K., Mitra, S.K., and Sanyal, D. 2001. Fruits: Tropical and Subtropical, Volume I. Third revised edition, Naya Udyog, Kolkata, India

Chadha, K.L. Handbook of Horticulture. Indian Council of Agricultural Research, New Delhi.

Chauhan, K.S. 2001. Litchi: Botany, Production and Utilization. Kalyani Publishers, New Delhi.

Ghosh, S.P. 2005. World trade in litchi: Past, present and future. *Acta Hort.*, 558: 23-30.

Holcroft, D.M. and Mitcham, E.J. 1996. Postharvest physiology and handling of litchi (*Litchi chinensis* Sonn.). *Postharvest Biol. Technol.*, 9:265-281.

Huang, H.B., and Xu, J.K. 1983. The developmental patterns of fruit tissues and their correlative relationships in *Litchi chinensis* Sonn. *Scientia Hort.*, 19: 335-342.

Jiang, Y.M. 2000. Role of anthocyanins, polyphenol oxidase, and phenols in litchi pericarp browning. *J. Sci. Food Agri.*, 80: 305-310.

Li, J.G., Huang, H.B., Gao, F.F., Huang, X.M., and Wang, H.C. 2001. An overview of litchi fruit cracking. *Acta Hort.*, 558: 205-208.

Menzel, C.M. 2002.The lychee crop in Asia and the pacific. RAP Publication: 2002/16. FAO. Pp. 115.

Menzel, C.M. and Waite, G.K. 2005. Litchi and Longan Botany, Production and Uses. Cabi Publishing, Wallingford, Oxfordshire OX10 8DE, UK. ISBN: 0851996965.

Mitra, S.K. 2002.Overview of lychee production in the Asia-Pacific region. In: Papademetriou, M.K., and Dent, F.J. (eds) Lychee production in the Asia-Pacific region. Food and Agriculture Organisation of the United Nations. Bangkok, Thailand.

Nath, V., Ahmad, F., Mir, H., Kundu, M., Sahay, S., Pandey, S.D., Srivastava, K., and Pongener, A. 2016. Litchi: Global Perspectives. Bihar Agricultural University, Sabour – 813210 (Bhagalpur), Bihar. Excel India Publishers, New Delhi. ISBN: 978-93-85777-71-4.

Paull, R.E., Chen, C.C., and Chen, N.C. 2004. Litchi. In: Gross, K.C., Wang, C.Y. and Salveit, M. (eds), The commercial storage of fruits, vegetables, and florist and nursery stocks. Agricultural Handbook No.66. USDA, ARS.

Singh, G., Nath, V., Pandey, S.D., Ray, P.K., and Singh, H.S. 2012.The Litchi. Food and Agriculture Organization of The United Nations. New Delhi, India.

Singh, H.P. and Babita, S. 2002. Lychee production in India. In: Papademetriou, M.K., and Dent, F.J. (eds) Lychee Production in the Asia-Pacific region. Food and Agricultural Organization of the United Nations, Bangkok, Thailand, pp. 106-113.

U.S. Department of Agriculture. 2013.Composition of Foods Raw, Processed, Prepared. National Nutrient database for standard reference. Release 26. USDA, Maryland, USA.

Underhill, S.J.R. and Critchley, C. 1993. Physiological, biochemical and anatomical changes in lychee (*Litchi chinensis* Sonn.) pericarp during storage. *J. Hort. Sci.,* 68: 327-35.

# 6

# Aonla

*Devi Singh and J.S. Bal*

## INTRODUCTION

Aonla or Indian gooseberry (*Emblica officinalis Gaertn.)* is an important indigenous minor fruit grown in tropical and subtropical regions of India. It is known with many vernacular names such as amla, aura, amlaki, nelli, amolphal, amlakamu, amlet, amlay. Aonla, cultivation is highly remunerative due to its high productivity even on marginal lands. The plant is quite hardy, heavy bearer giving better income to the farmers. It can be grown successfully in marginal soil and various kinds of wasteland conditions such as sodic and saline soils, ravines, dry and semi dry regions.

## NUTRITIVE AND CULTURAL SIGNIFICANCE

Aonla fruit is highly nutritive and popularly known as Amrit Phal. It is one of the richest natural source of vitamin C and contains 600-700 mg of ascorbic acid per 100g of pulp. Aonla is also rich in pectin, minerals such as iron, calcium and phosphorus and fair source of thiamine and riboflavin. The fruit contains 14 per cent carbohydrate 0.5 per cent protein, 30 mg/100g fruit vitamin $B_1$, 1.2 per cent iron, 0.7 per cent mineral matter and 3.4 per cent fibre.

Aonla fruit is valued high among indigenous medicines in India. It is the fruit to fill the gap of astringent food recommended by the Aurvedic and Unani system of medicines for balanced diet and sound health. Fruit is acidic, cooling, refrigerent, diuretic and laxative. Aonla dry powdered is useful in haemorrhage, diarrhoea, dysentery, diabetes, jaundice, dyspepsia, cough, piles and dermatitis. Aonla fruit increases food absorption, balances stomach acid, nourishes brain and heart, fortifies liver, strengthen lungs, enhances fertility, flushes out toxins, increases vitality and promotes healthier hair. It is the main ingredients of chyawanprash and one of the ingredient in Triphala and used in the treatment of headache , constipation and enlarged liver.

Aonla fruits are used as preserve (murabba), pickle, chutney, jam, jelly and candy. The fruits can be dried and powdered for off season use. It is commonly used in preparation of ink, hair oil and hair dyes. Aonla and its products are exported to different countries and importing countries are U.K., Germany, Italy, Belgium, Saudi Arabia, Bahrain, Hong Kong, Malaysia, Indonesia, Nepal, Pakistan, USA etc. Aonla fruit contains gallic acid which checks oxidation of vitamin C hence it has antioxidant property. Therefore, the fruit is rich source of vitamin C both in fresh and processed form.

Aonla leaves are offered to the Lord of Shri Satyanarayana Vrata, Shiva and Gowri on Nitya Somvara Vrata. The fruit and flowers are also used in worship. Aonla is regarded sacred by "Hindus" and has great mythological significance. Pind Dan is done near the base of the tree to get salvation of the elders.

## ORIGIN, HISTORY AND DISTRIBUTION

Aonla is native to tropical southeast Asia specially Central and Southern India, Sri Lanka, Bangladesh, Malaya and China. It grows wild in foot hills of northwest Himalayas to eastern Himalayas in Assam, Meghalaya, Manipur, Mizoram and Tripura and at elevation from 1250 to 1800 MSL in South India.

Aonla is known and grown in India for the last 3500 years. Sushruta mentioned its use in Ayurveda during 1500 BC-1300 BC. It finds a prominent place in ancient Indian mythological literature like Vedas Ramayana, Padmapuran, Kadambari, Charak Sanghita, Sushrut Shanghita, Askandhpuran etc.

Aonla is grown commercially in many countries of the world like India, Sri Lanka, Malaya, China, Pakistan, Singapore, West Indies, Thailand and USA (Hawaii and Florida). In India, its cultivation was first started in Varanasi. India ranks first in the world in area and production. Uttar Pradesh is the leading state followed by Madhya Pradesh and Gujarat. Other aonla growing states are Tamil Nadu, Jharkhand, Chhattisgarh, Assam, Bihar, Rajasthan, Haryana, Andhra Pradesh and Punjab.The total area under aonla cultivation in India is 95.09 thousand hectares with an annual production of 11.73 lakhs mt. The average productivity is 12.34 tonnes/hectare.

**Table 1:** Area, production and productivity of leading aonla producing states.

| Sr. No. | States | Area (000ha) | Production (ooo mt) | Productivity (tonnes/ hectare) |
|---|---|---|---|---|
| 1. | Uttar Pradesh | 34.34 | 374.28 | 10.90 |
| 2. | Madhya Pradesh | 13.98 | 373.00 | 26.68 |
| 3. | Gujarat | 9.67 | 95.63 | 9.89 |
| 4. | Tamil Nadu | 8.22 | 173.74 | 21.14 |
| 5. | Jharkhand | 7.90 | 33.82 | 4.28 |
| 6. | Chhattisgarh | 3.35 | 36.21 | 10.81 |
| 7. | Assam | 0.90 | 16.27 | 18.08 |
| 8. | Bihar | 1.00 | 15.00 | 15.00 |
| 9. | Rajasthan | 1.61 | 12.89 | 8.01 |
| 10. | Haryana | 2.23 | 11.02 | 4.94 |
| 11. | Andhra Pradesh | 0.85 | 9.17 | 10.79 |
| 12. | Punjab | 0.44 | 6.06 | 13.77 |

(*Source:* NHB, Indian Horticulture Database 2014-15)

## TAXONOMICAL AND BOTANICAL DESCRIPTION

Aonla was earlier categorised under genus *Phyllanthus* which comprises of about 500 species. As per nomenclature, aonla belongs to genus *Emblica* and family Euphorbiace. Chromosome No.X=7,2n=28. About 350 to 500 species have been reported under this genus. The important species are as follows.

***Emblica officinalis*:** This species is evergreen in tropical climate but deciduous under sub tropical conditions.Trees are spreading in nature and attain height of 10-20 meters under different soil conditions.

***Emblica fischeri*:** Trees are small bearing large leaves. The fruits of this species are used for making pickle and chutney.

***Emblica myrobalan*:** It is a small deciduous tree. Fruits are 1.5-4.0 cm. in size and are used for preparation of various products.

### Botanical Description

The cultivated species of aonla *Emblica officinalis* is a monoecious plant having male and female flowers separately on the same panicle. The inflorescence is racemose type. Flowers are minute, unisexual, with short pedicel. Male flowers appear first in clusters and bear in very high proportion, yellowish green to deep pink in colour. Female flowers have tiny green perianth and numbers of segments varies from 5 to 7 but commonly 6. These flowers bear on upper end of few branches. Ovary hypogynous, carpels 3-4, 3 chambered, placentation axile, 2 ovules per locule . Fruit is capsular, oval round to flattened round or triangular in shape, colour whitish green to yellowish green, skin smooth to rough, semi translucent, segments 6-8, surface smooth to raised. Stone hard, round to triangular in shape, seed 6-8, light brown to deep brown in colour and seed coat hard. Edible part is epicarp and mesocarp.

## CLIMATIC AND SOIL ADAPTABILITY

### Climatic Adaptability

Aonla is a hardy tree and can be grown under different agro-climatic conditions. It is a sub-tropical fruit but can be grown successfully in tropical climate The grown up plant can tolerate temperature as high as $46^0$ C. Aonla prefers distinct winter and summer for high productivity. In India, aonla is being grown from near sea coast to up to 1800 MSL. It is sensitive to much hot and frost but can tolerate mild frost under north Indian conditions. The tree can be damaged with the occurrence of severe frost. Vegetatively raised plants are more prone to frost injury as compared to seedlings. The young plants upto the age of 3-4 years should be protected from frost during winter and hot desiccating wind in summer. Low temperature of $7^0$C to $10^0$C is favourable for flower bud initiation.

Aonla is highly suitable for growing in arid conditions because soon after fruit set in springs, the fruits remain dormant throughout summer without any growth. Aonla tree can tolerate drought for longer peiod. There is varietal difference in tolerance to drought, Chakaiya is more tolerant to drought than Banarasi. The fruit growth in aonla takes place during July-August. The fruit develop good size under high humidity during growth and development of fruits.

### Soil Adaptability

Aonla is not very exacting in its soil requirement. It is a hardy tree and can be grown on variable soil conditions ranging from sandy loam to clay loam. But, well drained , fertile loamy soil is best for its cultivation. It is a potential crop for degraded lands and the marginal soils having soil pH 6.0 to 9.5. Aonla has good tolerance to saline, sodicity and moderately alkaline soils.

Aonla is an ideal plant of wastelands and road side plantation. Locations having high water table should be avoided for aonla planting. Stagnant water for a longer period particularly during rains is harmful for its growth and root system. It is more salt tolerant crop and has better potential for commercial cultivation in salt affected soils. Seedling plants show better tolerance than budded one. The cultivars Francis and Kanchan have higher tolerance to sodicity and salinity than cultivar Banarasi and Krishna.

## RECOMMENDED AND POPULAR CULTIVARS

In India most of the varieties have been developed through selection particularly from NDUAT, Faizabad, GAU, Gujarat and RBS College, Agra. The varieties like Banarasi, Chakaiya, Kanchan, Francis, Krishna, NA-7 and NA-10 are commonly growing in North India.

### Salient Features of Important Aonla Cultivars

**Banarasi:** It is a seedling selection from Varanasi. Trees are semi tall with spreading growth habit (P 1). Shy and slightly alternate bearer. Fruits are large (50g) in size, slightly conical at apex and light yellow at full maturity lobed and slightly fibrous. It is prone to heavy fruit drop. Keeping quality is poor.

**Plate 1:** Banarasi tree in dormancy

**Plate 2:** Chakaiya tree in dormancy

(See colour version on page 315 )

**Chakaiya:** It is also a seedling selection. Trees are medium tall with upright growth habit and regular prolific bearer (P 2, P 3). Medium size fruit, flattened at base and round apex, greenish colour and fibrous. Average fruit weight 28g. Keeping quality very good and suitable for pickle.

**Francis (Syn. Hathijhool):** It is a seedling selection. Trees are erect and tall with drooping branches, good and regular bearer. Fruits are large in size weighing about 60g, oval roundish, light green, slightly fibrous but susceptible to necrosis.

**Kanchan (NA.4):** A seedling selection from Chakaiya. Trees are tall with upright growth habit. It is heavy and regular bearer with medium size fruits having higher fibre content. Average fruit weight is 30g. Fruit skin smooth and light green matures in mid December. This variety is preferred by industries for pulp extraction and various product preparation. It is well adopted in semi-arid regions of Gujarat.

**Krishna (NA-5):** It is a chance seedling of Banarasi from Pratapgarh district of UP. Trees semi tall with spreading growth habit. It is early maturing variety having shy bearing. Fruit weight 40-50g, attractive pinkish green flesh and less fibrous.

**Amrit ( NA-6):** A selection from Chakaiya. Tree tall with semi spreading growth habit. It is prolific bearer and late maturing. Fruits medium to large in size (35-37g) with bright appearance. It is ideal for candy, preserve, jam, Chavanprash, fruit bar and toffee being very low in fibre content.

**Plate 3:** Chakaiya tree in blooming (See colour version on page 316 )

**Neelam (NA-7):** It is a selection from Francis. Tall, semi-spreading growth habit. It is a precocious, prolific and regular bearer. Average fruit weight is 40g, skin smooth, semi-translucent and yellowish green. Almost fibreless and soft flesh. It is mid maturing variety and suitable for processing purposes.

**NA-9:** A seedling selection from Banarasi. The trees are tall and semi spreading. Fruits large, smooth, semi-translucent, light green in colour. Early maturing and shy bearer. Keeping quality is moderate. It is susceptible to fruit necrosis.

**NA-10:** It is a chance seedling of Banarasi also known as Agra Bold. Trees are semi tall having semi spreading growth habit and dense foliage. Fruits are medium in size and flattened round in shape. Fruit skin is rough, Yellowish green with pink tinge. Highly astringent, good keeping quality. It matures in mid-November. Recommended with the name Balwant in Punjab.

**Gujarat Amla-1:** A seedling selection. Trees large with spreading habit having prolific bearing. Fruits bigger in size of good quality.

**Anand-1:** It was developed at GAU, Anand. A moderate bearer.

**Anand-2:** It was also developed at GAU, Anand. It is prolific bearer, fair keeping quality.

**Anand-3:** Developed at GAU, Anand. A moderate bearer yellowish green fruits having 6-7 segments.

**Laxmi-52:** Identified by CISH, Lucknow.

## PROPAGATION TECHNOLOGY AND ROOTSTOCKS

Both sexual and asexual methods of propagation are followed in aonla in different parts of the country.

### Seed Propagation

This method is commonly used in various parts of the country but does not ensure true-to-type plants due to cross pollination. The plants raised sexual comes into bearing late and do not produce uniform fruit size.

### Vegetative Propagation

Aonla is commercially propagated by patch budding method during June-September.Forkert method of budding is also used for its propagation which give high success from July to September. The plants can be propagated through T-budding method during rainy season from July to September. This method gives high success up to 80 per cent. Inarching or approach grafting can also be practiced with good success. Poly and net house, containers and rooting media have also proved very useful in propagation of aonla plants.

**Raising of Rootstock:** The fruits of aonla (desi) are collected during January-February and seeds are extracted. The seeds are soaked in water. Sinkers are used for sowing and floaters are discarded being embryoless. The seed should be sown in first fortnight of March in north India. The seeds can be sown on seed beds or in polythene bags just after harvesting as there is no seed dormancy. The seeds can be soaked in $GA_3$ @500 ppm for 24 hours to improve the germination percentage. The seeds germinate in 5-7 days and seedlings become ready for budding within 4-6 months. Patch budding is done from June to September when the seedling become 1.0 to 1.2 cm in diameter at 15 cm above the ground level. When the bud take has occurred, the rootstock should be cut above the bud union. Retain the vigorously growing shoot when the shoots from the bud attain a length of 15-20 cm and remove the weaker one. The sprouts below the bud union should be removed regularly to have a good size healthy plant.

## PLANNING AND PLANTING

Plantation of aonla can be done during rainy season and spring season depending upon the location and availability of plants, Normally aonla is planted during August-September in rainy season and mid-January to February-March in spring. Bare-rooted plants can be planted during January-February with good success. Pits of 1x1x1 m size are dug about a month before planting. Each pit should be filled up with the mixture of top fertile soil and well rotten FYM (30kg) and chlorpyriphos for better establishment of plants.

### Planting Density

The planting distance is kept 8-9 m in square system which accommodate 180-140 plants per hectare. Plantation of aonla is also made at 7.5 x 7.5 m in square system which accommodates 180 plants per hectare. Aonla is a cross-pollinated crop, therefore, plants of two or more varieties should be planted together to ensure proper pollination, fruit set and yield. The cultivar NA-7, NA-6 and Chakaiya may be planted as pollinizer. Hedge-row planting at a distance of 8x4 m and triangular planting system can also be adopted for effective utilization of space and solar radiation. Young aonla plants need protection from frost by erecting thatches during winter at early stage after transplanting.

## PLANT AND FRUIT PHYSIOLOGY

Aonla is a deciduous plant in subtropical conditions and shed its leaves during winter months. However, it behaves as an evergreen plant in tropical climate. In aonla flowering takes place on newly emerging determinate shoots begin to appear at the nodes in spring season. The flowering takes place during March-April in north India whereas in south India it is between June –July in addition to February - March. Aonla tree is sensitive to day length. Flower bud differentiation takes place in first week of March in north India. Male flowers appear in clusters in the axils of leaf all over the branchlet. Female flowers appear only on the upper end of few branchlets. Early flower initiation takes place in cv. Chakaiya and late initiation in cv. Kanchan.

Initial fruit set in aonla varies from 12 to 18 per cent. Fertilization takes place within 36 hours after pollination. Immediately after fertilization, zygote rests for 120-130 days and endosperm nucleus may rest for 70-80 days after fertilization in cv. Banarasi. Thereafter, fruits do not show any growth during summer months. Dormancy breaks as a result of division in endosperm nucleus in end July to realy August.

The fruit starts growing continuously and maximum growth takes place in the month of September.The fruit growth follows a pattern of double sigmoid. The fruit growth after dormancy is slow in first fifteen days. It show rapid growth from mid-August to early October. The growth of fruits slow down again between second week of October to early November and registered slight increase after that. It cease to grow after November. The seed length and weight increases rapidly during first rapid growth period of fruits when their size varies from 1 to 3.5 cm. The fruit is drupe. It has freshly exocarp and thick and massive endocarp. The fruits mature from end November to early December and continue ripening up to February. The fruits become dull and greenish yellow at maturity and ripening.

## SOIL CULTURAL PRACTICES TECHNOLOGY

### Water Need

Aonla is a hardy fruit tree and fully established tree can even withstand drought conditions. It is generally grown as a rainfed crop and little or no irrigation is provided. But for better establishment of orchard in early years few frequent irrigations are required. After planting, young plants should be irrigated at 3-4 days intervals during summer and at 10-15 days in winters. In bearing orchard, regular irrigation is needed for the development of embryo. The plants should be given irrigation after 10-15 days interval from March to mid-July in light textured soils. During rainy season, no irrigation is given to the aonla orchards.

The active fruit development period of aonla takes place in September and October. The plants need irrigation during growth and development period at monthly interval. The fruit maturity takes place from November to January, only one or two light irrigation can be given if needed. December to mid February are the coldest months in north India. During this peak winter period, if there are chances of occurrence of frost, light irrigation should be given to protect against frost. Avoid irrigation during flowering period. In general, 10-12 irrigations are required in a year..

The drip irrigation is well suited for widely spaced aonla crop in wasteland situation. This system can be adopted for undulating terrain and having shallow and porous soils and in water scarcity areas. The water savings with drip system is 40-70 per cent. It enhanced plant growth,check weed growth, improve fruit yield and quality.

## Nutritional Need

In aonla manures and fertilizers doses are fixed on the basis of soil fertility, age of plant and productivity. Nitrogen application encourages vegetative growth and female flowers whereas phosphorous improves sex ratio, initial fruit set, fruit retention, yield and quality parameters such as total soluble solids and vitamins. Potash improves fruit retention and quality of fruits. Aonla bear heavily and continue to bear up to about 50 years under proper conditions In general farmyard manure, nitrogen, phosphorous and potash should be applied @ 10 kg, 100g, 50g and 75g per tree per year age of the tree, respectively and quantities should be stabilized at 10 years of age. The entire quantity of farmyard manure, phosphorous and half of nitrogen and potash must be applied in February before flowering and remaining half of nitrogen and potash during August. The commonly followed fertilizers schedule in different parts of the country is given below in Table 2.

**Table 2:** Manuring and Fertilizers Schedule for Aonla

| Age of Tree | FYM | Urea | Superphosphate | Muriate of Potash |
|---|---|---|---|---|
| (years) | (kg/tree) | (g/tree) | ( g/tree) | ( g/tree) |
| 1-3 | 10-20 | 100-300 | 150-450 | 40-120 |
| 4-6 | 25-40 | 400-600 | 600-900 | 160-240 |
| 7-10 | 40 -50 | 700-1000 | 1050-1500 | 280-400 |
| 10& above | 60 | 1000 | 1500 | 500 |

The whole quantity of FYM,superphosphate , muriate of potash and half dose of urea is applied in January-February. The second half of urea is applied in August.

In Punjab, 15-20 kg farmyard manure is applied to young aonla plants and 30-40 kg to mature plants during July-August. In addition, 30 g nitrogen per plant is applied for each year of age of the plant up to 10 years and afterwards 500 g of nitrogen to full grown plants. In Rajasthan, five years old aonla plant is given 50-60 kg FYM, 600 g nitrogen, 300 g phosphorus and 375 g potash. FYM, phosphorus, potash and half nitrogen is given in February before flowering. Remaining nitrogen is applied during August.

## Micronutrient Deficiency

Boron deficiency causes fruit necrosis in aonla plants. For its correction, spray borax @ 0.6 per cent in September-October at fortnightly interval. Zinc and iron deficiency may occur in aonla in certain localities. These deficiency can be corrected by spraying of 0.5 per cent zinc sulphate and iron sulphate . Soil application of zinc sulphate @ 250-500 g per plant is beneficial to overcome the zinc deficiency.

## Inter Culture

Flowering in aonla takes place in February and after fruit set the fruits remain dormant during summer until monsoon. After this fruits begin to grow and harvested in December-January. Soils during summer months should not be disturbed and no cultivation is done. The basins should be cleaned with the beginning of monsoon and green manuring can be raised between vacant space in rows.

Intercropping in pre bearing stage is very beneficial particularly during the initial 3-4 years as sufficient space is available for raising these in between the tree rows. During November to February in winter months, when tree has sparse or no foliage, vegetable like pea, cauliflower, spinach, tomato, brinjal etc. can be grown. During pre-bearing period, leguminous crops like gram, horse gram, cow peas etc,can be intercropped which enrich the orchard soil. The fillers like karonda, phalsa and ber can be intercropped particularly in sodic and saline soils in each square or between the tree rows to supplement the income during prebearing period .

## Weed Control

Weeds grow luxuriantly in aonla field due to enough vacant space during initial years of planting. The weeds can be removed by doing regular hoeing or spraying herbicides for better root growth and moisture conservation. *Cyperus rotundus, Chenopodium album, Phyllanthus nineri, Bergia ammanoides, Calotropsis procera, Cenchus sp., Zizyphus sp., Prospis sp., Trigonella polycersta*, etc. are most common weeds grown in aonla orchards (Makhija and Singh, 2010).

Glyphosate @ 1.5 *l*/ha as post-emergence was the most effective herbicide in checking all types of weeds in aonla orchards. Pre-emergence herbicide diuron @ 3 kg/ha in first fortnight of March or glyphosate @ 3*l*/ha in second week of March in 500 litres of water can be sprayed to check weed flora in aonla orchards. If weeds grow continuously after rainy season, then give another spray of gramoxone @ 3 *l*/ha. It can be very helpful in controlling the weeds round the year.

## Mulching in Orchard

Mulching in aonla encourage tree growth,check weed growth, reduce moisture evaporation and regular soil moisture. The surface area or tree basin are covered with organic or inorganic mulches during active growth period. Black polythene and paddy straw used as mulch in aonla orchard are very effective in induction of early and uniform flowering which regulate fruit maturity and produce better coloured fruits. The practice of covering the soil with black polythene is very effective in conserving moisture, checking weeds and reducing the fruit drop.

Mulching with organic materials like paddy straw or sugarcane trash are helpful in improving the soil fertility in sodic soil in aonla plantation. Paddy straw, sugarcane trash, hay, banana leaves are generally used in aonla orchards. The thickness of organic mulch should be 20-30 cm and not be in close contact of the tree trunk. Paddy straw mulch is commonly used in north India due to its easy availability. Mulching in aonla should be done during December-January after addition of manures and fertilizers.

# PLANT CULTURAL PRACTICES TECHNOLOGY

## Training and Pruning

**Training:** Aonla plants should be trained to modified leader system for proper framework. There will be little or no limb breakage if tree is well trained.The plants should be encouraged to develop a medium headed tree. The plants should be trained to single stem up to the height of about 1 m. Then select 4 to 6 well spaced main branches on all directions around the trunk. Aonla bear heavily and due to fruit weight tender and lengthy branches tend to break.

**Pruning:** The trees does not require regular annual pruning. However, to give proper shape to the tree and to allow primary branches at regular space all around the trunk, corrective pruning should be done. After harvesting the crop, remove dead, diseased, intercrossed and infested twigs for further normal growth of the tree.The pruning of bearing trees can be done after the termination of the crop each tear.

**Rejuvenation:** Inferior old trees of aonla can be rejuvenated in to superior type by doing top working. Inferior trees can be headed back to a height of about 1.2m from the ground level in March and stump produced shoots can be budded in May-June with the scion of improved cultivars.

## Fruit Drop

The fruit drop is of great concern to the aonla growers as it reduces fruit yield to considerable extent. Flower and fruit drop in aonla occur at different stages.

Initially, about 70 per cent flowers drop down within three weeks of flowering due to unfertilized ovaries and degenerated ovules. The second drop consists of drop of the young fruit lets at the time of dormancy from June to August which may be due to lack of pollination and fertilization. Finally, the third fruit drop spread over a period of rapid growth from August to October and it may be due to embryological and physiological factors such as lack of auxins, improper nutrition and moisture stresses. Heavy fruit drop in aonla cultivars has been reported 12 to 38 per cent in August and 35 to 54 per cent in September.

The minor elements play an important role in aonla nutrition with regard to fruit size, yield and quality. Zinc @ 0.6 per cent as foliar application proved helpful in checking fruit drop in Banarasi cultivar. The minimum fruit drop up to 50 to 60 per cent was recorded with zinc application.

### Pollination

Aonla is a cross-pollinated crop. Honey bees are the main pollinating agent. Wind also play an important role in pollination process. Pollen viability and pollen grain germination varies from 82-96 and 16.7 to 37 per cent, respectively. Most of the aonla cultivars are self-incompatible and need pollinizer in order to improve fruit set and retention. Krishna (NA-5) and NA-6 are found to be best pollinizer for NA-7 cultivar.

## HARVESTING AND PRODUCTION OF FRUITS

The seedling trees of aonla starts giving commercial bearing after 10-12 years. But, the budded plants start bearing after 4-5 years of planting and the trees starts giving regular yield at the age of 10 years. Productive life of aonla tree is 50-60 years. The fruit are ready for harvest in November-December in North and in West India are harvested up to the month of February. In South India the availability of fresh aonla is from March to December and in East India from October to January. Aonla fruits should be harvested at full maturity.

**Maturity Indices**:Maturity is a stage when fruit attains full growth. Maturity can be judged by the change of seed colour from creamy white to black or change in fruit colour or by development of translucent exocarp. On the basis of maturity season, Krishna, Banarasi and NA-10 are early ripening cultivars. Francis,NA-7 and NA-8 are rated as mid season cultivars whereas Chakaiya and Kanchan are late ripening cultivars. Fruits are manually harvested individually. For harvesting shaking of tree is not a good practice and fruits should not be allowed to fall on the ground. The injured fruits cause spoilage to good fruits during packaging, transportation and storage. Ladder can be used for ease in harvesting. Use of plastic trays is good for collection and transportation of fruits. All the fruits on the tree do not mature uniformly. Therefore, two to three pickings should be done for harvesting of all the fruits on the tree. Peduncle should be properly clipped to check the spoilage during transportation.

**Yield:** The full bearing trees of different cultivars can yield 100 to 150 kg fruits per tree. Kanchan, Francis, Neelum and Balwant are high yielding cultivars which produce about 120 kg fruits per plant

## POST-HARVEST FRUIT TECHNOLOGY

### Grading and Packaging

The fruits should be graded into three grades viz A, B and C grades on the basis of size. In general, the fruits are large sized with diameter 4cm and above in A grade, small sized with less than 4cm in B grade. All the defective fruits should be placed in C grade. Both A and B grade healthy fruits are used for preserve and candy making. It is common practice in industry that small-sized fruits are used for making Chyawanprash. The fruits are commonly packed in gunny bags, nylon nets, perforated polythene bags, wooden crates and baskets. Baskets of 40-50 kg capacity lined with newspaper and aonla leaves as cushioning material are being used for packaging. However, the fruits should preferably be packed in CFB boxes for profitable marketing.

### Transportation of Fruits

In general, aonla fruits are transported loose in gunny bags or in the baskets. This cause damage to the fruits due to bruising and temperature build up.The fruits should be properly packed for sending to distant markets in transport vehicles. The aonla fruits are transported through tempo, three wheelers, trucks and rail. There is need to study the most suitable transport for this highly perishable fruit.

### Storage of Fruits

Aonla fruits are highly perishable in nature. The fruits should be handled quickly as period between harvesting and consumption is very limited. Shelf life of aonla fruits at ambient temperature is about one week but can be extended upto 12 days by storing in perforated polythene bags. The spoilage starts after a week at room temperature which is caused by blue mould. In zero energy cool chamber, the fruits of Chakaiya,Francis and Kanchan could be stored up to 18 days and of Banarsi and Krishna up to 12 days. The fruits can be kept at low temperature ($10^0$ C) for two weeks . Shelf life of fruits may be extended up to 75 days in 15 per cent brine solution. Pre-harvest spray of calcium nitrate 1 per cent + 0.1 per cent Topsin-M proved beneficial in extending storability of fruits up to 20 days at ambient temperature.

## Processing of Fruits

Aonla is processed into number of products which are easy available in the market. The various products like squash,RTS, jam, syrup, preserve, candy, toffees, pickle, sauce, churan, cosmetics are prepared from aonla. The major food product prepared from aonla is Chayawanprash .

## INSECT –PESTS AND MANAGEMENT

**Bark-eating caterpillar (*Indarbela tetraonis*** Moore): The pest is very common in neglected and poorly managed aonla plantation. It cause damage by eating the bark and boring holes into the main trunk, stem and branches. Due to bark eating, shoots and main trunk get girdled and dries up. Under severe attack tree growth,flowering and fruiting is affected. The full grown tree dries up. The infestation of this pest starts in April with the emergence of moths. To control this pest remove the webs and inject water emulsion of 0.05 per cent chlorpyriphos and plug the holes. The larvae of the pest are parasitized by gungus *Beauveria bassiana* in nature. This can be used as a potential bio-control agent.

**Leaf roller (*Garcillaria acidula*):** Larvae of moth bind the leaves together and feed there in. It reduce the photosynthetic capacity of leaves and causes leaf shedding. Collect the rolled leaves and destroy them along with the larvae. Spray 0.2 per cent malathion 0.05 per cent quinalphos.

**Aphids [*Cerciaphis emblica*** (Patel and Kulkarny), ***Schoutedonia emblica*** (Patel and Kulkarny) and ***Setaphis bougainvillea*** (Thunberg) **]**:The incidence occur from July to October.The leaves turn yellow and dry up. Infested shoots appear bended and twisted at the growing points. The nymphs and adult females suck the sap. It affect the tree growth and vigour and flowering and fruiting is suffered heavily. Remove the affected leaves and shoots and spray the plants with 0.06 per cent dimethoate

**Shoot gall maker ( *Betousa stylophora*** Swinhoe)**:** The pest remain active from June to December and cause gall formation on stem and shoots. Fresh galls are formed from June to August. However. The infestation lead to stunted growth of the trees. The terminal shoots swell and increase in size with passage of time.The galls can be visibly seen in October-November. It can be controlled by removing the galled shoots and destroy them. Spray the plants with 0.05 per cent chlorpyriphos and repeat spray at fortnightly intervals if need arises.

**Mealy bug** ( ***Nipaecoccus vestator*** Newstead)**:** The attack of mealy bugs is noticed from April to November. Both nymphs and adults feed on tree and cause drying and dropping of leaves and flowers. To control the pest prune the affected parts of the tree and destroy them. Spray 0.05 per cent quinalphos. Excessive excretion of honey dew is noticed. Prune the affected parts and destroy them at early stage of infestation.

## DISEASES AND MANAGEMENT

**Ring rust (*Revenelia emblicae* Styd.) :**The disease is common in aonla plantation in Rajasthan. On fruits, initially few black pustules appear on fruits which later develop in a ring. These postules join together and cover large area on the fruit.The affected fruits give dirty look. On leaves,pinkish brown pustules develop in August. It can be controlled by giving three sprays of 0.4 per cent wettable sulphur at monthly interval star from July. Repeat the spray at monthly interval. Dithane Z-78 @ 0.2 per cent from July to September after 2-3 weeks interval can also be given for its control. .

**Anthracnose (*Colletotrichum* state of *Glomerella cingulata*** (Stoenm.) Spauld and Shrenk): The disease appear during August-September. Initially minute,circular,brown to grey spots symptoms appears with yellowish margin on leaves. Depressed lesions are formed on fruits and later on turn dark in the centre. The affected fruits become shrivelled and rot. It is favoured by hot and humid weather. Spray the tress with Bordeaux mixture (2: 2: 250) or 50% copper oxychloride (0.3%) for the control of anthracnose.

**Blue mould (*Penicillium citrinum*):** The disease is common in aonla growing areas. It causes brown patches and water-soaked areas on the fruit surface. As the disease progresses, three distinct colour develop i.e.first bright yellow,then purple brown and finally bluish green . There is exudation of drops of yellowish liquid on the fruit surface and emit a bad odour. Bruising and injury at the time of harvesting should be avoided. Before packing of fruits discard affected fruits. The fruits may be treated with 0.2 per cent borax or 0.1 percent carbendazim after harvesting.

**Soft rot (*Phomospsis phyllanthi* Punith):** It appear during December to February. Smoke brown to black round lesions develop within 2-3 days of infestation. Later on, lesions show olive brown discolouration with water soaked areas and cover the whole fruit within 8 days. The fungus cause infection in both young and mature fruits. Fruit shape gets deformed. Avoid injury to the fruits during harvesting. Treat the fruits with Dithane M 45 or bavistin 0.1 per cent just after harvesting of fruits.

**Fruit rot:** It is caused due to the fungus ***Alternaria alternata***. The rot starts as a small brownish spherical necrosis spots which increases with the development of disease. The spots become dark brown to black in advanced stage. The centre portion of the infected tissues becomes soft and pulpy. Spray 0.1 per cent carbendazim 15 days prior to harvest. Avoid injury to the fruits during harvesting. The fruits should be treated with 0.2 per cent borex. Sanitary measures should be followed during transportation and storage of fruit

## Physiological Disorders

**Internal fruit necrosis:** The symptoms start with the browning of innermost part of the mesocarpic tissues at the time of endocarp hardening in the second and third week of September. Browning of mesocarp extends towards the epicarp resulting into brownish black areas on the fruit in the second and third week of October. In severe incidence, mesocarp of fruit turns black and become corky and gummy pockets develop. The deficiency of boron has been reported to be the cause of this disorder.

Francis and Banarasi cultivars are highly susceptible to this disorder. Chakaiya and NA-6 and NA-7 are almost free from this problem. To control the disorder, three sprays of borax @ 0.6 per cent in September and October at fortnightly interval should be given. Zinc sulphate (0.4%) + Copper sulphate (0.4%) and borax (0.4%) in September-October has also been found effective in controlling the disorder.

**Unfruitfulnes:** Unfruitfulness is a problem in aonla orchards. In aonla producing areas of Uttar Pradesh, trees are fraught with this problem. Banarsi cultivar is more prone to unfruitfulness problem, This happen largely due to absence of female flowers in the trees.The flowers in aonla are borne in leafy shoots which bear male flowers in upper end and female flowers in the lower ends. According to this pattern there are two types of shoots viz, i) shoots having both male and female flowers ii) shoots having only male flowers. In Banarsi cultivar,female flowers are inadequate,thus it is more prone to unfruitfulness.

Some precautions should be taken before doing vegetative propagation. Select scion shoots from a tree bearing maximum female flowers. Plant more than one variety in the orchard to avoid the male only on account of lack of female flowers

## MARKETING OF FRUITS

Most of the orchardists sell their crop to the contractors. The fruits are marketed in the wholesale market by the contractors. Trading between traders and commission agents is done through concealed auction. The traders from different parts of the country purchase aonla fruits from wholesale market and further sell the produce to the retailers. Traders purchase the premature fruit for chutney making and mature fruits for making preserve (muraba) and pickle. There are four different methods of sale of aonla prevalent amongst the growers. Contract sale of orchard, sale in village, sale to bulk consumers, and sale at market place are the common methods followed in aonla growing areas. There is no information dissemination system which guide the growers about emerging demand and price situation in the neighouring markets

Aonla is mainly consumed in processed form. Being astringent in taste it is not consumed as a table fruit. The major food product prepared from aonla is Chavanprash. Annual procurement for this product is estimated to be 25000 tonnes of fresh aonla. The industry is worth Rs 200 crores and marketed by few large companies like Dabur, Zandu, Baidyanath ,Charak etc.

The different cultivars of aonla yield 100 to 150 kg fruits per tree. The large sized fruits has better attraction in the market and fetch higher price. However, the small sized fruit are equally good for preparation of different products. The aonla fruit is sold in the retail market from Rs 50 to Rs 80 per kg during November to January depending upon the variety and market place. However, the farmers are getting only Rs 10-15 per kg in the wholesale market. Area under aonla is increasing rapidly in last few years. Intensive cultivation is being done in salt affected soils in many districts of Uttar Pradesh. It is also spreading into other states viz. Maharashtra, Rajasthan, Gujarat, Punjab, Haryana, Andhra Pradesh,Karnataka and Tamil Nadu.

## FUTURE STRATEGIES

- To develop ideal regular and high yielding cultivars.
- To ensure availability of better planting material.
- To standardize agro- techniques especially canopy management and high density planting.
- To develop most appropriate manuring and fertilization schedule.
- To strengthen post-harvest handling and processing aspects.
- To improve marketing infrastructure for better profitability to the growers.
- To develop efficient integrated insect-pest and disease management.

## LITERATURE CONSULTED

Anon, 2011. Fruits-Flowers-Vegetables-Production and Recommendations.CCS,HAU,Hisar.

Anon,2016. Package of Practices for Cultivation of Fruits. PAU,Ludhiana.

Bal, J.S.2014. Aonla, In: Fruit Growing. Kalyani Publishers, New Delhi, pp. 316-324.

Bal,J.S., Gill,S.S. and Sandhu, A.S. 2016. Raising Fruit Nursery. 2nd Edition. Kalyani Publishers, New Delhi.

Chandra, A. and Chandra, A.C. 1997. Aonla. In: Production and Post Harvest Technology of Fruits. NBS Publishers and Distributors, Bikaner, pp. 279-289.

Chundawat, B.S. and Sen, N.L. 2002. Aonla (Indian Gooseberry). In: Principles of Fruit Culture. Agrotech Publishing Academy, Udaipur, pp. 271-281.

Makhija,M and Singh,S.2010. Weed management studies in aonla. *Haryana J.Hort. Sci.*,39(3/4):185-87.

Maurya, I.B.. 2007. Aonla. In: Fruit Production Technology. ed. by Yadav, P.K. International Book Distributing Co. Lucknow, pp. 323-333.

Mitra, S.K.1997. Aonla. In: Tropical Horticulture vol.1.Eds (Bose et al),Naya Prokash,,Kolkata.

Pathak,R.K. 2002. Status Report on Genetic Resources of Indian Gooseberry-Aonla (*Emblica officinalis* Gaertn.) in South and Southeast Asia,NASC,New Delhi.

Shukla, A.K; Shukla, A.K. and Vashishtha, B.B. 2004. Aonla. In: Fruit Breeding – Approaches and Achievements. International Book Distributing Co. Lucknow, pp. 233-243.

Singh, D. 2009. Aonla (cultivation and Processing) AGROBIOS, Jodhpur, pp-202.

Singh, I.S.2012.Aonla-Production, Handling & Processing. West Willey Pub. House, New Delhi, pp-176.

## SECTION - 3

## CULTURE AND TECHNOLOGY OF MINOR SUB-TROPICAL FRUITS

# 7

# Loquat

*H.K. Bons and J.S. Bal*

## INTRODUCTION

Loquat (*Eriobotrya Japonica* Lindl.) is an evergreen fruit tree and grown extensively in subtropical regions of world which belongs to family Rosaceae. It is also known as Japanese Plum or Japanese Medlar. The word *Eriobotrya* has been derived from two greek words i.e 'erion' and 'botrys' meaning wool and cluster respectively which stands for wooly inflorescence The plant is often planted in parks and gardens due to its ornamental appearance. The fruit of loquat is available in the months of March - April when very few fresh fruits are available and fetches maximum returns.

## NUTRITIVE AND CULTURAL SIGNIFICANCE

The fresh fruit of loquat is very delicious and refreshing in taste. The fruit contains 87.4 % water, 10.2 % carbohydrates, 0.7 % protein, 0.3% fat ,0.9 % fiber, 0.03 % calcium,0.02 % phosphorus and 0.7 mg iron/100g of edible portion. Its fruit consists of 60-70% pulp and 15-20% seed. The principle constituents of ripe loquat are fructose, sucrose and malic acid. Out of total acids , malic acid accounts about 83% and remaining being citric acid. The loquat pulp also reported to contain 0.42% crude protein and 146 mg and 387 mg of essential and total amino acids per 100 g pulp. The fruit is good source of natural antioxidants like polyphenols, vitamin C, cartenoids and flavonoids but antioxidant content and capacity vary from cultivar to cultivar. These antioxidants lower the incidence of diseases like cancer, arthritis, heart diseases, brain dysfunction and early aging process. Loquat fruit is rich in dietary fiber known as pectin which is useful in holding moisture and getting rid of toxins from colon. Loquat is also good source of potassium (about 210mg/100g pulp)

Loquat is an excellent source of carotenoids or vitamin A which is responsible for skin and flesh colour that varies from yellowish white, yellow to deep orange colour. Total carotenoids found in peel are much higher than in pulp. The carotenoid content of yellow orange fruit was 5-10 times more than in yellow–white fruits. Beta carotene and β-crytoxanthin are principle pigments which contributes 44 and 27% respectively to total carotenoids (17.6 μg/g) were also main contributers to the vitamin A (175 RE/100g). The fruit has very high amount of antioxidant property.

Besides fruits, other parts of the plant are also reported to have high medicinal value i.e leaves are used to inhibit inflammation and fibrosis, to cure skin diseases, cough and tumors in china and Japan. Even loquat seed extract is reported to help in curing liver disorders. The flowers are used as expectorant. Honeybees are easily attracted to the fragrant white flowers, amber coloured honey with agreeable flavor is produced from its flora. An infusion of leaves is given in diarrhea. Loquat tea is suggested for treating diabetes

In addition to fresh consumption for dessert purposes, the loquat fruit can be processed into number of products like jam, juice, jelly, squash, syrup, canning, chutney etc. The tender branches of trees are used as a fodder. Its wood is hard and close grained. The wood is used for making ruler and drawing materials. In China, the loquat tree is projected as cultural symbolism for motivation and management. The Chinese Ganzhi system correlated loquat with luck under wealth and riches category (Ong Hean-Tatt, 2000).

## ORIGIN, HISTORY AND DISTRIBUTION

The loquat is indigenous to the hills of central- eastern china with mild cold and moist climate. The loquat has been cultivated in Japan for more than 1000 years. It was first learned by botanist Kaempfer in 1690.Then, Thunberg saw in Japan during 1712.The loquat was brought to Europe by French in 1784 which was planted at National garden in Paris. It was first imported to England from Canton, China and planted at Kew in Royal botanical garden in 1787. After that loquat trees was grown on Riviera and in Malta and Algeria and fruits appeared in local markets. In 1818 excellent fruits were produced in hothouses of England. It was introduced in St. Micheal island in 1823.

Loquat cultivation spread to India and South East Asia, East Indies, Australia, New Zealand and South Africa. Loquat cultivation was expanded in Israel since 1960 due to dwarfing character on Quince rootstock. Now it is widely distributed throughout subtropical regions of world and commercially grown in Japan, China, Pakistan, India, Madagascar, Mauritius Island, Egypt, the Mediterranean countries (Spain, Italy, Greece, Israel), Tunisia Australia, USA

(California and Florida) Argentina, South Africa, Brazil, Venezula, Chile, Equador, Hawaii, Armenia and Azerbaijan. China is largest producer of loquat in world covering about 1,20,000 hectares with an annual production of 4000,000 tons (Lin, 2007).

The exact date of loquat introduction in India is not known. However, it was grown for the first time in Saharanpur (U.P) at Government Botanical Garden (now Horticultural Experiment and Training Centre) from where it has been spread to others parts of country. Now commercial cultivation of loquat in India is confined to Uttar Pradesh (Saharanpur, Dehradun, Muzzaffarnagar, Meerut, Farrukabad, Kanpur, Bareilly, Gorakhpur) Delhi, Himachal Pradesh (Kangra) and to small extent in Assam, Maharashtra, Nandi hills of Tamil Nadu and Mysore. In Punjab, it is highly preferred fruit for kitchen garden and also being grown on a small scale in the orchards at Gurdaspur, Hoshiarpur, Amritsar, Pathankot, Jalandhar, Roopnagar, Patiala and SAS Nagar districts.

## TAXONOMICAL AND BOTANICAL DESCRIPTION

Common name : Loquat, Japanese Plum , Japanese Medlar , Japan Plum,Chinese Plum

**Classification**

| | |
|---|---|
| Kingdom : | Plantae- Plants |
| Subkingdom : | Tracheobionta- Vascular plants |
| Superdivision : | Spermatophyta- Seed plants |
| Division : | Magnoliophyta - Flowering plants |
| Class : | Magnoliopsida- Dicotyledons |
| Subclass : | Rosidae |
| Order : | Rosales |
| Family : | Rosaceae |
| Genus : | *Eriobotrya* |
| Species : | *Japonica* |

In China, fourteen species of genus *Eriobotrya* and more than hundred varieties of *Eriobotrya japonica* has been described (Ding et al 1995). Botanically, Japanese classified loquat into two groups i.e the Chinese type and Japanese type. The first type is characterized by large fruits, pear shaped and yellow fleshed .The second type is characterized by small fruits, round shaped and white or pale yellow fleshed. The most important species of genus *Eriobotrya* are as follow:

### *Eriobotrya japonica* Lindl.

*Eriobotrya petiolata,* Hook.f

*Eriobotrya latifolia* Hook.f

*Eriobotrya longifolia* (Decne.) Hook.f

*Eriobotrya dupia,* (Lindl .) Decne.

*Eriobotrya bengalensis (Roxb.)* Hook.f

*Eriobotrya elliptica* Lindl .

*Eriobotrya hookeriana* Decne.

*Eriobotrya augustissima* Hook.f

*Eriobotrya fragrans* Chump. ex Benth

The loquat trees are evergreen, spreading and erect growing in nature. It attains height of about 7-8 m.It has rounded, compact and dense crown, short trunk with wooly new twigs. The leaves are large, 12-30 cm long, 7.5-10cm wide whorled at the branch tips. The shape of leaves is elliptical, lanceolate to obovate with dentate margins, dark green and glossy on upper surface and lower surface is whitish or rusty hairy. Flowers are small, white, fragrant appear on inflorescence/wooly panicles which arises terminally on new shoots. Each flower comprises 5 petals, 5 sepals, 20 stamens and 5 pistils joined at the base. Ovary is inferior, 2-5 celled with 20 ovules, out of which only two to five ovules develop into seed in fruit and rest of ovules abort.

The fruit is true pome which is round, oval or pear shape appears in clusters of 4 to 30. The colour of skin and pulp varies from pale yellow to deep orange . Seeds are large, brown, ovoid or angular one to few number varies from cultivar to cultivar. The edible portion is fleshy thalamus The basic chromosome number of loquat is n=17.

## CLIMATIC AND SOIL ADAPTABILITY

Loquat grows successfully in subtropical climate where temperature does not fall below the freezing point. The foliage is heavily injured if temperature goes below 0°C for considerable length of time. The cool weather during some parts of year with well distributed annual rainfall of 60-100 cm is best suited for its cultivation. It can tolerate drought and heat but not severe frost. Loquat is grown throughout India up to elevation of 1525 m above sea level. It can also be grown up to an elevation 2000 m and well adapted to Mediterranean regions. Warm and dry climate is essential at the time of fruit ripening for development of desired flavour and sweetness. However, hot scorching winds before the ripening of fruits, results in small sized fruits or fruits ripens unevenly. The fruits are very susceptible to sunburn which makes them unacceptable in market. Prevalence of cool and foggy weather at fruit ripening reduce the fruit quality in terms of sweetness and flavor. However, thick and leathery leaves of loquat can withstand adverse conditions of climate without major damage as compared to many other fruit crops. For successful cultivation of loquat submontane region of north India or places with mild climate free from severe hot weather are considered ideal.

Loquat grows on wide range of soils. It is suggested that medium heavy soils with good drainage should be preferred instead of light soils. Loquat also grows well in fertile clay loam soils. The soil should be free from any hard pan, gravel and lime for its successful cultivation. It cannot tolerate waterlogged conditions which are harmful to even grown up trees.

## RECOMMENDED AND POPULAR CULTIVARS

The loquat cultivars have been classified on the basis of their maturity. Given below is the group of varieties ripened early, mid and late as per their ripening period.

A. **Early Maturing Varieties**: Ripens from mid-March to first week of April
   Golden Yellow, Improved Golden Yellow, Pale Yellow, Thames Pride, Pathankot

B. **Mid-Season Varieties**: Ripens from end-March to mid-April
   Fire Ball, Mammoth, Improved Pale Yellow, Safeda

C. **Late- Season Varieties**: Ripens during second fortnight of April
   California Advance, Tanaka

**Golden Yellow:** It has medium sized egg shaped fruits with golden-yellow colour and yellowish flesh. Each fruit contains 4-5 medium sized seeds. This cultivar ripens during third week of March .Recommended for cultivation in north India.

**Pale Yellow:** It has large fruits slightly conical to roundish in shape with pale yellow colour and white flesh .Each fruit contains 2-3 medium sized seeds. This cultivar ripens during second week of April. Recommended for cultivation in Punjab and adjoining states.

**California Advance**: It has medium sized conical to round fruits with yellow colour and creamy white flesh. Each fruit contains 2-3 medium sized seeds. This cultivar ripens during fourth week of April. The variety is grown commercially in northern Indian states.

**Improved Golden Yellow:** It has large fruits oval to pyriform in shape with golden yellow colour and orange flesh. Each fruit contains 3-5 seeds. It ripens in end of March.

**Improved Pale Yellow**: It has medium sized fruits oblong pyriform with pale yellow colour and creamy white flesh. Each fruit contains 3-4 seeds. It ripens in mid-April.

**Tanaka:** It has medium sized fruits ovate or round with orange or orange-yellow colour and yellow flesh. Each fruit contains 2-4 medium sized seeds. It is late variety and ripens in end of April.

**Pathankot:** The fruits are medium to large in size and ovate in shape. The average weight is about 40g.It attain orange colour at maturity. The fruit contains 3 seeds. Pulp is 61 per cent. It ripens during third week of March.

**Loquat Selection**: The fruits are medium in size and oval in shape. The average fruit weight is 37g. It attains orange colour at maturity. Pulp percentage is 53.Individual fruit contains 4-5 seeds. The fruit ripens during fourth week of March.

**Fire Ball:** The fruits are round and deep red in colour. The fruits are small to medium in size, oblong to ovate and good in quality. The variety is grown in Uttar Pradesh. It is mid season maturing variety and moderately seedy.

**Mammoth**:The fruits are small and oblong to pyriform in shape. Pulp is medium thick, pale orange in colour,coarse and granular. It has pleasant taste. Seeds are few in number. It is a mid season variety

**Thames Pride**: The fruits are medium in size with deep yellow colour at maturity .Pulp medium pale orange. Taste is mild sub-acid. It is moderately seeded variety. The variety is grown in Uttar Pradesh and Tamil Nadu on commercial scale.

## PROPAGATION TECHNOLOGY AND ROOTSTOCKS

As compare to other fruits, loquat is very difficult to propagate. Nurserymen usually follow propagation through seeds which has disadvantage of slow growth and provides plants which are not true to type and bear fruits of variable and inferior quality. There is urgent need to standardize method and time of vegetative propagation to procure true-to-type plants for growing loquat at commercial scale.

## Raising of Rootstock

Commercial loquat cultivars are generally used for raising rootstock seedlings. Pear, apple, quince and Mespilus have also used in many places as rootstocks for commercial loquat cultivars. Loquat seeds should be sown immediately after extraction from fruits as exposure to heat and light results in poor germination and stunted growth. Fresh seeds are sown in month of April-May in moist sand for germination . When the seedlings are 4-5 cm tall, they are transplanted in the nursery for further growth and grafting. The seedling grow rapidly and becomes fit for grafting in coming rainy season. In case the mother plants are high headed, then the seedlings are transplanted in the pots and brought in contact with mother plant by raising plant forms when they attain inarchable size.

## Propagation Techniques

**Stem cuttings:** The loquat can be grown from stem cuttings. The best time reported to prepare cuttings is after rainy season when spring season growth has become more or less dormant. Care should be taken that only wood of current season growth should be used.

**Air layering:** Air layering can be tried but success is very less. For air layering three month old shoots are ringed and layered. Treatment of ringed shoots with 3% NAA or 2500 ppm IBA has enhanced success of air layering. About 40% rooting success is reported in air layers with IBA treatment. Air layers of loquat treated with IBA and NAA (250 ppm ) resulted into maximum rooting (92.5%). Roots emerged after 44 days as compared to 73 days in control. The treatment of 250 ppm IBA also resulted in maximum number of primary root and greatest root length and thickness .

**Grafting:** The plants are generally propagated through inarching. The best time for inarching is July-August. Cleft grafting is used on large scale in China and still used in Japan for top working. An improved method of cleft grafting called 'Young stock cleft graft' was developed in china which is used for large scale multiplication. Veneer grafting is commonly practiced in Japan. Cleft grafting in early spring (1st March) results in better success rate (62.50%) in Pakistan. Side grafting resulted in 70% success in March-April, while 85% success was achieved in monsoon season under Pakistan conditions.

**Budding:** Shield budding gave more encouraging results in the months of January and February as compared to September and October, when buds from 3 month old branches were used. In Turkey, it was observed that March was most suitable time for budding with 95.48% bud take. Bud take was also > 90% between mid February and mid April. Patch budding was more successful (87.88%) than T(85.43%) budding.

**Micro propagation:** Protocol for micro propagation has been standardized by using shoot tips in China. Axillary shoots were successfully derived from shoot-tip culture.

## PLANT AND FRUIT PHYSIOLOGY

The loquat is an evergreen fruit tree. It is grown in subtropical regions of the world. Loquat bears flower at terminal end of current year's growth. The flowering period of loquat is very long which starts from mid July and continue up to January. In tropical climate, it gives three reproductive flushes but the intermediate flush gives the higher yield of the better sized fruits. In Uttar Pradesh, only one flush continued flower from September to February at Saharanpur. The opening of flowers continued throughout the day. In most of the varieties, dehiscence complete in more than one day. In Golden yellow and Pale yellow dehiscence complete only in one day. In Punjab, the flowering starts in the first week of October and continues up to third week of December. In general the number of flowers per cluster vary from 50 to 100 and about 15-20 fruits per cluster are set.

The loquat fruit develops very rapidly after fertilization. The fruit growth takes place by thickening of the toral rim immediately above the carpel level. The sepals grow towards the centre and cover over in 'hood' fashion, the distal portions of the carpels. The functional ovule develops into a fertile seed occupying the whole central region of the fruit. As the fruit is mature the carpel wall are more or less parchment like. The fruit skin is tough and leathery. Seed is very hard. The growth pattern of loquat cv. Tanaka is neither single nor double sigmoid but is exponential. The edible portion in loquat fruit is entirely toral in nature comprising of pith and cortical areas . The edible portion developed through uniform growth of tissue throughout the fruit.

## PLANNING AND PLANTING

The site should be properly laid out before the establishment of loquat orchard. The usual planting distance for loquat is kept 6.5 × 6.5 m in square system. This will accommodates 225 plants per hectare However ,in many localities loquat is planted even at 7×7 m and 8×8 m apart in square system. Before planting, the field should be leveled properly. The pit should be dug 1×1×1m size. The pits should be refiled with mixture of 3-4 baskets of farmyard manure, one feet top soil and 200 g of single superphosphate. To each pit add 15 ml of chlorpyriphos 20 EC mixed in about 2 kg soil to avoid the attack of white ants.

Loquat is planted in two planting seasons viz; Feb-March (spring season) and August-September (Rainy season). The plants should preferably be planted during August-September when the weather has cooled down sufficiently. The young plants should be given utmost care to grow healthy and straight .

## SOIL CULTURAL PRACTICES TECHNOLOGY

### Water Need

Water need of loquat trees is particular on account of its fleshy and leathery leaves. Enough moisture should be maintained in the soil for proper growth of the foliage and shoots. The loquat is more tolerant to drought. But for the better growth and productivity ,the orchard should be irrigated judiciously. In order to enable the shoots to develop properly and the mature terminal buds to fill out properly, sufficient moisture in the soil is required. The young plants should be irrigated at weekly intervals or at fortnight intervals from September onwards for better growth.

The loquat orchard should be irrigated before the swelling of blossom buds. Watering to the trees during growth and development period of fruits is very important. Therefore, three to four irrigations are recommended from November to March. To the bearing trees, irrigation should be applied at the end of November or early December to afford protection from frost injury.

### Nutritional Need

Loquat trees is known to exhaust the soil. Therefore, the plants need adequate manures and fertilizers to bear good regular crop over the year. Green manuring of loquat orchard is useful practice which also improve physical conditions of the soil. The nutritional schedule recommended for loquat is given below.

| Age of trees (years) | FYM (kg/tree) | Nitrogen (g/tree) | Phosphorus (g/tree) | Potassium (g/tree) |
|---|---|---|---|---|
| 1-3 | 10-20 | 75-250 | 30-80 | 90-240 |
| 4-6 | 25-40 | 300-375 | 100-190 | 360-600 |
| 7-10 | 40-50 | 400-500 | 240-320 | 660-900 |
| Above 10 | 60 | 500 | 320 | 900 |

The farmyard manure and whole quantity of phosphorus and potash should be applied in September. Apply nitrogenous fertilizers in two split doses. One half dose should be applied in October before flowering and remaining half in February – March i.e after fruit set.

### Weed Control

Weed population in loquat orchard reduce the fruit yield considerably by competing for moisture , nutrients, light and space. The most common weeds in loquat orchard are false mallow (*Malvastrum coromandelianum*), coat buttons (*Tridax procumbens*), wood sorrel (*Oxalis corniculata*), wild poinsettia *(Euphorbia geniculata*) and few grasses. Out of weed flora, 69% were annual and rest were perennials. To check the weed growth, thorough cultivation of loquat orchard should be done during summer and after rainy season. However, deep ploughing should be avoided in loquat orchards.

## PLANT CULTURAL PRACTICES TECHNOLOGY

### Training and Pruning

Modified leader system or open centre system is commonly followed to train the loquat plants at initial stage. At this stage, the main stem should be kept clean up to the height of 50 cm by removing all the branches. After that, pruning is confined to removal of dead and diseased branches in the full grown tree. Annual pruning in loquat may be attempted to regulate fruit bearing. Harvesting of mature bunch is a kind of pruning which encourages the new growth. Heavy pruning in loquat is not advocated. The best time of pruning is during summer months after the harvesting of fruits.

### Top Working

It has been observed that most of seedling loquat trees are not productive and are unprofitable. Such healthy trees should be top worked with some improved commercial cultivars. The best time for doing top working in loquat is summer months i.e May. Sufficient number of healthy shoots will emerge from the healthy headed back plants during the rainy season. Retain one or two healthy shoots for doing grafting and other should be removed. These plants can be propagated vegetatively by employing inarching , T- budding and bark grafting. Bark grafting was found more successful for top working as compared to other methods of grafting. The grafting plants will start producing fruits of superior quality.

### Fruit Thinning

Thinning of flowers and fruits is necessary to promote the growth of fruits. Flower thinning is done by removing about one third of the whole inflorescence. In addition, further thinning of flowers should be attempted from remaining inflorescence. The loquat tree has tendency to bear heavily and as such the fruits remained small in size. Thinning of fruits is done as soon as the fruits develop. While doing thinning, fruits of uniform shape and size are retained. Thinning of fruits in clusters or removing of some of the cluster may be practised for improving the size and quality of fruits. Paper bags can be used while doing fruit thinning to maintain their attractive appearance.

### Pollination

Most of the popular cultivars grown in northern India are either self-incompatible or partially self compatible. Therefore, it is advisable not to plant a solid block of single loquat cultivar. California advance is a good pollinizer for Golden yellow and Pale yellow and Tanaka cultivars and should be planted along with these cultivars. Under Saharanpur conditions, Golden yellow cultivar was found totally self- incompatible whereas California Advance , Tanaka and Pale Yellow cultivars were partially self-incompatible. It is reported that Tanaka pollinated with Pale Yellow resulted lower fruit set than the self one, thus indicating a partial cross-incompatibility. Self-incompatibility in loquat is reported due to gametophytic nature. *Apis dorsata* is the main flower visitor and pollinate the flowers. Honey bees also pollinate the flower when temperature rises above 14°C.

### Use of Plant Growth Regulators

Use of growth regulators play important role in improving fruit set and retention , fruit size, fruit thinning, induction of seedlessness and improve quality in loquat. Fruit set and fruit quality is obtained better with GA at 60 ppm. Both NAA and NAAM at 25 ppm application resulted optimum level of thinning in fruits. The most effective treatment found to thin loquat fruits and increase the average fruit size was with application of 20 ppm NAA 10-15 days after anthesis. GA has been applied to induce seedless fruits in loquat. Spray application of GA at 250 ppm applied after the emergence of floral buds or NAA at 20 ppm applied during full bloom produced seedless fruits. Seedless fruits were smaller, elongated and mature 4-5 weeks earlier than seeded fruits. Similarly GA 40 ppm sprayed at pea stage of fruit and again one week later improved the fruit retention, volume, weight, pulp content and minimised the seed content. TSS, reducing, non- reducing sugars and ascorbic acid increased with reduction in acidity was obtained with 40 ppm NAA.

## HARVESTING AND PRODUCTION OF FRUITS

**Harvest Maturity:** The loquat is a non-climacteric fruit and it is harvested when fully ripened on tree. The fruit reaches from flowering to maturity in about 140-150 days. The harvest maturity of loquat fruits is generally judged visually. Fully ripe fruits with golden pale or orange colour should be harvested. The taste and quality of fruit is superior at maturity as sugar content of fruit increases and acid content decreases during maturation and ripening. The quality of loquat fruit depends on flesh firmness and sugar: acid ratio. So, Hunter "a" value as well as visually assessed colour score in relation to sugar: acid ratio can be index of maturity.

**Harvesting**: Since loquat bear in clusters, all the fruits usually matures and ripens uniformly and entire cluster should be cut at one time. If in some cases, some fruits in cluster ripe earlier than they should be carefully harvested with help of clipper. The fruits should never be pulled from tree by hand. In this way, the stem would separate from the flesh and cause decay to set in at once. Harvesting of immature fruits should be avoided. The fully mature fruits should only be picked up. Cultivars such as California Advance, Thames Pride and Golden Yellow should be harvested at Total soluble solids of 11 per cent. The loquat tree generally starts bearing fruits after third year of planting. The tree becomes commercial at the age of 15 when it bears maximum fruits. The average yield of loquat tree is about 16-20 kg .However, healthy and well managed tree can also bear up to 40 kg fruits/tree.

## POST-HARVEST FRUIT TECHNOLOGY

### Grading and Packaging of Fruits

Loquat is classified as non-climacteric fruit. No respiratory peaks of ethylene is either from the attached fruit or after harvest. However, some metabolic changes in fruit occur which improve the colour development, texture and flavor, thus making the fruit acceptable to consumers. Whole clusters are not preferred in market. So, the individual fruits are clipped from cluster and graded for size and colour to provide uniform pack. In general, two grades of loquat fruits are considered. The large size fruits free from blemishes should be placed in one box while all the remaining marketable fruits should be packed separately. However, third grade can be made with undersized and misshapen fruits that can used for processing of Jams, Jellies or other products. All the superfluous stems should be clipped off and badly bruised, shriveled and scarred fruits should be discarded. The fruits require careful packing and paper is placed at the bottom of the boxes. The first grade (large size) fruits should be properly packed and must receive better handling. The paper cuttings can be used for cushioning purpose. The wooden boxes of 14 kg should be used for sending fruits to nearby markets. For distant markets small packages are used

In Japan 3-4 grades are prepared according to quality and 4-5 grades on the basis of size and packed in 300 g or 500 g bag and 1kg or 2 kg carton. In China, the fruit cluster is cut, packed in bamboo baskets or wooden boxes and send to market where it may be separated into 2 or 3 grades. In Turkey, fruits are carried to market in large containers up to 15-20 kg. Recently , some farmers are getting higher prices by grading fruits according to sizes and selling in small trays of 1 kg.

**Storage of Fruits:**. Loquats can be stored for 4-6 days at ambient temperature and for 2 weeks at 11°C temperature and 85-90% humidity. The firm ripe fruits are more suitable for storage as compared to fully ripe fruits. Loquat fruits stored at 10°C resulted in less spoilage and surface disorder than stored at 25°C. However, pre-cooling was found effective at both the temperatures in preventing softening and decreasing sugar and acid content of fruits. High quality with good taste was maintained at 10°C after pre-cooling .Low temperature conditioning at 5°C for 6 days has been suggested to facilitate storage for 50 days at 0°C. Chitosan coating (1 % w/V) could be used to extend the shelf life of loquat fruits. It improve the storability of fruits up to 21 days in cold storage. Chitosan coating 0.75%. significantly reduces weight loss and suppress flesh browning of loquat fruits during cold storage. Moreover, there was significant increase in TSS, TSS/Total acid ratio, pH, and vitamin C during storage at 7°C for 28 days.

**Processing of Fruits:**: The loquat fruits are mostly consumed fresh. It mixes well with other fruits for preparing fresh fruit salads or fruit cups. Firm but slightly immature fruits are best for preparing pies or tarts. Fully mature fruits can be processed into jams, jellies, juice, canned fruits and squash. Wine can also be made from loquat fruits. Tanaka cultivar is good for preparation of canned fruits while Saharanpur special is suitable for preparing dehydrated products. Loquat slices dried with Refractance Window drying (RWD) method have higher quality with colour and appearance closer to fresh loquat .

## INSECT-PESTS AND MANAGEMENT

**Bark eating caterpillar (*Inderbela quadrinotata* Walker)**: Caterpillars causes damage by boring holes into the stem and branches and feeding on the bark under the cover of its excreta. They destroy the conducting tissues. This incidence of bark eating is noticed during October-November.

The holes are plugged with cotton wool soaked in carbon disulphide and chloroform to kill the caterpillars inside the bark. Remove webbing and apply kerosene into the holes during October-November and again in January. All the alternate host plants in vicinity should be treated.

**Fruit fly *(Bactrocera dorsalis)*** : Sometimes the fruits are attacked by fruit fly during April-May. The maggots are found inside the fruits. The fruits fall and maggots come out of the fruits to pupate in the soil.

Orchard land should be ploughed in January-February to expose the pupae to natural enemies. This can reduce the population considerably . Affected fruits should be collected and destroyed. Fruit fly traps containing 100 ml emulsion of methyl eugenol 0.1% and 0.1% malathion should be hanged in April in the orchard. Spray 1.25 litre of malathion 50EC +12.5 kg sugar in 1250 litre of water per ha. If infestation still persists, repeat spray at 7-10 days intervals.

## DISEASES AND MANAGEMENT

**Root rot/white rot:** This disease is caused by *Polyporus palustris*.The bark and wood of the root and root collar is affected. The decay wood turn pinkish to dull violet in colour. In advance stage small, white, elongated pocket appear and form a mess of spongy white fibres. The affected tree shows wilt symptoms, early leaf fall and increase in fruit set. The fruiting bodies of about 30 cm diameter or more appears when disease is in advance stage. They are either hidden by the litter or lie exposed on the soil surface.

For its control examine the roots and collar regions of roots showing weakening signs. First dug out decayed roots and remove them completely from color region. Then, treat immediately cut ends with disinfectant solution, when dry apply Bordeaux paste. Drench the soil with Bordeaux mixture from where dead roots dug out. Irrigation water should not be allowed to come into contact with the stem. Deep hoeing and interculture should be avoided which cause injuries to the roots and results into attack of fungus.

**Fruit shoot blight and bark canker:** This disease is caused by *Botryodiplodia theobromae* and *Phoma glomerata*. The cankers appear on bud scars, wounds, twig stubs or in crotches. Small circular brown spots appears around a leaf scar. As the canker enlarges ,centers become sunken with healthy surrounding bark. The fungus multiples in bark cankers.

Cankers on trunk and crotches should be removed with 2 cm of healthy bark and destroyed. Then cover the wound with disinfectant solution and apply Bordeaux paste. After a week, apply Bordeaux paint. After this treatment, spray the trees with Bordeaux mixture (2:2:250) just before monsoon, during monsoon and till October.

**Crown rot:** It is caused due to fungus (*Phytophthora* sp).The fungus attacks the bark producing canker from ground level to shoots. The affected trees shows profuse flowering but fruiting is sparse and of low grade. The leaves becomes yellowish green and tree shows partial or complete wilt symptoms. With time whole tree dry up completely. The pathogen is soil borne and perpetuates itself in the dead cankers.

Remove and burn infested trees. First remove the diseased bark in dry season and treat with disinfectant solution and then apply Bordeaux paint after a week. After that, spray the trees with Bordeaux mixture (2:2:250) . Repeat the spray just before monsoon, during monsoon and till October.

## MARKETING OF FRUITS

Loquat fruits becomes available during end-March to mid-May in India. This is a period when only few fresh fruits are available in the market, Thus, it fill the gap between citrus, apple, ber and stone fruits of the season. Therefore, loquat fruit provides a decent return to the growers.

General marketing of loquat fruit is little variable as compared to other commercially grown fruits. It is sold mostly in the local or short distant markets on account of perishable nature of this fruit. The commercial bearing of loquat starts after four years from planting and on an average its tree from different varieties yield 20 kg fruits. But, a better managed orchard still can give higher yield up to 40 kg fruits per tree. Loquat is planted at 6.5m x 6.5m apart which accommodates 225 plants per hectare in square system. In this way, one hectare loquat plantation yield approximately 45 quintals fruits. Large sized fruits of A grade category fetch better premium in the market than the medium sized B grade fruits. Normally, A grade fruits are sold @Rs 30-40 per kg and B grade fruits @Rs 25-30 per kg. The growers who follow scientific recommendations and adopting latest techniques can earn net profit of Rs 1.5 lakh per hectare. The loquat orchard relatively needs less care than the other fruits but it can excel the other fruits in yielding net profit per unit area.

### Marketing Strategies

There is need to understand the marketing strategies of loquat fruit keeping into view the consumer's acceptance. When whole cluster of fruits are sold in the market, they are not attractive and fetch less rate of the produce. Therefore, individual fruits should be clipped from the cluster, stalk to be detached from each fruit, thus, attract high premium in the market. Harvesting and packing are highly labour intensive operations. This limit the area of loquat production for fruit growers.

### Trading and Export Potential

Trade in loquat is witnessed low even in the international market. Major trading of this fruit is done in Spain and they export 36-47% of their total production. In Turkey, loquat is consumed fresh in spring only small quantities are exported to the Middle East and Central and Northern Europe. Loquat fruits can be stored for about 40 days under CA storage. This suggest that loquat is the future fruit for export market if shelf life is extended further with CA storage.

## FUTURE STRATEGIES

- Identification, collection and multiplication of superior genotypes.
- Hybridization for high yield and quality characters.
- Production of quality planting material through vegetative propagation.
- Standardization of rootstocks for high density planting.
- Efficient use of growth regulators for quality regulation.
- Standardization of suitable pollinizers for different commercial cultivars.
- Technology development for protected cultivation.
- Extending shelf life of fruits at C A storage for extended marketing.
- To develop processing cultivars for consumer acceptance.
- To recommend technologies for more processing products.

## LITERATURE CONSULTED

Abassi, N. A., Hafiz, I. A., Qureshi, A. A., Ali, I. and Mahmood, S. R. 2014.Evaluating the success of vegetative propagation techniques in loquat. *Pak J Bot.,* 46:579-84.

Abbasi, N.A., Hussan, A. and Hafiz, I.A. 2011.Loquat production in Pakistan and strategies for improvement. *Acta Hort.,* 7: 117-121.

Agusti, M., Juan, M., Almela, V. and Gariglio, N. 2000. Loquat fruit size is increased through the thinning of naphthalene acetic acid. *Plant Growth Regulation,* 31: 167-171.

Bal, J.S. 2004. Loquat. In: Fruit Growing.3rd Edition: Pp: 272-79. Kalyani Publishers, New Delhi.

Bal, J.S., Gill, S.S. and Sandhu, A.S. 2016. Raising Fruit Nursery. 2nd Edition. Pp: 86-87. Kalyani Publishers, New Delhi.

Bose, T.K., Mitra, S.K. and Sanyal, D. 2002. Loquat. In: Fruits: Tropical and Sub-tropical. Pp: 449-474. Naya Udyog, Calcutta.

Caballera, P. and Fernandez, M.A. 2003. Loquat production and market. Ist International Symposium on Loquat, Zaragoza: CIHEAM. Pp: 11-20. In: Liacer, G. (ed) at Badeves, M.L. (ed).

Chaudhary, A.S. Singh, M. and Singh, G.N. 1994. Effect of napthaleneacetic acid 2,4,5 trichlorophenoxyacetic acid and gibberellic acid on loquat cv. Safeda batia *Indian J. Agric. Res., 28:* 127-132.

Ding, C.K., Chen, Q.F., Sun, T.L. and Xia, Q.Z. 1995. Germplasm resources and breeding of *Eriobotrya japonica* Lindl. in China. *Acta. Horticulturae,* 403: 121-126.

El-Safy, F.S. 2014. Drying characteristics of loquat slices using different dehydration methods by comparative evaluation. *World J. of Dairy & Food Sciences,*9 (2): 272-284.

Ito, H. E., Kobayashi, S. H., Li, T., Hatano, D., Sugita, N., Kubo, S. , Shimura ,Y.Itoh. , Tokuda, H. and Nishino, H. 2002. Anti-tumor activity of compounds isolated from leaves of *Eriobotrya japonica J. Agric. Food Chem.,* 50:2400-2403.

Jackson, D.I. , Looney, N.E., Morley-Bunker, M. and Thiele, G. 2011. Temperate and Sub-tropical Fruit Production, 3rd Ed.,www.cabi.org.

Karadeniz, T. (2003). Loquat (*Eriobotrya japonica* Lindl).

Khan, A.R.; Siddiqui, I.M. and Chattha, G.A. 2002. Propagation trials in loquat ( (*Eriobotrya japonica*). *Pakistan J. Agric. Res.*,17:198-199.

Kumar, R. 1976. Induction of seedlessness in loquat. *Indian J. Hort.,* 33: 26-32.

Kumar, S., Ritu and Pallavi, G. 2014. A critical review on loquat (*Eriobotrya japonica* Thunb/ Lindl.). *Int. J. Pharmaceutical & Biological Archives,* 5 (2): 1-7

Lin, S, Sharpe, R.H. and Janick, J. 1999. Loquat: Botany and Horticulture. *Hort. Reviews.,* 23: 233-276.

Lin, S.Q.2007. World loquat production and research with special reference to China. *Acta Horti.,* 750: II, International Symposium on Loquat: 750.

Loquat. 1997. http://crfg.org/pubs/ff/loquat.html (assesed on 29.3.2017)

Loquat. 2017. https://en.wikipedia.org/wiki/Loquat

Loquat–PurdueHorticulture–PurdueUniversity html Https://www.hort.purdue.edu/newcrop/mortan/loquat

Misra, K.K. 2011. Loquat. In: Subtropical Fruit Culture. Pp: 298-314. Kalyani Publishers, New Delhi.

Nishioka Y., Yoshioka,S., Kusunose,M.,Cui,T.L.,Hamada,A., Ono,M., Miyamura,M., Kyotani,S., and Cui,T.L.. 2002. Effects of extract derived from *Eriobotrya japonica* on liver function improvement in rats. *Biol. Pharma. Bull.*, 25:1053-1057.

Ong Hean-Tatt, 2000. Cultural significance of the Plant Symbolisns. In: Chinese Plant Symbolisns GUI World of Culture. Article C500/3.

Petriccione, M., Pasquariello, M.S., Mastrobuoni, F., Zampella, L., Patra, D.D. and Scortichini, M. 2015. Influence of chitosan coating on the quality and nutraceutical traits of loquat fruit during post-harvest life. *Scientia Horticulturae,* 197: 287-296.

Polat A.A. 2007. Loquat production in Turkey: Problems and solutions the European. *J. Plant Sci. & Biotech.,* 1 (2): 187-199.

Pujari, S. Loquat Cultivation in India – Production Area, Climate, Harvesting and Fruit Handling. http://www.yourarticlelibrary.com/cultivation/loquat-cultivation-in-india-production-area-climate-harvesting-and-fruit-handling/24701/

Roy, S.K., Khurdiya, D.S. and Waskar, D.P. 1995. Other Subtropical Fruits. In: Handbook of Fruit Science and Technology Production,Composition, Storage and Processing Pp:539-561.

Sakuramata, Y. H., Kusano, S. and Aki, O. 2004. Effects of combination of Caiapo Reg. with other plant-derived substance on anti-diabetic efficacy in KK-Ay mice. *Bio Factors,* 22:149-152.

Sadhu, M.K. and Chattopadhyay, P.K. 2001. Loquat. In: Introductory Fruit Crop. Pp: 353-356. Naya Prokash, Calcutta.

Singh, A. 1997. Loquat. In: Plant Physiology and Production. Pp: 380-384. Kalyani Publishers, New Delhi.

Singh, J. and Bal, J.S. 2011. Performance of loquat (*Eriobotrya japonica* Lindl.) varieties grown under Punjab conditions. Proc. "Climate Change and Fruit Production". Pp 29-36. Ed. by W.S.Dhillon *et.al.*

Singh, N. and Shukla, H.S. 1978. Response of loquat fruit to GA and urea. *Plant Sci.*,10: 77-83.

Wafae, M.A. and Nadir, A.S. 2012. Physiochemical and organoleptic characteristics of loquat fruits and its processing. *Nature & Science,* 10(6): 108-113.

Yoshioka S, Hamada A, Jobu K , Yokota J, Onagawa M, Kyotani S, Miyamura M, Saibara T, Onishi S, Nishioka Y. 2010.Effect of *Eriobotrya Japonica* seed extract on oxidative stress in rats with non-alcoholic steatohepati-tis. *J. Pharmacol.*, 62 :241-2.

# 8

# Jamun

*Rajbir Singh Boora*

## INTRODUCTION

Jamun (*Syzygium cumini* (L) Skeels Syn *Eugenia jambolana* Lamk.) is an important commercial indigenous minor fruit of India and is highly suitable for avenue and for social forestry plantation. It belongs to family Myrtaceae. The other major plants belong to this family are eucalyptus, bottle brush and cloves. It is known by different names like black plum, Indian black berry, Ram jamun, jambolan, Jambul and Phalinda in different parts of country. Trees are found growing throughout the tropics and subtropics as an avenue plantation on road side, school, religious places and as wind break around the orchards. It is also occurs in the lower range of the Himalayas up to an elevation of 1,300 meters and in the Kumaon hills up to 1,600 meters above mean sea level. It is widely grown in the larger parts of India from the Indo-Gangetic plains in the North to Tamil Nadu in the South. The importance of this fruit is increasing day by day due to increasing awareness of hypoglycemic properties of this fruit against diabetics. It is adequately rich in antioxidants.

## NUTRITIVE AND CULTURAL SIGNIFICANCE

Jamun is a rich source of antioxidant and every part of the tree has been used by both urban and rural dwellers. The fruit is a good source of iron, sugars, minerals, protein and carbohydrates. Fully ripe fruits are mainly eaten as fresh and can also be processed into many value added products like jelly, jam, squash, RTS, wine and vinegar. The jamun fruit has sub-acid spicy flavor a little quantity of fruit syrup is much useful in curing diarrhea. The vinegar prepared from juice extracted from slightly unripe fruit is stomachic, carminative and diuretic apart from having cooling and digestive properties. The fruits are used as an effective medicine against diabetes, heart and liver trouble. Glucose and fructose are the principal sugars in the ripe fruit, not even a trace of sucrose was detected. Malic acid is the major acid ie 0.50% of the fruit weight. In addition, a small quantity of oxalic acid is present. Gallic and tannins account for the astringency of the fruit. The purple color of the fruit is due to anthocyanin pigment. The blossoms are important source of nector for honey bees in North India during summer. Seed contains crude protein (8.5%), crude fiber (16.9%), ash (21.72%), Ca (0.41%), P (0.17%). Powdered seeds have also reputation of being useful in the treatment of diabetes. The seeds contained alkaloid jambosin and glycoside, jambolin or antimellin, which reduce the diastatic conversion of starch into sugars. Its wood is used for railway sleepers. The jamun timber is used in buildings, for agricultural implements and for well work as it resists the action of water.

**Table 1:** Nutritive value of jamun fruits

| | | | | | |
|---|---|---|---|---|---|
| Moisture (%) | 84.50-86.4 | Acidity (%) | 2.1-2.5 | Pulp (%) | 50-65 |
| Protein (%) | 0.53-0.65 | Pectin (%) | 2.3-3.7 | Calcium (%) | 0.02 |
| Fat (%) | 0.10 | Vitamin A (IU) | 74-100 | Iron (%) | 0.1 |
| Calorific value/ 100 g | 83 | VitaminC(mg/100 g) | 30-40 | Total sugars (%) | 5.8-6.9 |
| TSS (°B) | 11-17 | Phosphorus (%) | 0.01 | Anthocyanin (OD) | 1.8-1.9 |

*Source:* Bose *et al.*2001 and Singh *et al.* 1999

## CULTURAL SIGNIFICANCE

In Maharashtra, leaves are used in decorations of marriage pandal. There is a famous Marathi song "Jambhul pikalya zada khali ". The wood from Jamun tree is used to make bullock cart wheels and other agricultural equipment. The timber is used to construct doors and windows. Hindus use a sizable branch of the tree to inaugurate beginning of marriage preparations. Culturally, beautiful eyes are compared to the leaves of jamun. In the great epic of India, Mahabharata Sri Krishna's body color is compared to this fruit as well.

## ORIGIN, HISTORY AND DISTRIBUTION

The place origin of jamun is India. It is also found in Thailand, Philippines, Madagascar and some other countries. The jamun has successfully been introduced into many other tropical countries like West Indies, California, Algeria and Israel. In India, jamun trees are mostly found scattered throughout the tropical and sub-tropical regions including the lower ranges of Himalayas up to an elevation of 1300 meters and in the Kumaon hills up to 1600 meters. It is widely grown in the larger part of India from the Indo-Gangetic plains in the north to Tamil Nadu in the south. It is one of the most hardy fruit crops and can easily be grown in neglected and marshy areas, arid and semi-arid areas, resource- poor and wastelands where other fruits trees cannot be grown successfully.

Information regarding area and production statistic of this fruit in India is not available because it is not grown on a commercial scale but scattered plantation in patches are common in the villages parks, on roadsides, avenues and as windbreak. The production in India is mainly concentrated in the dry regions and the produce is collected by the villagers and sold in the local market. The production potential of jamun depends on the demand for the products. Its cultivation can be introduced in arid and semiarid, resource-poor and wastelands areas where other crop cannot be grown. In India, the maximum numbers of jamun trees are found scattered throughout the tropical and subtropical region.

## TAXONOMICAL AND BOTANICAL DESCRIPTION

Common name : Jamun, black plum, Indian black berry, Ram jamun, jambolan, Jambul, Neredu and Phalinda

**Classification** :

| | |
|---|---|
| Domain | Eukaryota |
| Kingdom | Plantae |
| Phylun | Spermatophyta |
| Subphylum | Angiospermae |
| Class | Dicotyledonae |
| Order | Myrtales |
| Family | Myrtaceae |
| Genus | *Syzygium* |
| Species | *Cumini* |

Jamun is Dicotyledonous plant belongs to the genus *Syzygium* of the family Myrtaceae having chromosome no. 2n=40. There are more than 400 species of which a few provide edible fruits. Some important species in this genus are described as below.

***Syzygium cumini* (L.) Skeels:** It is commercial species and fruits having medicinal properties. It has deep tap root system. The lateral roots are numerous, moderately long and distributed down the main root. It bears many flowers in a panicle. Flowers are in a few flowered panicles, hermaphrodite, light yellow in colour, borne in leaf axils on beanchlets, calyx lobes 4, calyx tube not extending beyond summit of ovary; petals 4, white spreading, stamens many, ovary inferior. Fruit is a berry, purplish red to black, ovoid and edible. The tree may reaches up to 20 to 35 m tall having broad crown. Bark is rough, cracked, flaking and discolored on the lower part of the trunk, becoming smooth and light-grey higher up, branchlets greyish white. The trunk has 3 to 5 m circumference with a semi-spreading crown up to 10 m in diameter. In jamun, new vegetative shoots emerge as terminal growth on the previous season's branchlets. The new shoots in jamun emerge in two distinct flushes, i.e. from February to May and August to October. The growth of the jamun remain continue even at low temperature. The flush which appears in the month of February produces maximum growth and flowering.

Leaves opposite simple and glossy, elliptic, pinnately veined with lateral veins close together. Leaves broadly oblong or oval, bluntish acuminate The leaves are evergreen and have a turpentine smell, and are opposite, 5-25 cm long, 2.5-10 cm wide, pinkish when young, becoming leathery, glossy, dark-green above, lighter beneath, with a yellowish midrib when mature.Flower bud differentiation has observed on 5 to 10 month old branches and it started from last week of January and continued for 43 days. The floral parts arose in acropetal succession. Flowers are fragrant and appear in clusters 2.5-10 cm long, each being 1.25 cm wide and 2.5 cm long, with a funnel-shaped calyx and 4-5 united petals, white at first, becoming rose-pink, shedding rapidly to leave only the numerous stamens. The flowering in jamun starts in the first week of March and continues up to the middle of April. The trees are in full bloom in the second week of April. The inflorescence in jamun is generally borne in the axils of leaves on branchlet. The flowers are hermaphrodite, light yellow in color. Jamun is a cross-pollinated fruit. The pollination is done by honeybees, house flies and wind. There is heavy drop of flowers and fruits within 3 to 4 weeks after blooming. The inflorescence is terminal or lateral and develops mostly on current year growth, one year old shoots and older branches. Flowers are regular bisexual with 5 free sepals, 8 stamens and a simple style.

Fruit appear in clusters of just a few or 10-40, are round or oblong, often curved, 1.25-5 cm long, turning from green to light-magenta, then dark-purple or nearly black, although a white-fruited form has been reported in Indonesia. The skin is thin, smooth, glossy, and adherent. The pulp is purple or white, very juicy, and normally encloses a single, oblong, green or brown seed, up to 4 cm long, though some fruits have 2-5 seeds tightly compressed within a leathery coat, and some are seedless. The fruit is usually astringent, sometimes unpalatably so, and the flavor varies from acid to fairly sweet. Three distinct phases of fruit growth in jamun are recorded. The color of jamun fruit changed from dark green at fruit set to light reddish color at partial ripening and dark or bright purple at full ripe stage. The fruit took 63 days for compete ripening from fruit set. The ripe jamun had 76 per cent edible portion.TSS and sugars followed an increasing trend, while tannin content followed a decreasing trend growth and development. Fruit is berry, purplish red, ovoid, edible and polyembryonic.

***Syzygium alternifolium* Syn. *Eugenia alternifolia* Wight.:** This species is endemic to the southern Eastern Ghats in India. *Syzygium alternifolium* is harvested for its timber and seeds The species was therefore assessed as

Endangered. A moderate-sized or large tree, 0.9 m in girth with 3 m bole, found in Assam, Andhra Pradesh and Tamil Nadu. Leaves alternate, sub-orbicular, thick, coriaceous, flowers in lateral, longish peduncle cymes; fruits sub-spherical of the cherry size.

***Syzygium aqueum* (Burm.f.) A *Eugenia aquea* Burm.** It is found in Assam and Meghalaya and cultivated in Bengal. Bark smooth, ash colored; leaves opposite, ovate-oblong or ovate-lanceolate, coriaceous, gland dotted; flowers in terminal or axillary cymes, large, white, red or purple; fruits as big as a small apple, turbinate, flattened at both the ends, white or pale rose colored, the fruit is eaten as such processed into beverages.

***Syzygium aromaticum* (L) Syn *Caryophyllus aromaticus* L.:** Clove, the dried unopened flower buds of the evergreen tree, *Syzygium aromaticum*, (Syn. *Eugenia caryophyllus*) is an important spices noted for its flavour and medicinal values. It is indigenous to Moluccas Island (Indonesia) and was introduced to India around 1800 A.D. by the East Indian Company in their spice garden in Courtallam, Tamil Nadu. The major producers of this spice today are Indonesia, Zanzibar and Madagascar. The important clove growing regions in India now are the Nilgiris, Tirunelveli and Kanyakumari districts of Tamil Nadu, Calicut, Kottayam, Quilon and Trivandrum districts of Kerala and South Kanara district of Karnataka. Food processing industry uses both whole and ground form of cloves in various preparations. Clove oil is used in perfumeries, pharmaceuticals and flavouring industries. Clove oleoresin is also increasingly used in the food processing industry. In Indonesia, the major part is absorbed for making KRETEK cigarette industry. The clove is an evergreen tree often reaching a height of 7 to 15 metres. Leaves posses plenty of oil glands on the lower surface.Clove oil, the spice determining factor, is about 16 to 21 % in the buds. The oil contains 70 to 90 % of free eugenol and 5 to 12 % of eugenyl acetate.

***Syzygium samarangense* (Blume.) Syn. *Eugenia javanica* Lam:** Common names include wax apple, love apple, java apple, A tree, 5-15 m tall, with low branched trunk and widely branched crown, found in the Andaman and Nicobar Islands. It is cultivated in many parts of India for its edible fruits. In Malaysia, two forms are recognized which are white fruited or red-fruited. The shapes of water apples are different according to the species. Its pulp is crisp and watery, hence the name watery apple. However, not much part of the fruit is edible: the centre is filled with woolly fibers, within which brown seeds are located, they may be 3 to 6 small seeds, but generally the fruits are seedless. The flavours are varies from a thin, slightly acid flavor to a full and fruity apple flavour .

***Syzygium jambos* (L.) Alston. Syn. *Eugenia jambos* L.:** The tree is ornamental. The fruits are light yellow-white and rose scented with persistent calyx. The trees are medium sized and evergreen. The leaves have very small petioles. The seeds are polyembryonic and it is being grown in Assam, Bihar, Andhra Pradesh, Tamil Nadu, West Bengal, coastal areas of Maharashtra and Gujarat.

***Syzygium fruitecosum* (Roxb.) DC.** The trees are medium sized and evergreen. The leaves are light green in color, fruit harvesting start after the jamun got harvested in north India It is suitable for windbreak. Fruits are edible and small in size.

***Syzygium unifora* L. (Surinam cherry or pitanga cherry).** The trees are medium sized and evergreen. It is suitable for use as a rootstock of *Syzygium cumini*. It is resistant to attack of termites It is also a small tree and bears small-sized fruits having bright red colour and aromatic flavor. The tree is found in South India.The leaves are spread on house floors in Brazil, so that when crushed underfoot, they exude a smell which repels flies. The strong, spicy emanation from bushes being pruned irritates the respiratory passages of sensitive persons The shrub or tree, to 25 ft (7.5 m) high, has slender, spreading branches and resinously aromatic foliage. The opposite leaves, bronze when young, are deep-green and glossy when mature; turn red in cold, dry winter weather. Long-stalked flowers, borne singly or as many as 4 together in the leaf axils, have 4 delicate, recurved, white petals and a tuft of 50 to 60 prominent white stamens with pale-yellow anthers. The 7- to 8-ribbed fruit, oblate, 3/4 to 1 1/2 inch (2-4 cm) wide, turns from green to orange as it develops and, when mature, bright-red to deep-scarlet or dark, purplish maroon ("black") when fully ripe.

***Syzygium zeylanicum* (L.) DC. Syn *Eugenia zeylanica* Wight.:** The tree is small and bears edible fruits. Leaves ovate or linear-lanceolate. Flowers white, in axillary and terminal, usually compact, many flowered cymes. Berries subglobose, white, the size of a pea. It is found in Western Ghats of India.

*Syzygium bracteatum and* Syzygium *malaccense* (L.) are other species which and ornamental type

## CLIMATIC AND SOIL ADAPTABILITY

It is adapted to a wide range of ecological conditions, reflecting its wide geographical distribution under tropical, subtropical and semiarid conditions. It is also occurs in the lower range of the Himalayas up to an elevation of 1300 meters and in the Kumaon hills upto 1600 meters above mean sea level. It is widely grown in the larger parts of India from North to South with an annual rainfall varying from 350-500 mm. It can be easily grown in neglected and marshy areas where other fruit plants fail to be grown successfully. The jamun requires dry weather at the time of flowering and fruit setting. In sub-tropical areas, early rain is considered very beneficial for growth and quality of the fruit i.e. ripening of fruit. The fruits show remarkable improvement in quality and size of the fruit after the first shower of rain.

The jamun can be grown on a wide range of soils. However, for high yield potential and good plant growth, deep loam and a well drained soil are needed. Such soils also retain sufficient soil moisture which is beneficial for optimum growth and quality of the fruits. Jamun can grow well under salinity and waterlogged conditions also. The west coast of India has lateritic soil, which is well drained and suitable for jamun growing. Likewise medium black soils of peninsular India are also suitable for jamun growing.

## RECOMMENDED AND POPULAR CULTIVARS

This crop is facing severe genetic erosion due to industrialization, urbanization and intensive agriculture. A number of seedling strains with lot of variation in respect or fruit shape and size, pulp color, TSS, acidity and earliness are available in the scattered population particularly in Uttar Pradesh, Gujarat, Maharashtra and Karnataka that provide a good scope for selection of better varieties. The great variability in physico-chemical characteristic of fruits offering possibility of selecting a variety of small seed size, high pulp percent, deep pulp color with better chemical qualities are considered ideal for fresh market and processing. There is a large genetic diversity in the present population for crop improvement. Following approaches may be followed for improvement of the crop:

1. Selection based on intensive survey and clonal variation.
2. Creation of genetic variability by mutation, their characterization and evaluation.
3. Hybridization.

### Commercial Cultivars

**Konkan Bahadoli:** This variety was released in 2004 having spreading growth habit, dome shape canopy, dark green foliage, big size fruits and more keeping quality and regular bearer. The ripening of the fruit start in the second week of the July and comparatively late maturing under Punjab condition. The fruit yield ranges between 125-150 kg per tree.

**Goma Priyanka:** It is selection from Ode village of Anand district of Gujarat. Peak period of flowering is in March. It ripens in the first week of June in

Gujarat condition. The average fruit weight 19.50 g, pulp (84.62%) TSS15.10%, 0.38% acidity.

**Rajendra Jamun 1:** This variety was released from Bihar Agricultural College Bhagalpur. It is early, high yielding cultivar with average fruit weight 12.86 g with 88.40% pulp, TSS, 18.20° brix, and acidity 0.31%.

**Rajamun**:The variety is very common among the people. It produces big sized fruit with average length of 2.5-3.5 cm and of diameter 1.5-2.0 cm. fruits are oblong in shape, deep purple or bluish black in color at fully ripe stage. The fruit is juicy and acidic sweet in taste. The stone is small in size. It ripens in the month of June-July and show alternate bearing rhythm.

## PROPAGATION TECHNOLOGY AND ROOTSTOCKS

The propagation of true to the type quality planting material is a basic foundation of success of an orchard. Since jamun has a long juvenile period. The utmost care is required in the selection of genuine planting material at the time of establishment of an orchard

### Sexual Propagation

The sexual propagation is mainly through seeds. The fully ripe fruits of superior quality genotype based on the fruit size, shape, pulp percentage and color of the fruits were selected. The seeds were extracted and sown immediately as jamun seeds have no dormancy; hence fresh seeds should be sown in polythene bags of size 25 cm x 15 cm at a depth of 3-4 cm. The bags should be filled with a mixture of soil and farmyard manure in 3:1 ratio. Small holes are made before filling the porous rooting mixture in the bottom and sides of polyethylene bag to drain out the excess water. The seedlings become ready for transplanting after one year.

### Vegetative Propagation

Presently the existing plant population is seedling. Seedling plant bears fruit of variable size and quality on the other hand having long juvenile period. The clonal multiplication of the jamun induces early fruiting. For propagation of true-to-true plant material, patch budding and wedge grafting are the commercial methods.

### Rootstocks Production

The fully ripe fruits of superior quality genotype based on the fruit size, shape, pulp percentage and color of the fruits are selected. The seeds are extracted and sown immediately as jamun seeds have no dormancy; hence fresh seeds should be sown in polythene bags of size 25 cm x 15 cm at a depth of 3-4 cm. Seeds can also be sown in the raised nursery beds in open but the rate of survival of the grafted plant from nursery bed is very poor due to deep root system and also labor intensive as compare to plants raised in bags. The bags should be filled with a mixture of soil and farmyard manure in 3:1 ratio. Small holes are made before filling the porous rooting mixture in the bottom and sides of polyethylene bag to drain out the excess water. The seedlings become ready for budding or grafting after one year. The healthy seedlings of uniform size having stem of pencil thickness are used as rootstock for budding and grafting. The plants raised in the polythene bags can easily be transported to distant places and give higher transplanting success.

**Patch budding:** The selection of the bud wood of the desired genotype is an important task. The bud sticks are available during the active growth period, well swollen and recently matured buds are collected, and the buds from the upper part of the new shoots are not suitable. The active growth period is indicated by sap movement judged by easy separation of the bark from the wood of scion sticks. The bud stick should not be stored as this will cause the shriveling of the buds. In this method, a healthy bud is selected from the axils of leaf. Leaf blade is removed with the help of a sharp budding knife leaving petiole intact. The upper cut is given about 2.0 cm above bud which goes downwards up to 1.0-1.5 cm below the bud without taking wood portion and then lower cut is given about 1.0 cm below the bud. The similar rectangle cut is made on the rootstock and bud is placed. The bud is pressed by hand to remove open space if any and tied tightly with white polythene strip of 200 gauge thickness and 2 cm wide. The rootstock is detopped just after budding, about 15 cm above the bud to facilitate bud sprouting. After union, the top of the rootstock is cut a little above the bud union and polythene strip is carefully removed. The bud start sprouting 15 to 20 days after budding and plant will be ready for transplanting within 90 to 100 days.

**Wedge grafting:** In his method about 15-20 cm long mature shoots are defoliated 10-15 days prior to grafting operation; detach these shoots from the mother plant with the help of sharp secateurs. For this, soft portion of seedling rootstock is cut at 20-25 cm height and the top portion is removed. With the help of sharp knife, 5 cm long vertical downward incision is made in the center of the rootstock. A sharp slanting cut of 5 cm is made on both the sides on the base of the scion shoot to make wedge shape. Thereafter, prepared scion is carefully inserted in vertical slit of the rootstock and tightly secured with the help of 2 cm wide polythene strips. The grafted plant is covered with the help of polythene tube and tide with the rubber band to maintain the temperature and proper humidity, this tube is removed 20 days after grafting when the buds start sprouting that visible from outside.

**Transplanting:** The plant propagated in the nursery beds are lifted with large earth balls 100-110 days after budding for transplanting in an orchard. A number of these plants may be lost due to damage while lifting, packaging, and transporting. These operations are also cumbersome and incur high costs due to the large size of the earth balls. The plant propagated in polythene bags can be easily transported and transplanted without mortality with high final survival rate.

## PLANNING AND PLANTING

The planning of the orchard should be made keeping in view the soil climate and marketing potential of the crop. The land may be prepared by proper ploughing, harrowing and leveling. The next step is to layout the orchard.

There should be a gentle slope to facilitate proper irrigation and drainage of excess water to avoid the harmful effects of water stagnation in root zone during the rainy season. Jamun may be grown under various cropping systems i.e. as an orchard and as an agro forestry species in mixed cropping systems. After marking the field, pits are dug out. The size of the pit should be 90 x 90 x 90 cm during summer months, while digging, it is necessary to keep the topsoil and subsoil separately in two heaps near each pit, this helps in exposing harmful soil organisms and solarization of the pit. The well-decomposed organic matter is mixed with soil and then pits are filled. Planting should be done during the rainy season when the soil in the pits has already settled. While planting, one should be careful that the earth ball does not break and graft union remains well above the ground level. The plants should be irrigated immediately after planting. The jamun is planted at the distance of 9 x 9 m thus accommodates 124plants per hectare. It is usually planted along field boundaries, canal banks, schools, as wind break around mango plantation and around religious places.

### Planting System

The main objective while following a particular planting system is to accommodate the maximum number of trees per unit area. The popular system of planting in vogue is square system. Jamun is an evergreen tree can be transplanted during February-March and but the trees planted in February- March have to pass through unfavorable weather conditions in May and June. However, the monsoon season i.e. August-September is considered better because the plants easily get established during this season. The plants are transplanted with earth ball and are given irrigation till they get established.

### High Density Planting System

High density planting system ensures better utilization of land, labor and solar radiation with high return per unit area, because of accommodating more number of plants. High-density orcharding appears to be the most appropriate answer to the problem of low productivity and ensure early economic returns from the orchards. For the high density planting the plants should be trained in the initial years this will be helpful in the harvesting of the crop. Jamun is the only fruit crop having maximum per kg harvesting cost and also accident due to breakage of branches can be avoided.

## SOIL CULTURAL PRACTICES TECHNOLOGY

### Water Need

Jamun plants are not grown on commercial scale in a systematic regular field so the scattered population depends on the natural rainfall even in the irrigated zones. Irrigation is normally not practiced in jamun cultivation, but it promotes better growth during establishment and the early stages of growth, especially during the dry seasons. In the early age, plants require 8-10 irrigations a year while, bearing trees require 4-5 irrigations during the time of fruit development and ripening. In early stages, the jamun tree requires frequent irrigations but after the trees gets established, the interval between irrigations can be decreased.

### Nutritional Need

The natural grown population of the jamun is not generally manured. However, an annual dose of about 20 kg farmyard manure (FYM) during the pre-bearing period and 50-80 kg per tree at bearing stage is considered beneficial for proper growth and development of the fruit. In rich soils, the trees have a tendency to put on more vegetative growth with the result that the fruiting is delayed, under such conditions the trees should not be manured and irrigation should also be given sparingly and withhold. This helps in proper fruit bud formation and blossoming and good fruit setting. In general, 5 kg FYM, 125 g N, 50 g $P_2O_5$ and 50 g $K_2O$ per plant per year may be applied to one year old plant. This dose should be increased every year in the same proportion up to $10^{th}$ year, after which the fixed dose should be applied each year. Under rainfed conditions, foliar feeding is very useful in supplementing nutrients, particularly nitrogen and micronutrients. The optimum concentration of urea for foliar application seems to be 1.0% and one spray in April is effective to improve fruit retention, its growth and productivity. Supplementing micronutrients through foliar sprays of 0.2-0.1 % zinc sulphate has been observed to improve fruit quality and colour.

## Mulching

Mulching is the application of any layer of plant material or other suitable organic material to the surface of the soil, without incorporation into the soil. It is an ancient technique and having three main benefits, viz. improves soil physical properties; water conservation and reduction of weed growth. Use of much is very beneficial for successful cultivation of *jamun*. Mulching can be done with black polythene or any suitable organic material. Mulching with grasses, paddy straw and rice husk reduces the weed population and conserves the moisture in the soil. Organic mulching material improves soil pH, physical properties etc. Microbial and earthworm population in the basin soil increases with the use of paddy straw and grasses as mulch. Leaf litter of jamun under the canopy is effective to retain soil moisture during summer.

## Weed Control

In the scattered population the weed management is not generally practiced but productivity of regular orchards can be increased by following proper weed management. Weeds affect plant growth and yield adversely. Most weeds although complete their life cycle in a shorter period but compete with plants for light, water and nutrients. In the orchards, hoeing, hand weeding and ploughing of the land 2-3 times in a year are done to suppress weed growth. Intercropping and mulching also help in controlling weeds. Pre-emergence application of Hexuron 80 WP (diuron) @ 1.6 kg/acre can be made during first fortnight of March for rainy season and during first fortnight of September for winter season when field is free from weeds and stubbles. Glycel 41 SL (glyphosate) @ 1.6 litre/acre as post-emergence should be sprayed when the weeds are growing actively, preferably before weeds flower or attain a height of 15-20 cm i.e. during second fortnight of March for rainy season and during second fortnight of September for winter season . Dissolve the herbicides in 200 liters of water which is enough to give complete coverage on weeds in one acre. Spray during the calm day to avoid spray drift to the foliage of the trees.

## Inter Culture

The growing annuals or relatively short duration crops in the interspaces during the formative years in jamun plantation is referred to as intercropping and the growing of another perennial crop is called mixed cropping. Intercropping is intended to maximize land and space use efficiency to generate supplemental income particularly during the initial year of the growth when a lot of interspace is available in the orchard. The nitrogen fixing legume crop peas, gram, lentil, blackgram, cowpea, cluster bean, cucurbitaceous crops and capsicum may be grown as intercrops in the jamun orchard. Fruit crops like, guava, kinnow, kagzi lime, phalsa and papaya can be grown as filler trees.

## PLANT CULTURAL PRACTICES TECHNOLOGY

### Training and Pruning

#### Training

Basically, training is a potential tool to manage the canopy architecture of the plant. Young plants should be allowed with 4-5 well spaced branches to develop into the main scaffold structure of the tree. Framework of branches is allowed to develop above 60-75 cm from the ground level. Jamun plants may be trained through modified training systems.

**Modified Leader:** It is also known as delayed open centre. In this system, the main central leader is allowed to grow for a few years, until 8-10 scaffold develop around the central leader. The central leader is then cut to develop the side laterals which in due course, will grow as a modified leader. In this type of training, the trees develop well spaced limbs with strong crotches. The top being open, allows more sunlight to penetrate deep inside the tree canopy. This system is appropriate for training of jamun plants.

#### Pruning

Pruning is a tool to regulate tree and shape to achieve a desired architecture of the canopy and also to reduce the foliage density by removing the unproductive branches to make the tree open. Regular pruning in jamun plant is not required. However, the dry twigs and criss- crossed branches are to be removed regularly. To maintain the dwarf framework of the jamun tree, toping of main stem (4-6 meters) is needed. It will facilitate easy harvesting of the fruits. It is also observed that pruning of 50% annual extension growth after harvesting was effective to reduce the plant canopy and to improve the fruit quality attributes. The old unproductive damaged trees or young plants can easily be rejuvenated by top working.

### Canopy Management

Canopy management of the crop deals with the development and maintenance of their structure in relation to the size and shape for the maximum productivity and optimum fruit quality. The basic concept in canopy management of a perennial tree is to make the best use of the land and the climatic factors for increased productivity in a three-dimensional approach.

## Fruit Drop

Flowering and fruit set take place in March-April. It is observed that fruit drop in jamun starts just after fruit set and continues up to maturity and hardly 15-30% fruits reach maturity. The first drop takes place during bloom or shortly thereafter, and this proves to be the harvest drop as about 52% of the flowers drop after 4 weeks from flowering. The second drop starts after about 30-40 days after full bloom. The third drop takes place 40-50 days after full bloom.

**Control:** The fruit drop in jamun may be reduced by two sprays of 60 ppm $GA_3$, one at full bloom and the other 15 days after initial setting of fruit.

## Fruit Thinning

Fruit thinning is not practiced in the jamun although there is heavy natural fruit drop to balance the crop load.

# HARVESTING AND PRODUCTION OF FRUITS

## Maturity Indices

Optimum stage of fruit maturity is based on physical and chemical properties of the fruit. The jamun fruit followed a sigmoid curve and growth recorded in three distinct phases, viz during the first phase, the rate of growth was slow 15-52 days after fruit set, in the second phase 52-58 days after fruit set, the rate of development was quite rapid and the third and last phase 58-60 days after fruit set comprised slower growth with little increase in fruit weight, moisture content and sugars.

## Harvesting Practices and Yield

The seedling plants starts bearing after 8-10 years of planting while grafted ones after 4-5 years. However, commercial bearing starts after 6 to 8 years of planting in grafted plants and continues till the tree becomes 50 to 60 years old. The fruit ripens in June-July. The main characteristic of ripe fruit is that it turns deep purple or black in color. The jamun fruit should be picked immediately when it ripens. It cannot be retained on the tree at the ripe stage. The ripe fruits are picked singly by hand and in all cases care is taken to avoid all possible bruising to fruits. For harvesting, the picker climbs on the tree with bag of cotton cloth slung on the shoulder. When the bag is full of fruit, the picker hangs the bag down on the ground with the help of rope and another person standing below the tree catches the bag and empties in the baskets. The fruits of jamun are generally harvested daily and sent to market on the same day. There are 4-5 picking per plant as the fruit setting time decide the ripening phase of the fruit. The average yield of fruits from a full-grown jamun tree varies from 100-150 kg per year.

## POST- HARVEST FRUIT TECHNOLOGY

**Grading:** Grading is an important component of the marketing of the fruit.Generally it has been observed that after the fruit harvest, contractor or fruit growers place best quality fruits (known as chhap) in the top of the containers but this practice in long term neither helps the grower nor the consumer. Considerable variation exists in the quality of harvested fruit due to genetical variation, environmental and agronomic factors and therefore requires grading to get suitable returns from the market. Jamun should be graded on the basis of fruit weight, shape and color to fetch better price in the market.

**Packaging**: Packaging is a vital component of post harvest management of fruit to protect it from deterioration during transit while marketing. The appropriate packaging protects the fruits from physiological, pathological and physical deterioration in the marketing channels and retains their attractiveness. Jamun fruits are highly perishable and are normally packed in bamboo baskets for transport to local market. Physiological loss in weight, shriveling and rotting can be minimized when packed in bamboo baskets and CFB boxes using leaves or news paper as cushioning material. Among different types of containers one kg CFB boxes can be used for direct marketing of the fruits to prevent bruising losses.

**Storage:** The jamun fruit is highly perishable and can be kept in good condition for about 2-3 days under ordinary condition.

**Processing:** Jamun exhibits heavy post-harvest losses. It is necessary to process the fruits into different value added products by employing different methods of fruit preservation to ensure the availability of the jamun products round the year. The fruits may be utilized for making jam, jelly, beverages, wine and vinegar. A ready-to-serve beverage (nectar) is prepared with 25percent juice, $18^0$ Brix and 0.6 per cent acidity, which has an appealing purple color and it has been found to be highly acceptable. Acceptable dry table wine may be prepared from the jamun juice.

## MARKETING OF FRUITS

Jamun is high value minor fruit crop having the specific commercial importance in the market because of its medicinal and nutritional properties to cure diabetic patient. In the present context farmers collect the fruits from the shelf grown and naturally growing trees and sale this fruits in the local and distant market. While marketing grower do not follow standard marketing practices like grading and proper packaging. This trade is characterized by the large number of intermediaries and there is exploitation of farmers, alternatively producers share in the consumer's rupee is very less. Marketing plays an ultimate role in the fruit production as a successful marketing is the index of successful fruit production. The existing fruit trade is characterized by long chain of middlemen which reduces the share of *Jamun* growers and increase the price paid by the consumers. The area under jamun is not systematically planned so the producers are scattered over the wider areas. There is lack of any collective organization among the producers, while fruit merchants, commission agents and retailers are well organized. The different marketing channels with their fruits from *Jamun* growers to ultimate consumer are the village traders, pre-harvest contractors, commission agents/wholesalers, retailers/ hawkers were involved as intermediaries. With these intermediaries the commodity passes through different channels as presented below:

## Different Marketing Channels of Jamun in India

| Channel-I | Channel-II | Channel-III | Channel-IV | Channel-V |
|---|---|---|---|---|
| Producer | Producer | Producer | Producer | Producer at farm gate |
| ↓ | ↓ | ↓ | ↓ | ↓ |
| Pre-harvest Contractor | Pre-harvest Contractor | Commission agent | Pre-harvest Contractor at farm gate | Consumer |
| ↓ | ↓ | ↓ | ↓ | |
| Commission agent | Harvester | Hawker | | |
| ↓ | ↓ | ↓ | | |
| Hawker | Commission agent | Consumer | Consumer | |
| ↓ | ↓ | ↓ | ↓ | |
| Consumer | Hawker | | | |
| | ↓ | | | |
| | Consumer | | | |

## Export Potential

Jamun possesses the commercial value in the national and international market due to hypoglycemic properties. The duration of availability of the jamun fruits is less and fruits are highly perishable in nature, therefore, it has tremendous scope for processing. The beverages like juice squash and vinegar may be exported to earn foreign exchange. But in the real sense there is no surplus production of jamun in India.

## INSECT-PESTS AND MANAGEMENT

Among the pests, white fly and leaf eating caterpillar cause great damage to the Jamun fruit crop and tree.

**Leaf eating caterpillar (*Corea subtilis*):** The caterpillars is serious pest that infest the tender leaves and tree becomes defoliated. As a result, the tree losses vitality and yield declines To control this pest, spray Rogar 30 EC 0.1 per cent or malathion 0.1 per cent.

**White fly (*Dialeurodes eugenie*):** White fly damages the fruits and fruits give wormy appearance on the surface due to attack of fruit fly. It can be controlled by maintaining sanitary situation in the orchard, which consist of picking up the affected fruits and burying them deep in the soil and area under the tree should be dug, so that the maggots and the pupae hibernating in the soil may be destroyed. In case of severe infestation, spraying with dimethoate 0.03 % is effective.

**Bark eating caterpillars (*Inderbela tetraonis* and *inderbela quadrinotata*):** It is a polyphagous pest of major importance of **jamun** especially in north-western India The larva feeds on the live bark tissues, shelters under the covering of silken webbing during the night later on makes a tunnel into the branch and stem and remain in the hole during day time. As a result of larvae feeding inside the branches and the trunk, the translocation of cell-sap is disrupted and growth as well as fruiting capacity of the tree is reduced the tree show yellowish appearance and drying of the branches. As a result, the tree looses vitality and yield declines. Remove the webbing and inject kerosene oil into the holes during September-October and again in January –February. In the hole and be plugged. Foliar sprays with Dimethoate (0.05%) triweekly interval control the pest effectively.

**Jamun leaf miner *(Acrocercops syngramma* and *Acrocercops phaeospora):*** This pest causes damage during reproductive phase i.e. from April –September. The newly hatched caterpillar mines a narrow thread like silvery gallery on the leaf along the mid-rib upwards or latterly. The larval mine later on is transformed in to a tubular blister like swelling on the dorsal surface of the (Lakra, 1997). The pest can be controlled by clipping and burning of affected leaves followed by spraying of dimethoate 30 EC 1.2ml/ l.

**Jamun leaf roller (*Polychorosis cellifera*):** The larvae web the leaves by folding the tip downwards on both the margins parallel to the mid-rib and feed inside. In case of severe attack 1/4$^{th}$ of the lamina is eaten up. The pest undergoes 3-4 generations between March- April and September-October in North India and second generations is most harmful (Lakra, 1997). Regular clipping and burning of affected leaves can keep the population under control. Spraying with chlorpyriphos 20 EC (2ml/l) is effective against control of Jamun leaf roller.

**Leaf Webbers *(Argyroploce aprobola* and *Argyroploce mormopa):*** The newly hatched larvae in large numbers web together the tender leaves at shoot tips and feed within (Lakra, 1997). Regular clipping and burning of affected leaves can keep the population under control. In case of severe attack, spraying with chlorpyriphos 20 EC (2ml/l) is recommended.

***Squirrels, parrots crows and Birds:*** The jamun crop is seriously damaged by pests like squirrels and birds like parrots and crows. These have to be frightened away by beating the drums or flinging stone.

**Fruit-borer *(Meridarchis reprobata)*:** The caterpillars bore into the fruits and feed within rendering fruits unfits for consumption. It be managed by spraying of spray dimethoate 0.05% 20 days before harvest.

**Termites:** The plant is damaged by the termites. It can be controlled by spraying with chlorpyriphos 20 EC (2 ml/litre) followed by soil drenching around the plant canopy.

## DISEASES AND MANAGEMENT

**Leaf spot and fruit rots (*Glomerella cingulata*):** Small spots appears on the leaves and fruit start rotting. Affected leaves show small-scattered spots, light brown or reddish brown in color. Affected fruits show small, water soaked, circular and depressed lesions. Fruits rot and shrivel. The affected fruits are not suitable for marketing. The disease can be controlled with the foliar application of the fungicides like Dithane Z-78 at 0.02% or Bordeaux mixture (2: 2:250).

## FUTURE STRATEGIES

The jamun is an important minor and medicinal fruit rich in antioxidants so having rich future potential for cultivation in India. Therefore, areas of research to be given priorities are suggested as below.

1. Promising genotypes having tolerance to the biotic and a biotic stresses should be selected. The crop improvement programme should be focused on screening of the available elite strain, mutation breeding to induce dwarfness for the ease of harvesting and to develop the varieties that have more shelf life and deep purple pulp color.

2. There is an urgent need to ensure easy availability of the grafted plants of elite strains.

3. The production technology like integrated nutrient management, high density planting, weed management, canopy architecture management and irrigation management should be standardized.

4. Jamun fruits are highly perishable in nature; therefore, varieties with prolonged shelf-life should be developed.
5. The cheap mechanical harvesting devices should be developed so that cost of the harvesting can be reduced and laborer accident during harvesting may be avoided.
6. There is need to develop value added products from jamun pulp and seeds and need to be popularized them in the common masses in rural and urban areas and possibilities may be explored in the global market also.
7. The research work on integrated nutrient management and pests' management needs to be initiated to improve the percent marketable fruits.
8. The plantation through social forestry system may be encouraged by creating the awareness in the youth and social institutions and children in the school.

## LITERATURE CONSULTED

Bose, T. K., Mitra, S. K. and Sanyal, D. 2001. Fruits: Tropical and Subtropical. Vol. 2: 643-656. Naya Udyog, Calcutta.

Chaturvedi ,M. D. 1956. Jamun is another of our prized trees. *Indian Farming,* 5:17-9.

Chovatia, R. S. and Singh, S. P. 2000. Effect of time on budding and grafting success in jamun (*Syzygium cuminii* Skeels). *Indian journal of Horticulture*, 57 (3): 255-8.

Chundawat,. B S. 1990. Arid Fruit Culture, pp 165-71. Oxford and IBH publishing Co. Pvt Ltd New Delhi.

Geetha, C. K., Babylata ,A. K., Methew, K. L. and George, S .T. 1992. Fruit development in jamun (*Syzygium cuminii* Skeels). *South Indian journal Horticulture* ,40 (6): 350-1.

Hebbara, M., Manjunatha, M. V., Patil, S. G. and Pati,l D. R. 2002. Performance of fruit spicies in saline-water logged soils. *Karnataka Journal of Agricultural Sciences* , 15 (1): 94-8.

Lakra, R.K. 1997. Insect pests of some under-exploited fruits and their management 1. Jamun (*Syzygium cumini* Skeels) A Lepidopterous pests. *Haryana J. Hort. Sci.*, 26(1-2):35-47.

Kadam, S. S. 2001. New products from arid and semi arid fruits. *Indian journal of Horticulture* , 4:117-20.

Khurdia, D. S and Roy, S. K. 1985. Processing of jamun fruit in to a ready-to-serve beverage. *Journal of Food Science and Technology ,22 : 27.*

Madalageri, M. B., Patil, V. S. and Nalawadi, U. G. 1991. Propagation of jamun *(Syzygium cuminii* Skeels) by soft wood wedge grafting. *My Forest,* 27 (2): 176-8.

Mishra, R. S. and Bajpai, P. N. 1971. Vegetative growth studies in the jamun. *Indian journal of Horticulture* ,28 (4): 273-7.

Shukla, J. P. and Prasad, A. 1980. Changing pattern of jamun frui during growth and development. III. Changes in respiratory activity. *Progressive Horticulture,*12 (3): 71-3.

Shukla, K. G., Joshi, M. C., Yadav, S. and Bist, N. S. 1991. Jamun wine making Standardization of a methodology and screening cultivars. *Journal of Food Science and Technology,* 28 (3): 142.

Singh, A. K., Bajpai, A,, Singh, A. and Reddy, B. M., C. 2007. Evaluation of variability in jamun (*Syzygium cuminii*) using morphological and physio-chemical characterization. *Indian Journal of Agricultural Science* ,77 (12): 845-8.

Singh, I. S. 2001. Minor fruits and their uses. *Indian journal of Horticulture*, 58(1-2): 178-82.

Singh I.S., Srivastava, A.K. and Singh, V. 1999. Improvement of some under utilized fruits through selection. *J. Applied Hort.*, 1: 34-37.

Singh, R. K. and Thakur, S. 1977. Seed germination and seedling growth of jamun (*Syzygium cuminii* Skeels) types. *Proceedings of Bihar Academy of Agricultural Sciences,* 25(1): 139-42.

Singh, S., Krishnamurthy, S. and Katyal, S. L. 1967. *Fruit Culture in India,* pp 255-9. ICAR, New Delhi.

Singh, Sanjay and Singh, A K. 2006. Standardization of metod and time of propagation in jamun (*Syzygium cuminii*) under semi arid environment of western India. *Indian Journal of Agricultural Sciences,*76: 142-5.

Singh, Sanjay, Joshi, H.K., Singh, A.K., Lenin, V., Bagle, B.G. and Dhandar, D.G. 2007a. Reproductive biology of jamun (*Syzygium cuminii* Skeels) under semi arid tropics of western India. *Horticulture Journal,* 20(2): 76-80.

Singh, Sanjay, Singh, A.K., Bagle, B.G. and Joshi, H. K. 2007b. Scientific cultivation of jamun (*Syzygium cuminii* Skeels). *Agriculture Update,* 2(2):37-40.

Singh,S., Singh,A.K., Joshi,H.K., Bagle,B.G. and Dhandar,D,G. 2007c. Jamun-A fruit for Future, ICAR, New Delhi.

# 9

# The Avocado

*M. Tamilselvan*

## INTRODUCTION

The avocado (*Persea americana* Mill) is the highly nutritive fruit. Its common names are alligator pear or butter fruit in English; Aguacate, Palta in Spanish and Avokado in Afrikaans. It is an evergreen sub-tropical fruit.

## NUTRITIVE AND CULTURAL SIGNIFICANCE

Avocado is the most nutritive among fruits. The pulp is rich in proteins and fat, but low in carbohydrates. Avocados have the highest energy value of 245 cal/ 100 g and reservoir of several vitamins and minerals. Avocado is rich in copper and iron, potassium and is associated with lower blood pressure as it is high in monounsaturated fat. It contains all amino acids. Another benefit is avocados shown to maintain good cholesterol while reducing bad cholesterol and triglyceride level.. Avocado is considered healthy for heart ,good for digestion and vision. It is a natural detoxification fruit and may help in preventing osteoporosis

Avocado is mainly used fresh, in sandwich filling or in salads. It can also be used in ice creams and milk shakes and the pulp may be preserved by freezing. The fat in avocado is similar to olive oil in composition and is widely used in the preparation of cosmetics.

## ORIGIN, HISTORY AND DISTRIBUTION

Avocado is a native of tropical America. It originated in Mexico and Central America, possibly from more than one wild species. The first written account of the avocado, so far as known, is contained in the report of Gonzalo Hernandez de Oviedo (1526), who saw the tree in Colombia, near the Isthmus of Panama. It was introduced into Jamaica in 1650 and to Southern Spain in 1601. It was reported in Zanzibar in 1892. It was first recorded in Florida in 1833 and in California in 1856. The early Spanish explorers recorded its cultivation from Mexico to Peru.

In India it was bought during first decade of nineteenth century by an American missionary, residing in Bangalore between the years 1906 and 1914 from Royal Botanical Gardens, Ceylon. No extensive propagation of these with an eye on their commercial possibilities was attempted till 1940, though the climatic conditions and soil requirements in the region were ideal for their large scale cultivation. In fact, on account of its close resemblance to butter and identical lack of any taste it is so named in these parts as 'Butter Fruit'. A few seedlings of their choice varieties were also occasionally brought from their home country by the American missionaries who came for periodical stays in India during 1912 and 1940. This is the reason that more than a dozen varieties of Avocado grown in many parts of this region on hill stations like Kallar near Nilgiris, Palni, Kodaikanal, Yercaud, Coorg, etc. The small avocado orchard in the Maharaja's Palace at Bangalore and in the bungalows of a few Britishers were planted during that period. The influx of Americans in very large numbers soon after the outbreak of the Second World War, renewed interest in the propagation of avocados. In the meantime the Government of Mysore opened a research station for non-citrus fruits at Hessaraghatta, Bangalore and about 150 avocado seedlings of different species were introduced in the research station. Later on these trees were removed but lots of seedlings were planted in Bangalore and parts of Karnataka and Kerala. Lot of promising varieties of avocado were brought for planting in Karnataka, Kerala, Tamil Nadu, Sikkim and other states time to time for planting. The avocado cultivation has gained an overwhelming popularity during last one decade due to nutrition properties of the fruits.

Mexico is the largest producer and exporter of Avocado (supplies 45% of the international avocado market) in the world and it is followed by Chile, Indonesia, United States, Dominican Republic, Colombia, Brazil, Peru etc. The Avocado production in Asia is limited but some of the countries like China, Vietnam, Korea are known for avacodo cultivation. The USA is the number one importer in the world, followed by the Netherlands.

The production in India is very limited and there are not commercial plantation of Avocado.

The agro-climatic conditions prevailing in various parts of the country appear to be favourable for bringing more areas under avocado. Presently, plantations are not well organized and they are scattered. Also, quite a good number of improved varieties are now available with higher yield potential. Avocados are grown scattered in southern tropical states like Tamil Nadu, Kerala, Karnataka, and Maharashtra. Also popular in the northeastern Himalayan state of Sikkim on hill slopes at elevations of 800-1,600 meters. Avocados are grown at higher elevations frequently, to prevent soil erosion. Avocado is grown successfully in neighboring Sri Lanka, where good-quality fruits are harvested during May to August and December to January at different regions. Similar agro-climatic conditions to Sri Lanka are available in the Andamans and Nicobar Islands and in the tropical southern India. With proper varietal selection it should be possible to exploit also the possibility of out-of-season production, thus enhancing the availability of fruits for a longer period during the year. Avocado fruits produced in the country can be marketed without much difficulty, particularly to meet the requirement of the growing tourist industry. The mainland India and the Andaman and Nicobar Islands are attracting foreign tourists in a large number of places, where avocado could find a good market access. Avocado has also a good export potential.

## TAXONOMICAL AND BOTANICAL DESCRIPTION

Three ecological races have been recognized i.e. Mexican race, Guatemalan race and West Indian race. Guatemalan and West Indian races are placed in group *of Persea americana.* Mexican race is placed in group *Persea drymifolia.* The avocado belongs to the family Lauraceae.

**Mexican Race**: The leaves are smaller in this race than those of Guatemalan and West Indian races Leaves sharper at the apex. The flowers are hairy pubescent. Fruits small in size weighing less than 250 g and tends to ripen at the apex within six to eight months after flowering. Fruits have thin, smooth skin with a large seed fitted loosely in the central cavity.This is a high land race. This race is the most resistant to cold temperature. It does not thrive in topical region.

**Guatemalan Race**: It is native to the highlands of Central America. Fruits are larger in size and weighing 600g. They requires a longer time to develop and ripen. The fruits ripen in 9 to 12 months after flowering. The fruits have thicker, harder and rougher skin as compared to Mexican race. Seeds held tightly in the cavity of the fruit,are small. The oil content ranges between 8-15 per cent. The trees are less resistant to cold than Mexican race.

**West Indian Race**: These are native to the low lands of Central America. Fruits are medium sized, skin is smooth but leathery and glossy. Fruits are borne on long stalks and require up to 9 months for ripening from flowering. Seeds are large, fitted loosely in the cavity. The oil content is 3-10 per cent. This race is least resistant to cold and is likely to be successful in most parts of tropical India.

## Botanical Description

***Growth habit:*** The avocado is a dense, evergreen tree, shedding many leaves in early spring. It is fast growing and can reach 24 m, although usually less, and generally branches to form a broad tree. Growth is in frequent flushes during warm weather in southern regions with only one long flush a year in cooler areas.

***Foliage:*** Avocado leaves are alternate, glossy, elliptic and dark green with paler veins. They normally remain on the tree for 2 to 3 years. The leaves of West Indian varieties are scentless, while Guatemalan types are rarely anise-scented and have medicinal uses. The leaves of Mexican types have a pronounced anise scent when crushed. The leaves are high in oils and slow to compost and may collect in mounds beneath the trees.

***Flowers:*** Avocado flowers appear in January to March before the first seasonal growth, in terminal panicles of 200 to 300 small yellow-green blooms. Each panicle will produce only one to three fruit. The flowers are perfect, but are either receptive to pollen in the morning or shed pollen the following after-noon (type A), or are receptive to pollen in the afternoon, and shed pollen the following morning (type B). About 5% of flowers are defective in form and sterile. Production is best with cross-pollination between types A and B. The flowers attract bees and hoverflies and pollination is usually good, except during cool weather. Off-season blooms may appear during the year and often set fruit. Some cultivars bloom and set fruit in alternate years.

***Fruit:*** Guatemalan types produce medium, ovoid or pear-shaped pebbled green fruit that turn blackish green when ripe. The fruit of Mexican varieties is small (170 g) with paper-thin skins that turn glossy green or black when ripe. The flesh of avocados is deep green near the skin, becoming yellowish nearer the single large, inedible ovoid seed. The flesh is hard when harvested but softens to a buttery texture. Wind-caused abrasion can scar the skin, forming cracks which extend into the flesh. "Cukes" are seedless, pickle-shaped fruit.

Off-season fruit should not be harvested with the main crop, but left on the tree to mature. Seeds may sprout within an avocado when it is over-matured, causing internal moulds and breakdown. High in monosaturates, the oil content of avocados is second only to olives among fruit, and some-times greater. Clinical feeding studies in humans have shown that avocado oil can reduce blood cholesterol.

## CLIMATIC AND SOIL ADAPTABILITY

Avocados cannot tolerate northern India's hot dry winds and frosts, typically grown in tropical or semitropical areas with rainfall in summer and humid, subtropical rainfall areas. The climate zone of avocados is from true tropical to warmer parts of the temperate zone. Depending on the race and varieties, avocados can thrive and perform well in climatic conditions ranging from true tropical to warmer parts of the temperate zone. In India, avocado is not a commercial fruit crop. It was introduced from Sri Lanka in the early part of the twentieth century. In a very limited scale and in a scattered way it is grown in Tamil Nadu, Kerala, Maharashtra, Karnataka in the south-central India and in the eastern Himalayan state of Sikkim. It cannot tolerate the hot dry winds and frosts of northern India. Climatically, it is grown in tropical or semitropical areas experiencing some rainfall in summer, and in humid, subtropical summer rainfall areas.

Avocados can be grown on a wide range of soils, but they are extremely sensitive to poor drainage and cannot withstand water-logging. They are intolerant to saline conditions. Optimum range of pH is from 5 to 7. This is a unique characteristic of avocado tree. The plants perform best on loam of medium texture overlying a porous sub soil. Loams and sandy loam of alluvial origin are also found suitable for cultivation of avocados.

## RECOMMENDED AND POPULAR CULTIVARS

All three horticultural races adapted to tropical and sub-tropical conditions i.e. West Indian, Guatemalan and Mexican have been tried in India. The cultivars of West Indian race are grown in localized pockets in Maharashtra, Tamil Nadu and Karnataka. In tropical and near-tropical areas, only West Indian race is well-adapted but its hybrids with Guatemalan (e.g. Both selection) perform well and are considered valuable for extending the harvest season. In less tropical regions, hybrids of Guatemalan with Mexican race predominate since they combine the cold hardiness of the latter with the superior horticultural traits of both and also bridge the two seasons of maturity. In the eastern Himalayan state of Sikkim, avocado has been introduced successfully in hill ranges with an altitude of 800 to 1,600 metres. Both the Mexican and Gautemalan races are grown successfully in Sikkim. In avocado-growing areas of Sikkim, temperatures range from 12 to 30°C with an average annual rainfall of 2,000 millimetres. The Mexican race is cultivated on mid-altitude hills (pH 5-6).

## Popular Varieties in India

Green Type, Purple, TKD-1, Nabal, Linda, Puyevla, Gott-Froid, Furete,Pullock and Waldin.

**Pullock:** Pullock originated in Miami before 1896. Commercially propagated in 1901, oblong to pear shaped, very large, up to 2.25 kg, skin smooth; flesh green near skin, contains 3 to 5% oil, seed large, frequently loose in cavity. Season: early July to Aug. or Oct. Shy-bearing and too large but of superior quality.

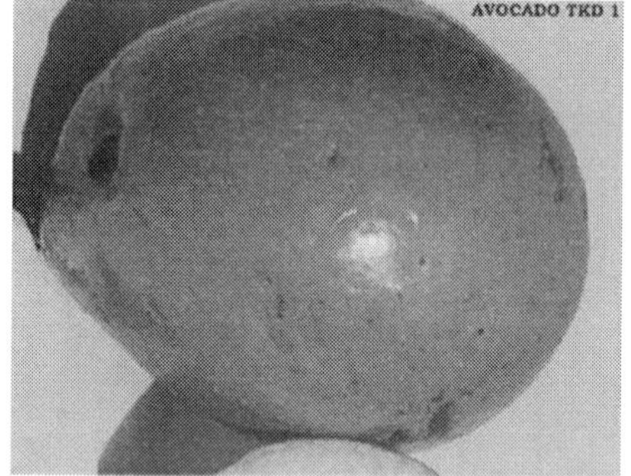


**TKD 1:** TKD 1 was developed at Horticultural Research Station, Thadiyankudisai of TNAU, Tamil Nadu. The fruits are medium sized and round. Trees upright and semi spreading hence suited for high density planting. Yield 264kg / tree. Fruits are sweet TSS 8° brix, fat 23.8%, protein 1.35%.

**Fuerte:** Mexican x Guatemalan hybrid collected by Carl Schmidt in 1911 from Atlixco, Puebla, Mexico; survived extremely cold weather in California, January 1913 and consequently named 'Fuerte' meaning 'strong' or 'hardy'. Flower group B; Pear shaped, medium to large size weighing 170 – 500g. Smooth green fruits of high quality with creamy, pale green flesh, easily peeled and available late fall through spring; 75–77% recovery.

**Purple**: This cultivar was introduced from Sri Lanka. It belongs to West Indian race. Fruit is pear shaped.12-15X6-9 cm in size, long necked and weighing 250 g. Its skin is smooth, deep crimson or maroon in colour and leathery in texture. Flesh is thick, smooth, fine textured and deep yellow in colour. Seed is set loose in the cavity. On ripening it gives rich nutty flavor.

**Green**: It belong to Guatemalan race. Introduced from Sri Lanka. Fruits are large and oval in shape. The colour of fruits is yellowish green but has slightly rough surface with thin skin. Seeds are large and sits tight in the cavity. Flesh thick, greenish yellow. It has nutty flavor at ripening.

## PROPAGATION TECHNOLOGY AND ROOTSTOCKS

In India, avocado is commonly propagated through seeds. The viability of seeds of avocado is quite short for about 2 to 3 weeks but this can be improved by storing the seed in dry peat or sand at 5°C. Removal of seed coat before sowing hastens germination. In India most of the trees grown are seedlings in origin.

### Seed Propagation

The seeds taken from mature fruits are sown directly in the nursery or in polyethylene bags. When 8-12 months old, the seedlings are ready for transplanting. The seedling trees took more time to start fruiting and the yield and fruit quality is highly variable. Due to cross-pollination, there is great variability in the seedlings produced from seeds, it is impossible to obtain genetically uniform plant as indicated for the formation of commercial orchards. These seedlings plants take long time to produce first crop and fruit quality in unrealiable.

### Vegetative Propagation

Vegetative propagation of superior clones of avocado by budding or grafting is essential to avoid this problem. Cleft grafting is the most suitable for the multiplication of avocado. It is done during the months of September and March. At the Fruit Research Station, Kallar, Tamil Nadu, layering as well as inarching gave up to 75 per cent success, while in West Bengal chip-budding is reported to be successful. In India, presently, there is no commercial nursery engaged in vegetative propagation of avocado, nor is there any initiative either at governmental or private level to undertake nursery production of avocado planting material.

### Rootstocks

West Indian rootstocks are preferred in warmer areas and under saline conditions. Purple avocado is generally used as a rootstock for avocado propagation. Guatemalan rootstocks are more sensitive to cold. They are susceptible to high pH, chlorosis and to *vertcillium* wilt.

## PLANT AND FRUIT PHYSIOLOGY

Avocado is an evergreen fruit plant. It starts bearing at 5-6 years after planting and has a marked tendency to biennial bearing which is prevalent in a number of other fruit trees. But there is specific problem in fruit set as far as avocado concerned. In avocado, the inflorescence is a compound panicle. The individual flowers are morphologically bisexual having fertile male and female organs. But they exhibit dichogamy viz., the male and female organs coming to maturity at different time thereby avoiding self pollination of an individual flower. In dichogamy, they are protogynous viz., the female parts coming to maturity before male organs. The type of dichogamy in avocado is a complicated one unique to avocado-the diurnally synchronous dichogamy. The female parts of all flowers that open at a time in a particular tree will mature simultaneously and hence behave functionally as female flower. The male parts of same flowers will come to maturity when the flowers open next time and hence all of them behave as male flowers during that period. By this the cross pollination between flowers of the same tree are also ruled out. The situation is further worsened by the fact that all the trees of a particular group will be exhibiting the same sex phase at a particular time and the opposite sex phase during the next opening of the same flower. So if the trees of single group are planted in mass, they will not set fruit and each group requires interplanting of trees of mother group the two groups, being compatible with one another.

The single tree of avocado is often unfruitful and mixed planting of varieties is desirable. Pollination is done mainly by honey bees. Avocado produces flowers from November to January in three main flushes at Poona. Three flushes of flower clusters do not raise from different branches but arise in succession from the extension of the same leafy shoots .November-December flowering fail to set fruit but January flowering set a successful crop. In foot hills of Nilgiris , March-April flowering set good crop.

The growth of avocado fruit follows the pattern of single sigmoid curve. The early period of fruit growth is characterized by rapid cell division. It continues during fruit development and even occurs in the mature fruit attached to the tree.

## PLANNING AND PLANTING

The avocado plantation in a relatively new area required care in selection of the varieties. The varieties of both A and B groups should be selected and their flowering must overlap. The proportion of A and B group varieties can be 1:1 or 2:1. Avocado is planted out to a distance of 6 to 12 meters depending on the vigour of variety and its growth habit. For varieties having a spreading type of growth, like Fuerte, a wider spacing should be given. In areas prone to excess water, they should be planted on mounds as avocados cannot withstand waterlogging. In Sikkim, a planting distance of 10 x 10 meters on hills slopes is preferred. While in south India, when it is planted with coffee the planting distance varies from 6 meter to 12 meters. The pits of 1 cubic meter size are dug during April – May and filled with farmyard manure and top soil (1:1 ratio) before planting. Planting is done in June-July or sometimes in September. In Coorg, a region of Karnataka, avocado trees are planted also as one of the mixed crops in a primarily coffee based cropping system (Tripathi and Karunakaran, 2013).

## SOIL CULTURAL PRACTICES TECHNOLOGY

### Water Need

In India, avocado is grown in those areas where rainfall is high and fairly distributed throughout the year. Therefore it is grown under rain fed conditions and irrigation is generally not given. Irrigation at intervals of three to four weeks during the dry months is beneficial. Sprinkler irrigation has been reported to improve the fruit size and oil percentage and advances harvesting time. To avoid moisture stress during winter season, mulching with dry grass/dry leaves is desirable. Flooding is undesirable as it promotes root rot incidence.

## Nutritional Need

Avocados need heavy manuring , and application of nitrogen has been found to be most essential. In general, young avocado trees should receive N, $P_2O_5$ and $K_2O$ in a proportion of 1:1:1 and older trees in the proportion of 2:1:2.

To young plant of avocado from 1 to 3 years age ,apply 25 kg farmyard manure and NPK @ 40,25,35 g per plant .To full grown trees of 10 years age, apply 50 kg farmyard manure and NPK @ 200,45,165 g per plant. Nitrogenous fertilizer improve fruit size and yield markedly. In Coorg area and the humid tropical region of Karnataka, Kerala and Tamil Nadu the fertilizers should be applied in two split doses in May -June and September-October. While in Northern India fertilizer may be applied in two split doses in March-April and September-October or just before and after the onset of the monsoon. Foliar application of zinc sulphate (0.5 per cent) and other micro nutrients may be undertaken in April-May or September– October. These micro nutrients may be applied as soil application along with other fertilizers.

## Inter Culture

In young orchards of avocado legumes or shallow rooted crops can be practiced to get income during pre bearing periods. However, inter cropping should be stopped when avocado starts bearing fruits after the age of 4-5 years.

## Weed Control

Deep cultivation in avocado orchards should be avoided because of surface roots.Intercropping with legumes or shallow-rooted crops can be done in young orchards which can smoother weeds also. The monoculture plantation of avocado may be maintain with sod culture. The weeds are major problems in high rainfall zones of south India. The use of gramoxone or glyphosate is recommended to control weeds. In coffee based plantation system, scruffling done for coffee is sufficient to control weed. Care should taken that the roots of avocado not disturbed during scruffling.

# PLANT CULTURAL PRACTICES TECHNOLOGY

## Training and Pruning

The plants need to be given light pruning in initial stages for developing an open centre canopy. After that pruning is rarely practiced. In upright varieties such as Pollock topping is done to reduce the tree size while in spreading varieties like Fuerte, branches are thinned and shortened. The dropping and ground touching branches need to be pruned for ease in cultural practices. Heavy pruning has been found to promote excessive vegetative growth, consequently reducing the yield.

## Avocado Quality Improvement

Avocados are a highly perishable product and susceptible to damage. Improving avocado quality at retail level is a major strategic objective for the industry. In 2006, detailed sensory research was undertaken to quantify consumer's preferences and tolerances of internal defects. This research established, amongst other things, that consumers will tolerate up to 10% internal defects with higher levels of damage adversely impacting future purchasing decisions.

A suit of R&D and extension projects have been implemented since 2006 to address the issues affecting avocado quality:

- Education materials developed and disseminated (hard copy and on-line) for each sector of the supply chain – growers, packers, transporters, wholesalers, ripeners, retailers.
- Weekly and quarterly forecasting and dispatch reporting to help industry to achieve more consistent crop flow through the supply chain across the year.
- National hands-on retailer training.
- Extension program to facilitate adoption of best practice for growers, packers, transporters, ripeners, wholesalers.
- Detailed ripening guidelines developed for ripeners.
- Research to improve field disease management.
- Research to identify and minimise causes of fruit bruising.

A monitoring program was also established to monitor changes in fruit quality at retail level over time. Fruit is sampled from 64 stores across 4 capital cities every month and assessed against key quality criteria. This data has provided insights into the main quality defects (rots and bruising) and has helped to direct future research.

The extensive data set provides hard evidence that the initiatives undertaken through the levy program to improve avocado quality at retail level are having a positive impact. The percentage of Hass avocados at retail level with unacceptable levels of damage decreased by 38% from 2008 to 2012. This improvement in quality is helping to drive increased consumer demand and returns to levy payers. Per capita consumption of avocados has increased 70% in the past ten years and returns to growers have also increased over this time.

## HARVESTING AND PRODUCTION OF FRUITS

Avocado plants raised from seeds start bearing five to six years after planting while grafted plants stars yielding in 3-4 years. Mature fruits of purple varieties change their colour from purple to maroon, whereas fruits of green varieties become greenish-yellow. Fruits are ready for harvest when the colour of seed coat within the fruit changes from yellowish white to dark brown. Mature fruits ripen six to ten days after harvesting. The fruits remain hard as long as they stay on the trees, softening only after harvest.

The yield ranges from about 100 to 500 fruits per tree. Purple avocado tree of 10-15 years age can bear 200-300 fruits .However, 20-25 years old trees can yield even up to 1000 to 1500 fruits per tree. In Sikkim, fruits are harvested during July to October is the usual harvesting time. In Coorg area fruits are available from June to October. In Tamil Nadu, July-August is the peak harvest time.

## POST-HARVEST FRUIT TECHNOLOGY

Avocado is a climacteric fruit. They should be harvested at the correct stage of maturity. The fruits are harvested at maturity and ripens in 4-5 days after harvesting. Ripening of fruits takes place at 15-21°C satisfactorily.

Avocados do not ripen on the tree, and fruits soften only after they are picked. Fruits need to be picked carefully. In India, fruits of 250 to 300 g in size are preferred. Most popular varieties are Hass, Fuerte and Green. Hard, mature fruits are harvested and allowed to ripen during transport and distribution. The temperature before and during the ripening phase is important. Zauberman et al. (1977) found temperatures above 30°C had adverse effects on avocado ripening. Eaks (1978) found the ethylene rise to occur earlier as the ripening temperature increased from 20°C to 25°C, while the peak value decreased at temperatures higher than 25°C. Low-temperature storage was also found to affect ripening. Unripe fruit held at temperatures at and below 5°C for 2 weeks showed greatly reduced ethylene peaks when ripened at 20°C, while storage for up to a week caused an earlier climacteric (Eaks1980).

The fruits are best packed in single layer in small, flat and well ventilated wooden boxes by providing sufficient padding with wood wool. The fruits of cultivars Fuerte and Nabal do not soften for 3 and 6 weeks, respectively at gradually decreasing temperature i.e. 2 days at 17 °C,2 days at 14 °C,4 days at 12and 4 days at 8 °C. Unripe avocados can be stored for up to four weeks at 5.5 to 8°C.

The exposure of fruit to ethylene before or during cold storage increases the risk of internal chilling injury. The combination of ethylene and low temperature accelerates skin injury and pulp browning in 'Ettinger', 'Fuerte' and 'Hass' fruit after just two weeks of cold storage.

## INSECT- PESTS AND MANAGEMENT

Mites, mealy bugs, scales are the important insect pests of avocado.

**Mites:** Three species of mites prey on avocado trees. The avocado brown mite lays eggs on the tree leaves, and in large numbers damages and destroys the plant's foliage. The persea mite also harms avocado tree foliage, leading to fewer fruits, though the six-spotted mite typically only causes leaf discoloration. Predatory mites will keep the population of all three mites under control, as will avoiding the use of chemical pesticides that kill their natural enemies.

**Thrips-** ***Scirtothrips perseae.*** Obvious feeding by thrips scars on fruit. These scars begin as scabs or leathery patches and spread across fruit. The adult insect is orange-yellow in color with distinct brown bands and reaches 0.7 mm in length. Insect thrives in cooler temperatures and may undergo 6 or more generations per year. Addition of coarse organic mulch about 6 inches thick below trees may help to reduce survival of thrips pupating in soil. Selective insecticide should be selected to control these insects so that minimum damage is done to populations of natural enemies.

**Mealy bugs -** ***Planococcus citri.*** Mealy bugs are not major problem in avocado but use of insecticides kills the natural enemies of mealy bugs causing major problem. Mealy bugs are sexually dimorphic. Yellow eggs are produced in a loose colony of waxy filaments. About 50-100 or more than 100 eggs covered in an ovisac are deposited by female. Mealy bug breeds continuously on different hosts like citrus, *Murraya koenigii,* coffee in Coorg region. Mealy bug observed on fruits during September-October months. It was found both on the immature and ripened fruits. The mealy bugs produced large quantity of honey dew, which attracts other insects and these insect lay egg on the fruits and deteriorate the quality of fruits. Mealy bugs are tend to be serious pests in the presence of ants because the ants protect them from predators and parasites. These may be control by release of lady bird beetle, *Cryptolaemus montrouzieri at* 10 beetles/ tree after fruit set. Spraying 150ml dimethoate + 250ml kerosene in 100 litres of water gives effective control of mealy bugs.

**Fruit fly –*Bactrocera dorsalis, B. caryeaea:*** Fruit fly infests the ripened fruits. Its infestation is more in southern states. The female fruit fly lays eggs on the mature fruits with the help of its pointed ovipositor. After hatching the maggots feed on pulp of these fruits and the infested fruits starts rotting and fall down. As a result brown patch appears around the place of oviposition. The maggots come out of the affected fruit and pupate in the soil.Pre-harvest IPM combined with sanitation (Collection and destruction of fallen/infested fruits) + Placing Methyl eugenol trap @ 4-6/acre + In severe infestation spraying of bait spray (Decamethrin (Decis) 2ml+ 100g of jaggery in 1 litre of water) is recommended.

## DISEASES AND MANAGEMENT

The Anthracnose root rot, leaf spot, Stem rot, scab are the major diseases affecting avocado.

**Anthracnose:** The disease is caused by *Colletotrichum gloeosprotioides.* It is becoming major problem but effecting the fruit yield and quality. The symptoms are developed in the fruits either or after harvest initially the symptom is large light, brown circular, lesions which turns into dark brown or black colour after sometime. Infection results in shedding of young fruits. Remaining fruits become deformed. The copper based fungicide copper oxide, copper trioxide may be used to control this disease in the initial stage. The orchard sanitation like burning of fallen leaves and fruits help to reduce inoculum. Postharvest treatment of disease is also recommended in many countries. Controlled atmospheric storage of fruits in 2% O2 at 7.2°C for 3-4 weeks helps to prevent the development of the fungus in storage.

**Phytophthora root rot:** It is serious disease of avocado caused by *Phytophthora cinnamoni*, leading to death of plant. The disease affects the roots and they became black and eventually die which effects the overall growth and yield of the plants. The disease situation is aggravated by ill drained and waterlogged conditions. Metalaxyl (Ridomil) mixed with soil before planting or applied as a soil drench controls root rot at least for four months after treatment. Soil drenching of Ridomil (1g a.i./10 lit) controls root rot.This disease may be minimized by using tolerant varieties, avoiding water logged areas for planting, soil solarization, use of metalaxyl, potassium phosphonate may be used.

**Avocado scab:** Avocado scab is caused by *Sphaceloma perseae.* Oval or irregular brown or purple spots on fruit with rough texture are formed. High humidity encourages scab growth and spread. Use of tolerant varieties and spray of copper containing fungicides helps to control the scab problem.

## MARKETING OF FRUITS

Marketing system for avocado is not properly organized as the production is small and production areas are scattered. Major constraint in avocado marketing is due to the consumer acceptability. The avocado fruit is not liked by the common people in the domestic markets due to the nature of its taste. But, on account of its high nutritive value and now awareness in the society , it is expected that avocado will find good place in the fruit markets of India in near future. The fruit is able to provide household nutritional security to the Indian population thus, avocado can be a potential fruit of India.

The fruits of avocado are available in Indian markets from July to October. Avocado sold in fruit markets between Rs 80-160 per kg depending upon the season and its availability in particular market and region.

## FUTURE STRATEGIES

**Strategic Planning Process:** Through the strategic planning process we scrutinized intelligence drawn from HAB's first industry-wide avocado forum, global stakeholder survey, a series of one-on-one interviews and a planning meeting including influential avocado players from around the world. In May 2016, HAB finalized and approved the plan.

**Purpose:** The end benefit of a product or company to people or society.

Hass avocados improve lives through a unique, flavorful eating experience and health benefits.

**Mission:** An organization'score business

HAB exists to support the global avocado industry stakeholders in our collective efforts toward market expansion in the U.S.

**Vision:** A vivid description of a specific destination for the Hass Avocado Board

HAB is the catalyst for fresh avocados being the No. 1 consumed fruit in the U.S. and industry stakeholders being successful.

### Guiding Values

- Think future
- Be inclusive
- Respect one another's point of view
- Follow through on commitments to
- Action

- Be transparent
- Operate with an open mind
- Build on agreements
- Celebrate success
- Collaboration
- Learning.

## LITERATURE CONSULTED

Adato, I. and Gazit,S. 1974. Water-deficit stress, ethylene production, and ripening in avocado fruits.*Plant Physiol.,* 53:45-46.

Adato, I. and Gazit,S. 1977. Role of ethylene in avocado fruit development and ripening. 1. Fruit drop. *J.Expt. Bot.,* 28:636-643.

Adato, I., Gazit,S. and Blumenfeld,A.1976. Relationship between changes in abscisic acid and ethylene production during ripening of avocado fruit. *Austral. J. Plant Physiol.* ,3:555-558.

Bal,J.S.2014. Avocado .In: Fruit Growing.3rd Edition. Kalyani Publishers, New Delhi.

Bergh, B.O., and Ellstrand, N.1986. Taxonomy of the avocado. Calif Avocado Soc. Yearb. 70: 135-145.

Blumenfeld, A., Gazit, S. and Argaman, E. 1983. Factors involved in avocado productivity. Special Publ. No. 222. Department of Subtropical Horticulture, Center, Bet-Dagan, Israel. pp. 84-85.

Durand, B.J. 1971. Introduction to Avocado Growing in South Africa: Institute for Subtropical and Tropical Crops.

Eaks,I.L.1978. Ripening, respiration and ethylene production of "Hass" avocado fruits at 20-40$^0$ C. *J. Am.Soc. Hort. Sci.,*103:576-78.

Eaks,I.L.1980. Respiratory rate, ethylene production and ripening response of avocado fruit to ethylene or propylene following harvest at different maturity. *J.Am.Soc. Hort. Sci.,* 105:744—47.

Garner, LC. 2004. Characterization and manipulation of flower and fruit abscission of avocado (*Persea americana* Mill.). Univ. Calif., Riverside, PhD Diss., UMINo. 3130259.

Ghosh, S. P.2000. Avocado Production in India. In*:* Avocado production in Asia and the Pacific. FAO Corporate Document Repository.

Jayanthi Mala B. R., Tripathi, P. C., Sunanda Sanganal and Sankar, V. 2014. Pollinators diversity and their role in pollination of Avocado (*Persea americana* Mill*). International symposium on conservation and management of pollinators for sustainable agriculture and ecosystem services"* at Pusa, New Delhi September 24-26.

Tingwa, P. O. and R. E. Young. 1975. Studies on the inhibition of ripening in attached avocado. (*Persea americana* Mill.) fruits. *J. Am. Soc. Hort. Sci.,*100:447-449.

Tripathi, P.C, and Karunakaran, G. 2013. Bharat Mai Navaneet Phal (Avocado) ki Kheti:Varthaman sthith ievam Sambhavanaye (In Hindi) (Avocado cultivation in India: Present status and possibilities) Bhumi Nirman (Bhopal) 16th Jan -15th Feb : 11

Visit the Voice of Horticulture website>http:voicefhorticulture.org.au/

Zauberman, G., Schiffman, M. and Yanko,U.1977. The response of avocado fruit to different storage temperature. *HortScience*, 12:353-54.

# 10

# Phalsa

*Deepa H. Dwivedi and Munni Gond*

## INTRODUCTION

Phalsa, ***Grewia subinaequalis*** DC (syn. ***Grewia asiatica*** L.) is native to the Indian sub-continent and South-East Asia and belongs to family Tiliaceae, to which jute, our major money earning fiber crop, also belongs.Phalsa has been mentioned in our Vedic books and is native to India and other parts of southeast Asia, including Pakistan, Sri Lanka and Bangladesh. It is known by many names viz., Hindi (Phalsa, *Shukri, Tadachi, Dhaman, Parusha*); English (Indian phalsa, phalsa); Bengali (Phalsa, *Shunkri*); Kannada(*Phulsa*), Telegu (*Phutiki*), Tamil (*Unnu*), Malayalam (*Chadicha*) Gujarati (*Shukri*), Filipino (*Bariuangulod*) and Thai (*Potaohai,Yapkheethao,Malai,Laikhon*). The plant is commonly known as Gangeran. *Grewia* species is one of the best examples of multipurpose fruit species which are useful source of food, fodder, fiber, fuelwood, timber and a range of traditional medicines which cure a number of diseases (Sharma and Patni, 2012).

## NUTRITIVE AND CULTURAL SIGNIFICANCE

The species is known for its edible fruits which are nutritionally balanced and rich in iron and calcium. The drupes also contain amino acids, mineral elements (K, Ca, Mn, Fe, Cu and Zn), tannin and pectic substances useful in a range of traditional medicines that cure various perilous diseases and have mild antibiotic properties. The plant preparations are used for the treatment of bone fracture and for bone strengthening and tissue healing. The fruits are used for promoting fertility in females and are considered in special diets for pregnant women and anaemic children.

The plant parts are rich in amino acids and mineral elements and contain some pharmacologically active constituents. In traditional folklore medicine, the fruit has been used as astringent, stomachic and cooling agent. When unripe, it has been reported to alleviate inflammation and is administered in respiratory, cardiac and blood disorders, as well as in fever. Root and bark has been prescribed for rheumatism and its infusion is used as a demulcent. The leaves are applied on skin eruptions. The plant possesses antioxidant, anti diabetic, anti hyperglycaemic, radio protective, antimicrobial, hepato protective, anti fertility, antifungal, analgesic, anti pyeretic and antiviral activities.

Fruits are delicious, sour too sweet in taste with attractive colour and are good source of phosphorus and iron, contain 50 to 60% juice, 10 to 11% sugar and 2.0 to 2.5% acids. They are excellent for making juice and squash and are mostly used as fresh fruit and have cooling effect. Medicinally it works as a digestive tonic and the fruits are astringent which may help in curing inflammation, heart and blood disorders, fever, heat troubles and constipation (Abid et al., 2012). It is one of the hardiest fruit crop and is rarely affected by any insects, pests and diseases. It is an exotic bush plant considered horticulturally as a small fruit crop but also used as a folk medicine. The ripe phalsa fruits are consumed fresh, in desserts, or processed into refreshing fruit and soft drinks enjoyed during hot summer months in India (Salunkhe and Desai,1984). However, phalsa fruit has a short shelf life and is considered suitable only for local marketing.

**Table 1:** Nutritional quality of phalsa fruits (Yadav, 1999).

| Nutrients Values/100 g. | |
|---|---|
| Calories | 329/1b (724perkg) |
| Moisture | 81.13% |
| Protein | 1.58% |
| Fat | 1.82% |
| Crude fiber | 1.77% |
| Sugar | 10.27% |

The plants play a wide role in maintaining the ecological balance. The shrubs are used in apiculture since bees visit the flowers for pollen and nectar. The young leaves are consumed by livestock and have fairly good feed value. The plants have an aggressive root system which holds fast to the soil protecting it from water and wind erosion. Leaf litter from the shrub improves soil physical and chemical properties (Orwa et al.,2009). Ecologically, it can withstand environmental stress more easily than annual crops and thus, makes an important contribution to sustainable production without needing expensive inputs of water or fertilizer. It is also said to regenerate well and is traditionally protected by farmers for this reason (Akundabweni et al., 2010). Despite this the persistent medicinal plant species is on the verge of extinction due to overgrazing, debarking by animals, encroachments, unsustainable utilization and other developmental activities (Jain et al., 2009).

## ORIGIN, HISTORY AND DISTRIBUTION

Phalsa is cultivated on a commercial scale mainly in the northern and western states of India (Hays, 1953;Chundawat and Singh, 1980). Phalsa originated in India in the Baroda region, and other parts of Southeast Asia, including Pakistan, Sri Lanka and Bangladesh. In the early 20th century, the fruit was introduced to Indonesia and the Philippines, where it has since naturalized. The Luzon province displays an abundance of the small, purple fruits in its lower elevations in dry zones. It is an exotic plant in Algeria, Botswana, Chad, Djibouti, Ethiopia, Iran, Kenya, Mali, Mauritania, Morocco, Namibia, Niger, Nigeria, Saudi Arabia, Senegal, Somalia, South Africa, Sudan, Tanzania, Uganda, and Zimbabwe. Thailand, Vietnam, Cambodia, and Laos. Few countries in the west cultivate phalsa, though some gardeners and research laboratories in Florida and Puerto Rico grow the fruit out of interest or for educational purposes.

In India, phalsa can grow in the Himalayan regions and thrives at elevations up to 3,000 feet throughout greater part of India in Punjab, Western Himalayas and up to 1000 m, in north Bengal, Bihar, Chhota Nagpur, Orissa, Gujarat, Konkan, Deccan and South India. It is spread over Andhra Pradesh, Bihar, Gujarat, Haryana, Karnataka, Kerala, Madhya Pradesh, Maharashtra, Punjab, Rajasthan, Tamil Nadu and West Bengal.

The phalsa is found growing in the wild form, all along the foot hills of the Himalyas, U.P., Punjab, Rajasthan, M.P., Maharashtra, West Bengal, Bihar and in south India, in the Western Ghats and Malabar Coast. In the cultivated form, it is quite commonly grown near urban centres in Punjab, Himalyas, Uttar Pradesh and Andhra Pradesh.

## TAXONOMICAL AND BOTANICAL DESCRIPTION

**Taxonomy:** *Grewia* is a genus of approximately 150 species of family Tiliaceae which include small trees and shrubs, distributed in subtropical and tropical regions of the world. The name *Grewia* was given due to Nehemiah Grew, one of the founders of plant physiology. Different species of *Grewia* are small trees or shrubs which grow to 4 m or more in height. *Grewia* is the only genus in family Tiliaceae with edible fruits. Extensively cultivated species for their fruit values are *Grewia subinaequalis* DC. (syn. *Grewia asiatica.*) and *Grewia tenax* (Frosk.) (Youngken, 1951). *G. asiatica, G. hirsuta, G. damine, G. lasiodiscus*, *G. optiva*, *G. biloba*, *G. bicolor*, *G. tiliaefolia*, *G. flavescens* and many more species are important for their medicinal and economic values.

Most of the genus of Tiliaceae family are wild and known for their fodder, fuel wood, craft works, timbers and therapeutic values viz. *Grewia flavescens* A. Juss, *G. villos* and *G. hirsuta. Grewia tenax* is highly drought resistant and occurs in the driest savannas at desert margins and regions of higher rainfall, where it grows in thickets on termite mounds in otherwise seasonally flooded country. *Grewia tenax* has often been cited as a prime candidate for domestication as a useful horticultural plant (Gebauer et al., 2002). A wild species *Grewia elastic* (*G. asiatica* var.*vestita*) grows on lower hills all over India. (Singh, 2015)

## Botanical Description

The phalsa plant is a large, shaggy shrub or a small tree reaching 4 m or more in height (Sastri 1956). The phalsa plants grow to become straggling tall shrub with rough bark on the stem and have numerous long, slender, drooping branches where the young branchlets are densely covered with a coating of hair. The alternate, deciduous, widely spaced, thick and large leaves are broadly heart-shaped or ovate, pointed at the apex, oblique at the base, measure up to 20 cm in length and 15 cm in width and coarsely toothed, with a light, whitish blush on the underside (Paul, 2015). Bark is greyish green, internally reddish brown, sometimes creamish in colour, thick, fibrous, tough and leathery. Leaf is shortly petioled, heart shaped, 5 to 7 nerved, main nerves connected by parallel venations, margin serrate, upper surface stellately pubescent, lower surface tomentose (Dey and Das, 1995). Fruits are lobed drupes. The individual bright orange-yellow flowers have a diameter of about 2cm with five large (12 mm) sepals and five smaller (4 to 5 mm) petals and are borne in dense cymes in the leaf axils in late spring (Sastri, 1956). The small fruits, almost round fleshy, fibrous drupes like blueberry are greyish purple at maturity which turn crimson or cherry red in color when ripe. They are borne on a 2- to 3 cm long peduncle and are produced in great numbers in open, branched clusters. Individual fruits measure from 1.0 to 1.9 cm in diameter, 0.8 to 1.6 cm in vertical height and 0.5 to 2.2 g in weight. Fruits often have surface having black circular depressed spots with large stellate covering trichomes and rest of the surface with small stellate covering trichomes.

Seeds are 1 or 2 in number, pointed at one end and grooved on the surface. Seed coat is stony and hard. Large fruits have two hemispherical, hard, buff-colored seeds up to 5 mm in diameter while small fruits are generally single-seeded. They are 1 or 2 chambered and endosperm is oily (Dey and Das, 1995). Fruits ripen gradually on bushes during the summer months (Yadav, 1999).

## CLIMATIC AND SOIL ADAPTABILITY

The plant grows in both tropical and subtropical climates but will tolerate other climates, except at high altitude. However, it does best in regions having distinct summer and winter seasons. It is very hardy and capable of existing under severe conditions like short periods of light frost and drought and is therefore suitable for arid region but requires protection from the freezing cold temperatures. In the absence of a prominent change of seasons the plant does not shed its leaves, flowers erratically throughout the year and fruits poorly. This limits its distribution in Southeast Asia.

The phalsa is a warm climate fruit plant. In India, this plant grows satisfactorily and produces well up to an elevation of 1,000 m. It can tolerate light frost although at the cost of defoliation. The plant is deciduous and normally loses its leaves slowly in those areas with mild winter season. Adequate sunlight and warm or hot temperatures are required for fruit ripening, development of appropriate fruit color and good eating quality. The average monthly temperatures for quality fruit production in phalsa is 44°C (Yadav, 1999). Phalsa can tolerate drought conditions fairly well after fruit harvest. For getting high fruit yield it requires irrigation at regular intervals of 20-days during April to June. No irrigation may be applied during rainy season and in dormancy.

Phalsa can grow with good success in almost all kinds of soils.The plants grow well and yield bumper crop even in heavy and lighter soils.However, phalsa perform best in rich loamy soils. In water logged soils plants become chlorotic and make poor growth. Proper drainage of soils is very important for normal plant growth. Low lying areas and water logged conditions should be avoided for commercial cultivation of phalsa.

## RECOMMENDED AND POPULAR CULTIVARS

No distinct varieties have been developed. Even if plants are raised from seed and flowers are cross pollinated, no varieties have come up.Basically two types tall and dwarf are described but they have not shown much difference between them when planted in the same situation except for its habit.

Two distinct types, tall and dwarf, have been developed in India that differ with respect to various chemical and physical characteristics (Table 1). The juice yield is slightly higher in the tall type because it is directly related to edible portion, while more total sugars and non-reducing sugars were observed in the dwarf type (Samiksha, 2016). Tall type had more reducing sugars and titratable acidity and a greater amount of seed protein than the dwarf types (Dhawan *et al.*, 1993).

**Table 2:** Nutrient content in fruits of tall and dwarf types of Phalsa (Dhawan *et al*., 1993).

| Content (%) | Tall cultivars | Dwarf cultivars |
|---|---|---|
| Edible portion | 91.30 | 90.79 |
| Seed | 8.70 | 9.21 |
| Juice yield | 67.50 | 65.90 |
| Pomace | 32.50 | 34.10 |
| Moisture | 76.80 | 74.83 |
| Total sugars | 5.73 | 7.95 |
| Reducing sugars | 1.24 | 0.99 |
| Non-reducing sugars | 4.49 | 6.96 |
| Titratable acidity | 1.48 | 1.12 |
| Fruit protein | 3.13 | 1.89 |
| Seed protein | 8.75 | 7.00 |
| Pulp protein | 1.40 | 7.00 |

The crop is mostly propagated through seeds and thus, there are no standard cultivars which are available. It is mostly the seedling populations which are found growing. This could be one of the major reasons why there are no planned orchards of the crop. However, one variety called "Sharbati" or "local" is found mentioned in literature.

The tall-growing wild plants bear acidic fruits which are not relished. The dwarf, shrubby type, with a blend of sweet-and-acid is the best fruits, is generally cultivated.

## PROPAGATION TECHNOLOGY

### Sexual Propagation

The phalsa is usually propagated from seed. For raising the seedling, large sized purple black coloured fruits are collected when the crop is ready in the end of May or beginning of June. After extracting from the fruits, the seeds should be washed and dried under shade. These seeds are sown on raised beds in lines which are 10-15 cm apart and 1.5-2.0 cm deep. The seeds should be covered with sand or light soil mixed with well rotten farmyard manure. The bed should be kept free from weeds. Initially, one hand weeding is recommended, otherwise the roots of small plants get damaged. The freshly extracted seeds should be used for raising seedling but these seeds lose their viability in 90 to 100 days under ordinary storage and 175 to 185 days under cold storage. Seeds germinate within 10-15 days and may be transplanted in the field within 3-4 months (Bose et al, 1985).

When the seedling come out and have made 5-7cm growth, light dressing of calcium ammonium nitrate or ammonium sulphate at the rate of 50g/m$^2$ of nursery area is applied to encourage rapid growth of the seedling. Seedling is ready for transplanting in winter i.e. January to Febraury (Shastri 1956). Bold seeds give 90 per cent germination during July. Seeds may be sown on raised beds 2cm deep in lines 10cm apart and are covered with a mixture of sand + F.Y.M in 1:1 ratio.Watering is done immediately after sowing. Flooding of seed beds is avoided, failing which root rot may appear due to the fungus *Pythium spp.* Apply 1% Bavistin solution after the seed germination and a solution of Dursban 20EC (chlorpyriphos) @ 10ml/L of water after 30days of seed sowing to check the attack of white ants.

Seedlings are transplanted from seedbeds into well-prepared holes and are usually spaced 10 to 15 ft (3-4.5 m) apart, though some experiments have favored 6 x 6 ft (1.8 x 1.8 m) or 8 x 8 ft (2.4 x 2.4 m) to maximize efficiency in harvesting. Fruiting will commence in 13 to 15 months. Annual pruning to a height of 3 to 4 ft (0.9-1.2 m) encourages new shoots and better yields than more drastic trimming.

### Vegetative Propagation

**Cuttings :**The phalsa plant is also propagated by rooting of hardwood cuttings as well as layering (Samson 1986). Cuttings are difficult to root due to the presence of mucilage. Some reports say that IBA treated cuttings prepared in December and left for callusing root successfully when planted in January. Wood type and planting date influence rooting of phalsa (Singh et al., 1961) but treatment with auxins (IAA, IBA, NAA) improve rooting of difficult-to-root hardwood cuttings of phalsa, ground layers and air layers (Mohammed and Chauhan 1970). However, even today, seed is the preferred method of propagation in phalsa.

## PLANT AND FRUIT PHYSIOLOGY

Flowers develop only on the new shoots of the current growing season. The period between flowering and fruiting maturity is 45-55 days. Under subtropical conditions, there is a brief period of dormancy lasting 4-6 weeks when the plant sheds its leaves.Phalsa begins to bear fruit after only 2 years, but does not produce marketable fruit until the 3rd year. The individual flowers are yellowish in colour with five large (12 mm) sepals and five smaller (4 to 5 mm) petals. The flower has a diameter of about 2 cm and are arranged in dense cymes, (Paul, 2015). The flowers do not mature uniformly.

Flowering in phalsa starts from February-March and continues till May. The first flower to open is at the base. Flowers are borne in the axil of leaves. The flowers are mostly cross pollinated and honey-bee seems to play major role in pollination. The flowers buds become plumpy before anthesis. The first sign of anthesis is the appearance of a slit in sepals at the base of the bud. The slit widens and at first only one sepal falls apart. The other sepals fall one by one and the whole process of flower opening is complete within half an hour. The distance of anthesis in phalsa takes place before the flowers are completely open.

The fruits are small berry like and deep reddish brown in colour. This subtropical fruit flowers in February and the fruits ripen in second fortnight of April and continues up to middle of June. Phalsa is deciduous in habit in northern India and sheds its leaves during winter season which makes it capable of withstanding the frost. Phalsa produces fruits in cluster in axil of leaves of the young shoots. It is one of the hardiest fruit crop with regard to the attack of insect-pests and diseases (Kumar et al., 2014). Fruit are lobed drupes. Fruit changes colour from green to purple-red and after they are fully ripened, to dark-purple. There are two kinds of fruits of phalsa, large fruits containing two hemispherical, tough, buff coloured seeds and small fruits containing single-seed. Phalsa is a self-pollinated crop usually. (Malik et al., 2010). The main problem in the phalsa cultivation is the uneven ripening and small berries which are to be picked individually. So the cost of harvesting is too high which became a major constraint, to phalsa growers. (Kumar et al., 2014).

## PLANNING AND PLANTING

Phalsa seedlings are generally planted by the onset of the monsoons for successful establishment of the plants. Seedlings are transplanted from seedbeds into well-prepared holes when a year old and are usually spaced 10 to 15 ft (3-4.5 m) apart. Some experiments have favoured 6 x 6 ft (1.8 x 1.8 m) or 8 x 8 ft (2.4 x 2.4 m) to maximize efficiency in harvesting. Fruiting commences in 13 to 15 months. The square or rectangular system of planting may be followed. Since the phalsa plant is a small tree/bush hence, hedgerow or double hedgerow system of planting can also be explored for better plant performance, training and pruning operations and improved quality of fruit since the fruits are more exposed to sunlight under these systems of planting.

## SOIL CULTURAL PRACTICES TECHNOLOGY

### Water Need

Seedlings must be irrigated immediately once they are transplanted in the orchard. However, phalsa is a hardy plant and can survive under low irrigation schedules. The plant performs well in wastelands and is drought-tolerant, but occasional irrigation during the fruiting season and in dry periods, is profitable for growers.(Yadav, 1999). Phalsa performs well even under semi-arid conditions and is capable of growing under neglected and under conditions of water scarcity where only a few other crops would normally survive (Chundawat and Singh 1980; Israr et al., 2012).

### Nutritional Need

Phalsa is commonly not fertilized by growers. Moreover, reports on its manuring are also scanty. The crop is borne on new growth and application of manures and fertilizers will encourage vegetative growth. The plant under N, P or K deficiency showed stunted growth with less number of branches and leaves.High yields have been recorded upon application of N, P and K at 100, 40 and 25 kg, respectively, per hectare. Trials at Rajasthan Agricultural University, Udaipur also revealed that application of 100:40:25 kg of NPK per hectare gave higher yield (Bose et al., 2001).

Phalsa is basically a wasteland crop because of its hardy nature. It is able to survive under varying soil conditions and thus, fertilizer schedules for the crop have not been worked out. The phalsa plant is reported to shows good response to nitrogen applications. High levels of phosphorus supply increase sugar content in the fruit while higher potassium suppresses sugar and promotes acidity. Phalsa is considered stress-tolerant and is commonly grown under neglect (Hayes 1953).

### Inter Culture

Removal of weeds and general hygiene of the orchard should be maintained for better performance of the plants. Since phalsa is grown at a close spacing of 2- 4 m and comes into bearing by one year hence it is not possible to undertake any intercropping with vegetables, etc., in the orchards. However, since phalsa is a small tree/bush it may be considered as a filler crop with mango but since it is being cultivated as a wasteland crop on marginal lands which have poor fertility levels it would not be possible to obtain satisfactory performance of the main and intercrop unless assured irrigation and manuring in the orchards is ensured.

Green manuring with *moong,* cowpea or *urd* may be done in the initial one-two years of orchard establishment which should be turned in towards the rainy season.

## PLANT CULTURAL PRACTICES TECHNOLOGY

### Training and Pruning

Phalsa bears fruits on the current season growth and is an axillary bearer. Thus, training and pruning is a vital operation to increase yields and also improve fruit quality. Annual pruning to a height of 3 to 4 ft (0.9-1.2 m) encourages new shoots and better yields than more drastic trimming. In north India the plants slowly begin to shed their leaves towards the middle of winter and this is the appropriate stage for pruning. They start regrowth soon after pruning.

One-year-old seedlings are usually spaced 2 to 4.5 m apart. Bushes flower progressively during the spring months. Since phalsa bears fruit on current season's growth and it is an axillary bearer, there is a need for regular but moderate annual pruning before the on-set of spring. Annual pruning to a height of about 1 m encourages new shoot sand higher yield of marketable fruit than does more drastic trimming (Singh and Sharma 1961).

Phalsa is usually trained as a bush. It can also be trained on head system with a single stem. The height of the single stem should be kept at 90cm to 1.0 m. The plant produce shoots above this height just as in bush system. In North India, phalsa is pruned every year during January. When it is trained as a bush the shoots are cut from the ground level by leaving 2- 3cms stubs. From here many sprouts emerge on the stubs. Flowering occurs on these shoots in March-April, which results in improved fruiting.

### Use of Plant Growth Regulators

Application of growth substances has proved effective in increasing the fruit set and yield in phalsa. Treatment with $GA_3$ at 10 ppm and 2, 4, 5-T at 5ppm increased the setting of the fruit and also the yield. Studies at Allahabad agricultural institute showed that the use of cycocel at 250 ppm sprayed twice at an interval of 7 days after 50 per cent fruit set increased the fruit size . It also caused delay in fruit maturity by 3-5 days. Application of 1000 ppm ethrel resulted in maximum (71.1 %) ripening in 5 days after application. An increase in fruit size and improvement in fruit quality by spraying of $GA_3$ (60 ppm), 2, 4, 5-T (5 ppm) and 2, 4-D (2.5 ppm). Spraying of $GA_3$ at 60 ppm once at the beginning of flowering, another after 15 days and ethrel at 1000 ppm when ripening of berries had just started, increased the fruit retention per cent and yield. The treatment also reduced the harvesting span and increase the total soluble solids contents of fruits (Bose *et al.,* 2001)

## HARVESTING AND PRODUCTION OF FRUITS

Summer is the fruiting season for phalsa. Only a few fruits in a cluster ripen at any one time, so continuous harvesting is necessary. The fruits are highly perishable and must be marketed within 24 hours. Average yield per plant is 5.0-7.0 kg per plant in a season.

Phalsa is basically a non-climacteric fruit (Dave et al., 2015) and thus can be harvested only after it has ripened. Phalsa is green when unripe, slowly turning red while ripening and changes to a deep purple when ready for harvest. At final maturity of fruit the thin skin of the fruit should be pliable and tender.

For uniform ripening of phalsa fruits, apices of shoots may be pinched in mid-May to check further shoot growth. Fruits start ripening in the first week of June and continue for a month. Fruits should be harvested twice a week and are packed in small baskets of size 2kg or in packs. Phalsa fruit is highly perishable in nature, hence should be transported to the market soon after harvesting.

## POST –HARVEST FRUIT TECHNOLOGY

Storage of phalsa fruits is a major challenge because of the highly perishable nature of the fruit. It is perhaps, for this reason that it is marketed in local markets by the small vendors. However, processed products which may be developed from the fruits have great potential markets. Harvested fruits are pre cooled to reduce the field heat can be stored for 2-3 days.

The phalsa fruits should be graded according to size and fruit colour to fetch higher premium in the market. The fruits are packed in bamboo baskets and provided cushion by spreading paper shreds or placing old newspapers. The shelf life of fruits is determined by the stage of harvest. Fruits picked at turning stage can be stored for 2-3 days. However , red ripe fruits remained in good condition only for single day.

## INSECT PESTS AND MANAGEMENT

**Bark eating caterpillar (*Inderbela tetraonis*):** Young larvae feed inside the branches and trunk. With the result translocation of cell sap is disrupted. Tree growth is reduced and fruiting capacity is affected.Remove the webbing and inject kerosene oil into the holes in December-January after pruning.

**Psylla (*Diphornia grewia*):** It suck cell sap from young leaves. Fruit growth and yield is reduced. Spray Rogor 30EC (dimathoate) 1000 ml in 500 litres of water in first week of March.

**Leaf eating caterpillar (*Euproctis fraterna*)**: The caterpillars segregate and gnaw the leaves. The entire plant may be denuded in severe attack.The egg mass and gregariously feeding young caterpillars should be collected and destroyed. Spray Dursban 20EC (chlorpyriphos) 200 ml in 100 litres of water when caterpillars are noticed.

## DISEASES AND MANAGEMENT

**Pinspot of phalsa:** It is caused by *Phyllanthus grewiae*. The disease cause damage to the foliage. Any time in a year ,it can attack the phalsa plants. Small brown to dark brown circular to irregular pinspot like lesions appear on the leaves. Spray Bordeaux mixture 2:2;250 for the control of this disease.

**Brown spot of phalsa:** It is caused by the fungus *Cercospora grewiae*. The disease appear on the phalsa plantation from June to August. There is pre mature leaf fall under severe attack. Tiny lesions appears on upper and lower surface of the leaf. These lesions are covered with a white mass of fungus at early stage and enlarge gradually. Lesions attain reddish brown to dark brown colour. Sometimes these lesion formed clear big spots on the leaves and covered entire surface. For its control, leaves and pruned material should be collected and destroyed to check further infection to the new leaves appearing during spring season. The phalsa plants needs to be sprayed with Bordeaux mixture 2:2:250 or 0.3 per cent Dithane Z-78 as soon as disease appear on the foliage.

## MARKETING OF FRUITS

Transportation of ripe phalsa fruits over a long distance and their storage for a long period is a major challenge for further utilization of the fruit. The marketability of this fruit is lost very rapidly after harvesting due to quick discoloration, fermentation as well as spoilage of fruits due to its highly perishable nature. These fruits are generally sold by the roadside vendors, under an unhygienic and uncontrolled environment in the extreme heat of the summer months of May- June, which cuts short its shelf-life drastically. The quantitative and qualitative losses in this seasonal fruit are tremendous and efforts need to be made to minimize losses by developing appropriate storage and packaging technologies or by developing protocols for preparation of processed products from the fruits since it has immense potential for use as functional foods which would have export potential.

## FUTURE STRATEGIES

Phalsa is a hardy fruit crop with immense neutraceutical potential and thus, can be utilised as functional food in food supplements, health beverages or as other processed products. However, it has been truly neglected since cultural practices for the crop are lacking and not even one variety has been proclaimed suitable for commercial cultivation till date. Hence, every aspect of the crop, from it's ethenobotany to its commercial exploitation, has potential for scientific studies. Vegetative propagation techniques need to be standardized for the crop for mass multiplication of superior cultivars which need to be developed through systematic breeding programs.

Shelf life of the fruit is dismally poor and technologies for harvesting and its storage need to be developed. Processed products of phalsa in the form of beverages, fermented foods or even freeze-dried powders need to be explored to improve the marketability of the crop since it is a rich repository of phytometabolites. Thus, it can well be exploited as a functional food for fortifying already developed processed products in the market prepared from fruits accepted more for their palatability rather than nutraceutical value.

## LITERATURE CONSULTED


Abid, M., Muzamil, S., Kirmani, S. N., Khan, I. and Hassan, A. 2012.Effect of different levels of nitrogen and severity of pruning on growth, yield and quality of Phalsa (*Grewia subinaequalis L.*). *African J .Agri. Reseach*,7(35):4905-4910

Akundabweni, L.S.M., Munene, R..W., Maina, D.M., Mangala, J.M. 2010. Mineral micronutrient density characterization using energy dispersive X-ray fluorescence (XRF) analysis in four on-farm Kenyan wild African fruit tree germplasm. *African J. Food Agriculture Nutrition and Development*, 10(8): 2901-2938.

Bose, T.K.,Mitra,S.K. and Sanyal, D. (eds.) 2001.Fruits: Tropical and Subtropical: Volume I ,Kolkata, Naya Udyog.

Bose, T.K., Dhu, R.S., Mitra, S.K. and Sen, S.K. 1985. Synergism of phenolic substance in regeneration of root from litchi cuttings. *Indian J. Hort.*, 42: p153-155.

Chundawat, B. and Singh, R. 1980. Effect of Growth Regulators on Phalsa *(Grewia asiatica* L.) I. Growth and Fruiting. *Indian Journal Horticulture* ,37: 124-131.

Dave R., Rao,T. V. R. and Nandane, A. S. 2015. RSM - based optimization of edible-coating formulations for preserving post-harvest quality and enhancing storability of phalsa *(Grewia asiatica* L.), *Journal of Food Processing and Preservation,* ISSN doi:10.1111/ jfpp.12630, 1-12.

Dey, D., Das, M..N. 1995. Pharmacognostic analysis of leaf, bark and fruitof *Grewia asiatica* Linn. *Int. Conf. Cum. Prog. Med. Arom. Plant Res.*Calcutta, India 131-132.

Dhawan, K., Malhotra, S., Dhawan, S..S., Singh, D.and Dhindsa, K.D. 1993. Nutrient composition and electrophoretic pattern of protein in two distinct types of phalsa *(Grewia subinequalis* D.C.). *Plant Food Hum. Nutr.*, 44:255-260.

Gebauer, J., El-Siddig, K. and Ebert, G. 2002. The Potential of underutilized Fruit Trees in Central Sudan. Jens Gebauer, Humboldt 104 University Berlin, Department of fruit science, Albrecht-Thaer-Weg 3,14195 Berlin, Germany.

Hayes, W.B. 1953. Fruit Growing in India.2[nd] revised edition. Kitabistan, Allahabad India.

Israr, F., Hassan,. F, Naqvi, B.S., Azhar, I. and Jabeen, 2012. Report: Studies on antibacterial activity of some traditional medicinal plants used in folk medicine. *Pak J Pharm Sci.,* 25: 669-674.

Jain, S.C., Jain, R. and Singh, R. 2009. Ethnobotanical survey of Sariska and Siliserh regions from Alwar district of Rajasthan, India. Ethnobotanical Leaflets,13: 171-88.

Kumar ,M., Dwivedi, R., Anand ,A.K. and Kumar, A. 2014. Effect of Nutrient on Vegetative Growth, Fruit Maturity and Yield Attributes of Phalsa (*Grewia subinaequalis* D.C.) *Global Journal of Bio- science and Biotechnology*, 3(3): 264-268.

Malik, S.K., Chaudhury, R., Dhariwal, O.P. and Bhandari, D.C. 2010. Genetic resources of Tropical underutilized fruits in India. Director. NBPGR. Pusa New Delhi.

Mohammed, S. and Chauhan, J.S.1970. Vegetative propagation of phalsa (*Grewia asiatica* L.) *Indian J. of Ag. Sci.,* 40: 581-586.

Orwa, C., Mutua, A., Kindt, R, Jamnadass, R. and Simons, A. 2009. Agroforestree Database: a tree reference and selection guide version 4.0. 2009.

Paul, S. 2015. Pharmacology action and potential use of *Grewia asiatica*: A review. *International Journal of Applied Research,* 1(9): 222-228.

Salunkhe, D.K. and B.B. Desai. 1984. Phalsa. p. 129. In: Salunkhe and Desai (eds.), Postharvest Biotechnology of Fruits. Vol. 2. CRC Press, Boca Raton, FL.

Samiksha,S.2016.http://www.yourarticlelibrary.com/cultivation/phalsa-cultivation-in-india-production-area-climate-harvesting-and-fruit-handling/24703.

Samson, J.A. 1986.The minor tropical fruits. p. 316. In: Tropical fruits. Longman Inc., New York.

Sastri, B.N. 1956. The Wealth of India Raw Material Number 4 *Grewia* Linn. Council of Scientific and Industrial Research. New Delhi, India.

Sharma, N. and Patni, V. 2012. *Grewia tenax* (Frosk.) Fiori.- A Traditional Medicinal Plant with Enormous Economic Prospective.*Asian Journal of Pharmaceutical and Clinical Research,* 5 (3): 0974 -2441.

Singh, J. P. Godera, P.S. and Singh R.P. 1961. Effect of type of wood and planting dates on the rooting of phalsa. *Indian J. Hort*., 18(1):46-50.

Singh, J.P. and Sharma, H.C. 1961. Effect of time and severity of pruning on growth, yield and fruit quality of phalsa (*Grewia asiatica* L.). *Indian J.Hort.,*18(1): 46-50.

Singh, R. 2015. Fruits. National Book Trust Publications.

Yadav, A. K. 1999. Phalsa: A potential new small fruit for Georgia: Perspectives on new crops and new uses.Edited by J. Janick, published by ASHS Press, Alexandria 348-352.

Youngken, H.W. 1951. Pharmaceutical Botany. McGraw-Hill Book Co. New York.

# 11

# Date Palm

*Anil Kumar, J.S. Bal and Nirmaljit Kaur*

## INTRODUCTION

The date palm (*Phoenix dactylifera* L.) is considered to be oldest among the domesticated tree fruits. It considered a symbol of life in desert because it tolerate high temperature, drought and salinity more than many other fruits. Date palm fruits are eaten as raw dates (Khalal stage), Soft dates (Tamar stage) and dry dates (Chhuhara). Different processed products viz sugar, starch, vinegar, juice, toffies, wine, chutney, jam, pickels etc. are prepared from date fruits. Leaves of date palm trees are utilized for preparation of baskets, brooms, ropes, building materials, fuel and paper. The plantation is also useful for stabilization and maintenance of ecological balance in the desert regions

## NUTRITIVE AND CULTURAL SIGNIFICANCE

The date fruits have been recognized for their highly nutritive and calorific food values. Date fruits provide abundant quantity of iron, potassium, calcium, nicotinic acid and small amount of protein, copper, magnesium, chlorine, sulphur, vitamin A, B1 and B2. Date pulp contains moisture (20%), sugar (60-65%), fiber (2.5%), protein (2%) and less than 2% fat, mineral matter and pectin substances in fresh fruits can supplement the dietary needs of dessert people and provide about 3550 calories per kg.

The tree of date palm provide food, shelter and timber materials. In other words every part of the date palm plant is used for different purposes. The leaves of palm have great significance in paper manufacturing. Large number of products are prepared from dates for human consumption.

Date palm leaves hold a very important role in the history of Middle East. In ancient Mesopotamia and ancient Egypt, palms represented fertility and long life. Paintings from long lost civilizations bear images of people harvesting dates.

Date palm fronds are also used in the Jewish holidays of Sukkot. Dates have a very special place in the Muslim faith (Cheng and Tsui, 2014).

## ORIGIN, HISTORY AND DISTRIBUTION

It is believed to be indigenous to countries around Persian gulf. Date cultivation has historical records dating back to 6000 BC along the Tigris and the Euphrates rivers in Iraq and to 2000 BC at Mohanjodaro along the river Indus. Several thousand years before its cultivation, pre historic man used the fruits from wild palm and carried the seed over the wide area from India through middle east and later to North Africa. The date palm *(Phoenix dactylifera* L.). was known to ancient people as the tree of life and is probably one of the first fruit trees to be cultivated. Representation of the date palm dating back the 3000 BC. Appear on Sumerian temple walls in present day in Iraq. In Egyptian hieroglyphics, the date palm is used as the symbol for a year and its frond is the symbol for a month. Date palm logs were used to roof royal tombs as early as 2700 BC.

Commercial cultivation of dates in the world lies $20^0$ and $35^0$ North and South latitude. Beyond it, from the equator commercial cultivation is rare owing to occurrence of high rainfall during fruit growth and ripening period . Iraq, Saudi Arbia, Iran, United Arab Republic and Algeria are the main date palm producing countries in the world. Besides these countries Pakistan, Morroco, Libya, Tunisia, Sudan, Muscat, Oman. The Aden, United States of America and Beharain also producing dates in substantial quantities. The dates are also producing in small quantities in Spain, Chad, Mexcio, Yamen, Israel and Somalia. These countries have a population of 60 million palms out of which Iraq alone has 20 million palms. Apart from the above a few date plantation exist in Peru, Brazil, Argentina, South Africa, Australia, Cyprus, Ethiopia and India , Before partition of India, the major date growing area were Multan, Muzaffarpur, Jhang, Deragazi Khan ,Bhawalpur, Kherpur which are now in Pakistan. In India, it is believed that date palm has been introduced by the soliders of Alexander in the $4^{th}$ Century BC in Indus valley. Babar has mentioned in his memories that date was grown in India in the early sixteenth century. Dates were also grown in India during the regime of Akbar (AD1555-1605) (Randhawa, 1980). Grooves of seedling date palm are found on the coastal belt from Anjar to Manadvi in Kutch district of Gujarat.

In India date palm is cultivated in coastal Saureshta,i.e Jamnagar, Bhavnagar, north Gujarat, Jaisalmer, Barmer, Bikaner, Nagaur, Jodhpur, Churu, Kutch and Sri Ganganagar districts of Rajasthan, Hisar (Haryana), Abohar (Punjab).

## TAXONOMICAL AND BOTANICAL DESCRIPTION

Date palm is related to the Order Palmae, family Palmaceae, genus *Phoenix* and the species *dactylifera*. So the scientific name is *Phoenix dactylifera* L. The date palm is monocotyledonous plant with strong straight unbranched stem growing to the height more than 30 ft. It is characterized by pinnate leaf palms by upward and lengthwise folding of the pinnae and peculiarly furrowed seeds. *Phoenix dactylifera* L the commercially grown date palm species produces superior quality fruits. This species has characterized feature viz. *Phoenix sylvestris* L Roxb and *Phoenix canariensis* Chaubaud ( Canary island palm). Canary island palm is mainly planted for ornamental purpose. *Phoenix sylvestris* L Roxb is found growing wild in almost all parts of India and produces fruit quality. This palm is used for production of crude sugar (date palm gur) and drink called neera.

Date palm being dioecious in nature, bearing male and female flowers on separate trees has chromosome 2n=36. Intra specific crosses of *Phoenix dactylifera* (L) with *P. sylvestris* (L) Roxb and *P.canariensis* Chabaud are successful and have shown metaxenia effect . Staminate and pistillate flowers are produced on separate palm in the axils of leaves of the previous year growth in all the species of date palm. The date palm does not possess a well developed tap root system. It has a large number of secondary roots more or less equal thickness, which arise in the form of a dense cluster from the base of the stem. The secondary roots outside the relatively well marked central portion (stele) have visibly great number of large air passages in peripheral (extra stelar) region resembling to the roots of the plants adopted for growing in soils which contains excessive amount of water such as rice. The large air passages in the date root structure indicate its requirement for plentiful air supply.

## CLIMATIC AND SOIL ADAPTABILITY

### Climatic Adaptability

As per Arabic saying "date palm should be grown with its feet in running water and head in fire. It requires prolonged hot and dry summers, moderate winter and almost rain free period at the time of fruit ripening (July-August). The date palm grows well in sand but it is not erinaceous. It has airspaces in its feet and may grow well where soil water is close to surface but it is not aquatic. It grows well in saline soil conditions but is not true halophyte and does better in quality soil and water. Its leaves are adapted to hot dry conditions but it is not xerophytes and requires abundant water. The date palm adapted to areas with long very hot summers with little rain and low humidity but with abundant underground water. The mean temperature required at the time of flowering and fruit ripening is 25° C and 40°· C respectively. The pattern of maximum temperature at the time of fruit maturity determined the rate of maturity and fruit quality

It needs practically dry rainy season for attaining pind stage. The excess rain during fruit ripening period may even rot the fruits. The fruit have to be harvested at whatever stage of maturity it may be, before the onset of monsoon. Late monsoon, lesser total rainfall and number of rainy days are the primary requirement of the date palm cultivation as these conditions enable good quality and quantity of fully ripen soft dates (pind khajoor). Heavy rainfall followed by clear weather is less harmful as compared to light shower followed by prolonged drizzle or followed by prolonged drizzle of cloudy weather. Similarly heavy rain during early stage of ripening is less harmful than light rains at later stage of ripening.

In the country, on the basis of base temperature at 10$^0$ C the requirement of the heat summation units for date varies 1951 to 3650 depending upon cultivars. However, the rate of maturity of the fruits and development of its quality depend on the pattern of daily maximum temperature and heat units during the fruit ripening period. The situation is beneficial in the Indian Thar desert due to early ripening, the fruit escape spoilage from the ensuring rains. The quality of dates depends upon the heat summations units. In India four districts of Rajasthan , Jaisalmer, Barmer, Bikaner and Jodhpur located in western part of Rajasthan are ideal suited for date palm cultivation and have vast potential. The annual average rainfall of Jaisalmer, Barmer, Bikaner and Jodhpur is 160 mm, 310 mm, 305 mm and 370 mm, respectively. In these areas rainfall distribution is erratic. The average maximum temperature during summer months generally remains above 40$^0$C and vary often reaching around 45-48$^0$C. The accumulation of heat summation at Jaisalmer and Bikaner is sufficient enough for date fruits to attain dang stage on the tree itself , which are required between flowering to fruit ripening period.

**Table 1:** Heat summation unit required for promising date palm varieties at Bikaner (from Spathe initiation to full dang stage

| Varieties | Heat summation units (Base temperature 10$^0$ C) | |
|---|---|---|
| Halawy | 1951 | 3101 |
| Medjool | 2648 | 3650 |
| Zahidi | 2322 | 3479 |
| Khalas | 2357 | 3281 |
| Barhee | 2323 | Not attained |
| Sewi | 2244 | 3281 |
| Zagloul | 2227 | 3631 |
| Hayani | 2460 | 3168 |
| Shamran | 2411 | 3567 |
| Khadrawi | 2213 | 3342 |

*Source :* Chandra et al (1992) Performance of date palm cultivars in Thar Desert.

## Soil Adaptability

Date palm thrive well in deep sandy loam soil having adequate facilities of subsoil drainage for optimum productivity .The soil should be fertile and having good water holding capacity. The soil pH in the date palm growing area generally ranged between 8-10 in Rajasthan . Date palm can tolerate more alkali or salt (mainly sulphate of sodium and magnesium and chlorides of these metals) as compared to other fruit plants.

## RECOMMENDED AND POPULAR CULTIVARS

There are more than 1000 cultivars of date palm in the world but only few are extensively are grown. The commercial cultivars grown in different countries are given below.

**Table 2:** Commercial cultivars of date palm in different countries

| Country | Cultivars |
|---|---|
| Algeria | Ghars, Deglet Nour |
| Bahrain | Murzaban, Khunaizi |
| UAE | Angal |
| Egypt | Hayani, Saidy, Zagloul, Samani |
| Saudi Arabia | Irzeiz |
| Iraq | Zaidi, Hillawi, Khadrawi, Sayer |
| Iran | Ustaumran (Sayer) |
| Libya | Bikraari,Taasfirt, Murzabad, Saidy |
| Oman | Mabsaly, Fardh |
| Morocco | Bufaguus, Medjool |
| Tunisia | Fatumi, Deglet Nour |
| Pakistan | Mozawati, Begum Jangi, Dhakki |

The cultivars depending upon flesh consistency are divided into three groups, namely soft, semi dry and dry in soft cultivars like Hillawi, Khadrawi, Medjool, Shmran (Sayer), Saidy and Hayani, almost all cane sugar is converted to invert reducing sugars during the process of ripening. These are also called invert sugar dates. The dry and semi dry cultivars like Thoory, Deglet Nour, Zaidi and Dayri retain a good amount of sucrose on full ripening and therefore called cane sugar dates. As much as one third of the total sugars may be sucrose in the fruits of the semi dry cultivars Deglet Nour and the dry or bread type Theory.

In India 40 commercial date cultivars came from the Middle East countries and North Africa directly or via California but only Hillawi, Barhee, Khadrawi, Zaidi, Medjool, Shamran and Kahalas have shown promise (Pareek, 1984).

## Characteristics Features of Important Date palm Cultivars

**Hillawi:** It is one of the most suitable cultivar for date palm cultivation in Rajasthan and Punjab. The trees of this variety are tall and vigorous in growth. The fruit colour is light orange with yellow shade. The fruit is oblong elliptical with obtuse apex and almost horizontal base. The average fruit weight is 9 g and fruit size of 3.56 cm x 2.10 cm having 31.0 TSS at doka stage. The fruits are sweet at full doka stage. The dates may be eaten raw .The fruits have a tendency to shrivel during the ripening stage particularly in case of insufficient irrigations. It is early ripened and reach at full doka stage by second fortnight of July. Its average yield 100kg/tree at doka stage.

**Barhee**: The trees of this variety are semi vigorous and medium in height, having good canopy stage. The fruits are golden yellow in colour at doka stage and oval in shape. The average fruit weight is 7.5g and fruit size 2.9 cm x 2.3cm having 32 per cent TSS at doka stage. The fruits are sweet and non astringent at doka stage. The fruits are most suitable for raw eating dates. However, it is not good for soft dates. It is late ripened variety, fruit reach at full doka stage by mid of August. It is moderately damaged by rains and high humidity. Average yield potential is 100 kg doka fruits per year.

**Khadrawi**: The tree of this variety are dwarf in growth. The fruits at doka stage are greenish yellow in colour and oblong in shape with broad apex and a bit slanting base. The average fruit weight is 7.7g and fruit size of 3.1cm x 2.1cm. It has pulp thickness of 0.66 cm having 30 per cent TSS at doka stage. The fruits are highly astringent at doka stage and badly spoiled due to rains at the time of ripening. The variety is suitable for preparation of both soft dates and as well as dry dates. It is medium in ripening as the fruit of this variety reach at full doka stage by end of July. The average yield potential is 80 kg doka fruits per year.

**Medjool:** It is popular variety and has the potential of commercial cultivation. The tree are vigorous and tall with erect growth habit. The fruits are large but variable in size having attractive golden yellow colour more than Barhee and Khadrawi. The irregularities in shape are common and are associated with ridges on the seed. The average fruit weight is 18g and fruit size 3.9cm x 2.8cm having 29 per cent TSS at doka stage .The fruits at doka stage are highly astringent. It is suitable for dry dates because of its large size fruits and good pulp thickness.It is tolerant to rainfall and high humidity conditions . It is late ripened as the fruits of this variety reach at full doka stage in Mid of August. The average yield potential is 75kg doka fruits per year.

**Zaidi:** The tree of this variety are semi vigorous in growth. The fruits are yellow in colour at doka stage with smooth and hard surface, hence least affected. The fruits are ovate in shape with broader end towards the apex. The average fruit weight is 10 g and fruit size 3.0cm x 2.2cm, having 30% TSS at doka stage. The fruits are astringent at doka stage. The partial dang stage fruits are eaten as raw dates. It is suitable variety for preparation of soft dates. It is late in ripening as the fruit of this variety reach at full doka stage in mid of August at Abohar. Its average yield potential 110kg doka fruits per year.

**Khunaizi**: This emerging as potential date variety because of quality fruits. The trees are semi vigorous having spreading growth habit. The fruits are dark red in colour with crispy pulp and dark brown at pind stage. The fruits are oval in shape with pointed apex. The average fruit weight 10.2g and fruit size 5cm x2.2cm having 43% TSS at full doka stage .The fruits are sweet at full doka stage and thus suitable for raw eating. These are also not good for soft dates. This variety is also early as the fruits reach at full doka stage by the end of first fortnight of July at Bikaner. Average yield potential is 60kg doka fruits per tree per year.

**Zagloul**: The trees of this variety are vigorous in growth having erect growth habit. The fruit are red in colour and large in size. These are oblong, oblique and asymmetrical in shape. The calyx is small in size and slightly depressed at the base. The fruit bunches are bigger in size. The average fruit weight 9.5g and fruit size is 3.9cm x 2.2 cm having 28% TSS at doka stage. The fruits are astringent at doka stage. The dang and pind fruits are relatively less sweet but have pleasant taste. It is suitable for soft dates and for preparation of fruit products such as date juice and preserved products. It is medium in ripening as fruit reach at full doka stage by second fortnight of July at Bikaner. It is prolific bearer and average yield potential is 125 kg doka fruits per year.

**Sewi**: It is vigorous in growth. The trees are tall and erect with moderate canopy. The fruits are yellowish green in colour at doka stage and oblong in shape. The average fruit weight 6.1g and fruit size 2.9 cm 1.9 cm having 32 per cent TSS at full doka stage. The fruits are sweet at full doka stage and suitable for raw eating as well as soft dates. It is late in ripening as the fruit of this variety reach at full doka stage by the first fortnight of August at Bikaner. The full grown tree of Sewi has average yield potential 50 kg doka fruits per year.

**Shamran:** The tree of this variety are semi vigorous. The fruits are yellow in colour with slightly pink shaded base. The fruits are medium in size and oblong in shape with obtuse apex .The average fruit weight is 6.5g and size 3.50 cm x 2.18cm having 30 per cent TSS at doka stage. The fruits are astringent at doka stage. It is suitable for dry dates and soft dates. Its soft dates having a typical aroma. It is medium in ripening as fruit reach full doka by end of July. It is prolific bearer and has an average yield potential of 100kg doka fruits per year per palm.

## PROPAGATION TECHNOLOGY

There are three techniques to propagate date palm, seed propagation, off shoot propagation (traditional method) and the recently developed tissue culture techniques

### Seed Propagation

Also called sexual propagation although used for breeding purpose. This is not a proper method of date palm multiplication. Because date palm is dioecious species. Seed propagation is discouraged because half of the progeny will be males and half will be females with no certain way to determine at an early stage the sex of progeny, nor fruits or pollen quality prior to flowering. The female plants are originating seedlings usually produce late maturing fruits a variable and generally inferior quality compared to established clonal palms. Date palm is heterozygous and thus there will be much variation within the progeny and desirable characteristics of the parent palm may be lost. In other words it is not true to type propagation and two seedlings palms are alike. Seedling differ considerably with regards to production potential, fruit quality and harvesting time , making them very difficult to market at harvest. Thus, seed propagation is by far the easiest and quickest method of propagation. Because of diversity the seed propagation could be useful for breeding purpose.

### Offshoot (Vegetative) Propagation

Offshoot propagation also called asexual or vegetative propagation.Offshoot plants are true to type to the parent palm. The offshoot develop from axillary buds on the trunk of mother plants and consequently the fruit produce will be the same quality as mother palm and ensure uniformity to produce. The offshoot plant will bear fruits 2-3 years earlier than seedling. The life span the date palm is divided into two distinct developmental phases: vegetative in which bud forming in the leaf axils develop into offshoots, and generative, in which buds form inflorescence and offshoots cease. From the time that the axillary bud of a leaf has differentiated into an offshoot until the time it grows outwards, take up to three years, with another 3 to 4 years before it reaches the desired size for its separation and planting (Hilgeman, 1954). Off shoot mainly produced in a limited number (20 to 30 at most) during the early life of palm (10-15 years from the date of its planting) depending on the variety and on prior fertilization treatment , irrigation and earthing up around the trunks. The offshoots selected for the removal must be disease and pest free and at least three to five years old with a base diameter 20-30cm, weighing around 10kg but not more than 25kg.

Vegetative propagation is mainly done by offshoots, which after 3-5 years of attachment to the parent palm produce root and by then start producing second generation of offshoots. Aerial offshoots is not used for planting as these are either devoid of roots or have poor root development. Ground offshoot of 8-10kg average weight and well developed root system are selected for commercial plantation.

### Tissue Culture Propagation

The application of tissue culture techniques for date palm also called in vitro propagation has many advantages that include propagation of elite female cultivars, large scale multiplication, absence of seasonal effects, production of genetically uniform plants and fast exchange of plants materials. In a commercial plantation 2 to 3 % male palm are sufficient to meet the pollination, hence main emphasis should be on increasing the production of female plants. There are many reports on in vitro multiplication of date palm using various explants shot tips, inflorescence, immature fruits, leaf segments, stem and root section. The area under date palm cultivation is stagnant due to non availability of planting materials, thus depriving the country to achieve the production potential. The tissue culture technology has potential to produce maximum number of plants in a limited time and space that are true to type.

## PLANT AND FRUIT PHYSIOLOGY

The date palm can adapt to extreme drought, to heat, and to relatively high levels of soil salinity. Flowering in date palm starts by spathe emergence. Spathe emergence takes place from late winter to spring which is influence by various factors such as growing site, climatic conditions mainly temperature fluctuations cultivars and sex of the plant etc. At Bikaner, it has been observed that generally spathe emergence take place much earlier in male palm as compared to female trees. The spathe emeregence commenced early in cultivars Zaidi, Nagal, Khadrawi and Khunaizi (first week of march) while it was delayed in Khasab, Barhee, Umshook (last week of March) at Abohar

Date palm being dioecious in nature bearing male and female flowers on separate trees requires artificial pollination for good fruit setting, as pollination through natural means such as wind and insect is negligible. Hand pollination is commonly practiced. It is done by dusting pollen through cotton balls on freshly opened female spathe in the early morning hours at least 2-3 days followed by placing pollen impregnated cotton balls in the female flowers or by hanging strands of male inflorescence in an inverted condition on female spathes. For collection of pollen from male spathe, the mature male spathe are swept on the paper. These collected pollen (6 hrs in sunlight and 18 hrs under shade) can be stored in airtight glass vials in a cool place at normal temperature for eight weeks or in refrigerator at $9^0$ C for a year.

On the loss of two unfertilized carpels (habbouk stage) after fertilization fruit set take place. Thereafter fruit growth take place. There are four fruit development stages. The Arabic names for these four stages are Kimri, Khalal, Rutab and Tamar. In Punjab these stages are termed as Gandora, Doka, Dang and Pind respectively. Kimri is stage when fruits attains pea stage and remain hard green. The Khalal stage is marked by when fruit attains full growth remain hard and their color changes into yellow or red, finally spotted with having characteristics of its variety. In this stage fruit has attained the full size. The third stage known as Rutab or dang characterized by the date becoming more or less translucent and softening start from distal end by darkening of the skin from yellow chrome or scarlet to brown or to nearly black. The final stage in the ripening of date is called Tamar or pind when the fruit get fully mature and has lost much of water, enough to make the sugar to water proportion sufficient high to prevent fermentation. This stage equivalent to that of raisin in the grape.

## PLANNING AND PLANTING

The optimum period of planting date palm offshoots in rainy season (July-September). During this period the relative humidity is high and temperature is favorable for establishment of suckers in the field. The spring season planting of date suckers is not favorable in the view of high temperature in following months which require much care for protecting newly planted suckers. The planting of suckers in early June is also not advisable because of very high temperature and low relative humidity which create a problem in establishment of suckers. The date palm offshoots are planted in the field in square system of planting at a distance of 8 x 8 meters. The distance may even be increased to 10 meters in case of soil is very fertile and irrigation water is abundance . Sufficient distance between rows and within rows is very essential as date palm remains in bearing for at least 50-60 years and close distance may create a problem in later year of bearing due to over crowding and shading. Proper distance also facilitate inter culture operations. 156 suckers can be planted in one hectare area at 8 meter planting distance in square system. In date palm orchards 10 percent male plants are required to get adequate pollen for pollinating female date bearing trees. The male suckers planted in outer row of facing towards the planting of offshoots is done in pits in thoroughly prepared field.

The pits are dug of one cubic meter size preferably 15 days before planting. The bottom and walls of the pit should be treated with chlorpyriphos for the control of termites. A mixture of 4 to 6 baskets of well rotten farmyard manure, top fertile soil 50 g Captan and 50-100 g insecticidal dust is used for refilling the pits. The pits are watered to enable the soil mixture to settle before planting. The offshoots having 8-10kg weight and well developed root system are selected for plantation. Aerial suckers are not used for planting because of poor root development. At the time of planting the offshoots are dipped in solution of carbendazim (0.2%) and chlorpyriphos, for the root initiation the offshoots may be dipped in IBA solution (1000ppm) for 2-5 minutes to ensure better root development and it should be planted 10-15cm above the ground level. The windbreak should be planted in North and west directions of the orchard to protect the date palm trees against hot and cold wind breaks for date palm. The newly planted offshoots are irrigated every day or alternate day depending upon the soil structure for initial three months period.

## SOIL CULTURAL PRACTICES TECHNOLOGY

### Water Need

For maximum growth and fruit production of date palm requires an abundance of water. The Arabs have a proverb that the date palm must have "its feet in water and its head in fires of heaven " the date palm is drought tolerant and is able to survive for long periods without irrigation. Drought however retards the growth. It is highly tolerant to excessive irrigation and floods but permanent water logging is injurious. To maintain the maximum growth the root zone up to 2-3 m should be kept moist and not allowed to dry up. Immediately after planting, light and frequent irrigation must be given. In sandy soils irrigation may have to be given every day or on alternate days. The frequency of irrigation is reduced after the offshoots established. The fully grown trees in the lighter soils must be irrigated at 15 to 30 days interval during winter 7 to 10 days interval during summers depending upon soil texture. One irrigation prior spathe initiation is critical for normal fruiting. Irrigation after fruit set should be given regularly. Scarcity of irrigation during fruiting may result premature fruit drop, shriveling of fruit and reduction in fruit size, weight and quality. The drip irrigation method in date palm orchard is useful as it provides efficient utilization of irrigation water. If drip irrigation is used from an early age, the root distribution of date palm will be within the area of moist soil surrounding the tree.

## Nutritional Need

Palm tree require proper nutrition particularly in sandy and gravely soils to achieve good growth and economical production. The amount of fertilizers needed by date palm tree depends upon soil type, kind of intercrops grown under as well as variety and age of the tree. The date palm trees respond to manure very well. The results of studies conducted at Abohar show that 25 to 50 kg farmyard manure and 1-2 kg ammonium sulphate per palm should be added. The quantities may be reduced if some leguminous crop has been grown in date palm. Farmyard manure is applied during December–January. The nitrogen dose should be given a fortnight before flowering in first week of February.

To get better yield a full grown date palm tree should be given 20-40kg FYM, 500-1000 g N, 250-500g $P_2O_5$ and $K_2O$ each annually. Farmyard manure should be applied after monsoon in September to December. The whole quantity of phosphorous and potassium and 60% nitrogen should be given three weeks before flowering .The remaining 40% nitrogen should be applied during fruit development period in March-April.

## Weed Control

Time to time removal of weeds is vary necessary in date orchard. Hoeing of newly planted date orchard should be done in the basin regularly. It not only helps in destruction of weeds but also provides better aeration to roots and moisture retention in the soil. In the well established orchards also, removal of weeds is very essential to avoid competition of weeds for nutrients and moisture with date palm trees. Thatai et al. (1994) conducted a trial on the effect of different herbicides to control the weeds in date palm at Abohar. The herbicides were sprayed twice as pre and post emergence stages at six months interval. All the herbicides reduced the weed population considerably as compared to control. Treatment with diuron (2kg/acre) followed by glyphosate (1.5 l/acre) gave the minimum population of monocot and dicot weeds. No phytotoxic effects were recorded on any of the trees with all the herbicides.

## Inter Culture

Date palm trees being unbranched, it provides plenty of space between the rows to be utilized for intercropping particularly during the initial year of establishment. It supports farmers with some income when date palm trees yet to start fruiting. However intercropping should not be done at the cost of date palm plantation. Therefore, the requirement of organic manures fertilizers irrigation water etc. for raising intercrops should be given separately . Selection of intercrop is also very crucial. The intercrop in any way should not be detrimental for growth and development of date palm trees. The leguminous crops such as gram, pea and guar, vegetables oilseeds crops such as mustard and sesame; fodder crops such as lucerne, berseem and oat can be grown as intercrops in date palm orchards. Small to medium sized fruit trees such as pomegranate, phalsa, papaya and citrus crops can also be grown as intercrops with date palm. Turmeric in the shade of date palm also performed well in the Abohar conditions.

## PLANT CULTURAL PRACTICES TECHNOLOGY

### Training and Pruning

Date palm is monocotyledonous trees have single stem and normally do not branch. Therefore, the training is minimal except to keep single stem growing without excessive leaves and removal offshoots from the palm trees as soon as they attain proper growth. Keeping offshoots attached with date palm trees for long even after attaining separable stage adversely affect their growth and productivity. The excessively dried leaves are removed from time to time. A fully grown date palm trees should have approximately 100 leaves. Very old leaves must be removed as these are less efficient in photosynthesis. The best period of pruning is during winter months (November to December). Besides removal of leaves, dehorning is also essential close to fruit bunches as it helps in pollination, thinning of bunches, spraying and harvesting of fruits. After fruit set, twisting of bunch is done during April and May which helps in withstand of bunch breakage due to high wind velocity. Removal of one third of their central strands after fruit set has shown better development of fruits and hasten ripening.

### Pollination and Fruit Thinning

Date palm is dioecious in nature, bearing male and female flowers on separate trees, requires artificial pollination for good fruit setting as pollination through natural means such as wind and insects is negligible. The spathe are generally emerge during February – March in north western India and flowers opening start during March-April. Receptivity of flowers in some date cultivars decreases rapidly after spathe splitting. The pollination is generally carried out during first 2-3 days after spathe open. Hand pollination is very common. It is done by dusting pollens through cotton balls on freshly opened female spathes in the early morning hours followed by placing pollen impregnated cotton balls in the female flowers or by hanging strands of male inflorescence in an inverted position on female spathes. For collection of pollen from male spathes are swept on paper. These collected pollen are dried (6 hours in sunlight followed by 18 hours in shade) and can be stored in air tight glass vials in cool place at normal temperature for 8 weeks or in refrigerator at $9^0$ C for a year. Fruit set is closely related to percentage of viable pollen and temperature condition during pollination. The poor fruit set resulting from low temperature may improved by placing paper bags over flower clusters. Heavy load of fruit may cause shriveling of berries, breaking of spate stalk, more damage due to rain and humidity, delay in ripening and alternate bearing. Fruit size is reduced and produce poor quality fruit. It is therefore necessary to keep only optimum quantity of fruit and thin out the rest. This is usually accomplished either by reducing number of fruits on each bunch and or by removing some of bunches. The number of fruits that a palm can safely depends on the cultivars, age , size and vigour of the palm and number of green leaves on it. Under normal condition 1-2 bunches in $4^{th}$ year and 3-4 bunches in $5^{th}$ year left. Normally 8-10 bunches per palm retained in India. Small, defective and broken bunches should be removed.

Bunch thinning either by removal of strands or shortening of strands or by combination of the two depends upon the cultivars and other conditions. In short strands varieties like Khadrawi, the strands are generally cut back to even, up the bunch from the top. Most of the fruit thinning is done by removal of half to two third of strands from the center. In long strands varieties like Deglet Noor, one third to half strands are cut in similar way as in Khadrawi, in addition strands are also cut back to remove about one third of the flowers. The desirable number of fruits to be left is between 1300 to 1600 per plam depending on the variety. The per cent of thinning is generally done 40-50 in Khadrawi, 50-55 in Hillawi, 50-60 in Zaidi and Barhee. Ethephon 100-400 ppm after 10 to 30 days from fruit set in Hayani was found effective in fruit thinning and advance ripening. The biennial bearing habit of treated plants was found to reduce by ethephon treatment.

## HARVESTING, PRODUCTION AND PROCESSING OF FRUITS

The fruit of most of the date palm varieties should be allowed to ripen on the tree until it reaches the stage of maturity at which it is to be consumed. The fruit maturity is influenced by the application of plant growth regulators. The fruit size increased and ripening delayed with gibberellic acid 50 ppm. Ethephon at 500-1500 ppm hastened the fruit maturity in date palm. NAA at 30-90 ppm applied two weeks after full bloom increase fruit size and weigh. Date palm trees come into bearing after six years of planting . Initially yield is less and it goes on increasing with the age of trees. Fruit yield differs from variety to variety and also depend upon the age of tree and orchard management. On an average 50 kg doka fruit per year can be harvested up to the age of 5 years and thereafter 75 -150 kg per tree per year by using recommended package of practices. Hillawi and Barhee yielded 50-100 kg per palm at doka stage. The average yield from Khadrawi and Shamran is obtained 40-70 kg fruit per palm. The yield in Medjool palm is up to 6o kg but fruit are very large in size.

The dates are harvested at different stages of maturity depending upon the varieties , demand from consumers and climatic conditions. The bunches are harvested at full doka stage for fresh dates in those cultivars which are not astringent in taste and edible at doka stage. The fruits are separated from strands and graded, shriveled, diseased and undersized fruits are discarded to make produce more attractive for the market. For preparation of dry dates (chhuhara) the fruits are harvested at full doka stage and for soft dates (pind khajoor) partial to full dang satge. At least two varieties of dates Hillawi and Barhee are liked most for eating in Khalal stage. The dates are hand picked at stage of maturity. All the dates in the same bunch do not ripened at the same time. It has been the practice to make several picking to harvest the fruit during a season. Some time season is favorable and more than 70 per cent fruit is ripe, the entire bunch is harvested

**Curing of dates :** Studies conducted on curing of dates at Abohar have shown that the rain and high humidity in the atmosphere at the time of ripening ( July and August) do not allow the ripening process on the tree to proceed satisfactory beyond doka stage. If the fruit is retained on the tree, there is checking, splitting and rotting. High humidity and rainfall also cause the fruit to drop. Therefore the crop has to be harvested at doka stage. Though in comparatively drier seasons partial crop on same trees may become dang (Mellow and soft) but the quantity of such fruit is limited. The berries are removed and graded manually on the basis of size and color.

A technique has developed to transform satisfactorily the date fruit at doka stage into dry dates (Chhuhara) of good quality at RFRS Abohar. At least four varieties Hillawi, Khadrawi, Shamran and Medjool yielded very good product. Chhuhara obtained from Medjool which is large size variety, compare very well with high quality Chhuhara imported from Middle-East countries. The techniques developed involves immersion of fruit at doka stage in boiling water for 6-8 minutes and then drying either in temperature controlled oven (air circulation type) for 80 to 120 hrs at 48 to 50$^0$ C or in the sun for 10-15 days if weather is dry. Thus an average 45 per cent product is obtained. Fruits at advanced doka stage or when or when they attained one fourth or one half or full dang (the berries become mellow and soft starting from distal end ) can be converted into soft dates (Khajoor) of good quality by drying either in the oven at 40$^0$C or in the sun. Thus for soft dates, only drying the berries at partial or full dang stage is required with no other treatment and this way a final product of soft Khajoor ranging from 50-60 per cent is obtained.

**Ripening of Dates at Doka Stage into Dang Stage.** In transformation from doka stage to dang stage berries were treated with 0.5% to 2% common salt (sodium chloride) and similar concentration of acetic acid in combination with 1 per cent salt at RFRS Abohar. The fruits were treated with salt were spread on polythene sheet and requisite quantity of salt was applied by rubbing and smearing uniformly on the berries. The fruits which were given acetic acid plus salt treatments, were first dipped in solutions of desired acetic acid concentrations for 2 minutes followed by salt application by the method described above. Each treated lot was packed into wooden boxes lined with old newspaper and packed in laboratory at room temperature. The boxes were open after 24 hours. From these studies it was found that with 2 per cent salt, 60 to 70 per cent doka fruit were transformed into dang. However, the dang obtained by the treatment was not as good in taste as that of naturally ripened on the tree, but still it was edible and generally acceptable on account of its having lost the astringency. But such product cannot be stored more than 24 hours and as such should be consumed to as early as possible.

## POST- HARVEST FRUIT TECHNOLOGY

The post- harvest losses in date palm fruits in India are as high as 32-40 per cent due to heavy rains during its fruit maturity. The post harvest losses in date fruit can be minimized to great extent by harvesting of date palm fruits at the right stage of maturity. Determining the maturity standards of different varieties of dates in therefore important for proper management, handling, harvesting, drying, packaging and storage .

Harvesting stage influences the fruit weight, acidity, TSS, organoleptic rating and spoilage percentage of fruits. The weight of fruits in all cultivars increased up to doka stage slightly decreased at dang stage. The total soluble solids in all the cultivars increased from gandora to dang stage, whereas acidity decreased. The study also revealed that for raw consumption of dates as well as for its better upkeep in terms of quality, fruit should be harvested at doka stage. A positive correlation was observed between TSS and organoleptic rating , hence, it is suggested that TSS may be taken as an index of maturity of dates.

The shelf life of fresh dates can be enhanced for two weeks in cold storage (5-7 °C and 85-90 % RH). Fresh dates of khalal stage can be stored up to two months at 0-1°C. Dried dates (Chhuharas) can be stored for one years at ambient conditions after packing in gunny bags. Storage studies of fresh dates fruits at Bikaner have revealed that at room temperature fruit could not be stored for more than four days, whereas these could be stored up to 30 days under refrigeration and up to 50 days at freezing temperature. All fruits of Hillawi turned to dang stage at freezing, whereas only 75 per cent doka fruits turns dang in refrigerator. The juice and sugars can be successfully utilized as sweetening and falvouring agent in ice cream in bakery products. A satisfactory pickle can also be obtained from six weeks age of picking of green fruits (kimri) treated with 50 per cent sodium chloride and 2 per cent acetic acid . There is possibility of utilizing date fruits, harvested at doka or khalal stage for preparation of alcohol, vinegar and concentrated juice.

## INSECT- PESTS AND MANAGEMENT

In north-western India, insect- pests like termite in young date plantation and thrips, scale insect, rhinoceros beetle (*Oryctes rhinoceros*) and Indian palm weevil (*Rhynchophorus ferrugineus*) in bearing trees have been observed to cause some damage. Stored dates are attacked by cigar hoeing beetle (*Lasioderma testaceum*

**Parlatoria date scale (*Parlatoria blanchardi* Targ):** In Abohar conditions, the Parlatoria date scale was first observed infesting the date pinnae in 1970. It was found to cober the entire surface of the foliage and also feed on the white succulent tissue at the base of the leaf stalk. As the population increased, they moved to the pinnae which ultimately withered and dried up. The date stone beetle was first observed infesting date fruits during June-July 1969. The adults beetle penetrate the unripe fruits and reaches the stone by constructing a direct circular hole through pulp. Both adults and grubs cause extensive damage by feeding inside the pulp. The infested fruit ultimately drop. The Hillawi and Khadrawi ripened early in Abohar conditions, infestation noted 22 and 17 per cent, respectively. However, Medjool and Thoory cultivars which ripened late, escaped infestation by this pest. It was found to remain active from June-August. Spraying the plant with fenvalerate 1ml/lit or imidacloprid 0.3ml/li or dimethoate 2ml/lit has proved effective for its control.

**Nitiduled beetle (*Heptoncus luteolus* Eri):** It was first observed feeding on date palm fruits during July-August 1969 at Abohar. The larvae penetrate the fruit and eat the inner portion of the pulp. The attack is followed by fungal decay. They also carry with fungal spores, thereby resulting in increased proportions of diseased fruits. Initially the dropped fruits are attacked. The pest spread to fruit bunches on tree. The pest over winters as pupae in soil. The adults are also found to infest decaying vegetation or other fruits. Spraying with fenvalerate l ml /lit gave satisfactory results. Regular removal of dropped fruits and their destruction is also recommended.

**Termite (*Odontotermes obesus* Rambur):** Termite workers occurs on all the tree parts all the year round and considerably affect the palm. Termite workers attack on root zone. Feed on the mine into the main trunk. The young establishing tree under its severe attack, wither and die. The root zone of the tree should be treated with chlorpyriphos 5 ml/lit water periodically. The galleries of termites on the tree trunk should be scrapped and treated with above mentioned insecticides.

**Black headed caterpillar (*Nephantis serinopa*):** In Kutch, black headed caterpillar causes serious damage ,while red palm weevil and rhinoceros beetle have also been observed. The black headed caterpillar feed on the leaves hiding inside the tunnel in the folds of leaves. It can be controlled by cutting and burning the badly affected frounds/leaflets. In severe infestation, removal and burning of fully dried 2-3 outer whorl of fronds helps in removing the pupae and other pest stages. For more effective control, population of its predator, *Goniozus nephantidis* should be built up in the plantation. The red palm weevil grubs enter the palm near the ground and bore into the trunk upwards. Clean cultivation is very useful. If grubs have already entered the trunk fumigation should be done and the holes should than be plugged.

**Lesser Date moth (*Batrachedra amydraula*):** It was observed for the first time in Bikaner. Larvae of moth attack on flower and fruits. Infestation occurs during reproductive period i.e between March to August. To control this pest decamatherin (3ml/ 10 lit water) or malathion (1ml/lit of water) should be applied at about one week after fruit set and repeated again 15 days after the first spray if required.

**Bird damage**: The period of fruit growth and development from May to July in date palm coincides with the period when there is almost no field crop available for birds to feed in the areas where date can be grown in Punjab and Rajasthan. Thus birds can cause severe damage to date fruits. The date bunches at the time of ripening are very much damaged by crow, parrot, sparrow etc. The maximum damage caused at full doka and dang stage. Therefore, protection from bird attack is very much needed during this period. The protection from the bird damage can also be done by covering fruit bunches with wire net gauge (3x3mm mesh) at the start of **doka** stage till harvesting.

## DISEASES AND MANAGEMENT

Graphiola leaf spot, alternaria leaf spot, rotting of main shoot of offshoots and fruit rotting are major diseases of date palm in Punjab and western Rajasthan. Graphiola leaf spot is only disease which causes economic losses by leaves to reduce their number on the palm affecting date fruit production.

**Graphiola leaf spot**: It has been observed to be severally affected by false smut of dates caused by *Graphiola phoenicis* (Moung). It is the most wide spread disease of date palm and probably occurs whenever the palm is cultivated under humid conditions which prevail in many subtropical to marginal date growing area. Leaves of date palm are spotted with yellowish dots with small black scales on worts having grey brown to dark brown horny outer surface and long sterile flexuous hyphae protruding from and inner membrane containing powder yellow of light brown spots. In case of heavy infestation leaves dry and consequently die. Thus it causes great damage to the plant growth and reduces the fruit yield. Studies on epidemiology, varietal screening and disease management were carried out in Punjab and Rajasthan, Jaisalmer, Barmer and Jodhpur which have hot arid climate. The disease progressively developed on older leaves as compared to newly one. Highest infection was recorded in the month of December due to low temperature and high humidity in the atmosphere. Incidence of disease has been recorded to be varying from 2.5 to 76.25 per cent in the field. Thirty four varieties were screened under natural as well as artificial conditions, out of which the variety Hillawi and Khalas found to be moderately susceptible and variety Zaidi are found to be resistant. Two spray of copper oxychloride (0.3%) were effective to minimize the Graphiola leaf spot followed by carbendazim (0.1%) .

The studies on changes in green pigments (chlorophyll and carotenoides) at different level of severity of the disease indicated that as the disease severity increase the green pigments also decreased proportionately in all the varieties of date palm. The per cent reduction in chlorophyll and caretenoids in diseased plant was obviously much more in highly susceptible varieties as compared to resistant ones.

**Fruit rots**: Pre- harvest fruit rot is another problem since their incidence is governed by occurrence of rain and high humidity during khalal stage and later stage of ripening. Considerable loss observed by fruit rots caused by *Aspergillus niger, Aspergillus flavus, Rhizopus sp. Penicillium sp.* and *Botryodiplodia sp* when humid and rainy weather occur during the ripening season. Out of 34 varieties screened against fruit rot before and after the rains, none of the variety was found immune, while varieties Khunaizi, Hillawi, Medjool, Khalas and Barhee were susceptible to the disease only one cultivar i.e Zagloul was found resistant i.e below 5 per cent disease, incidence. Two sprays of Trichodermaviride 1%+ Azadiractin (0.3%) at an interval of 15 days coupled with date leaf cover of fruits bunches were found effective to control of fruit rotting up to 62 per cent.

**Alternaria leaf spot**: This disease of date palm is caused by *Alternaria alternatea* (Fr.) Keissler, is also a most destructive disease of date palm. Symptom of the disease were observed on leaves in the form of small isolated pale or dark gray to black circular spots measuring 2-8 mm in diameter. Later on spots increased in size and become irregular straw colored and coalesced. On severity the disease leaves turned yellow and defoliated. Two sprays of carbendazim + mancozeb ( 0.2%) at an interval of 15 days were found effective to control the disease 52.69 per cent.

**Diplodia suckers rot**: This disease is caused by a fungus *Diplodia phoenicum* (sac), a fungus that some time affects leaf stalks and offshoots of date palm. This disease is very serious and become a limiting factor for further propagation of date palm. This disease is observed to characterized by the death of offshoots, either while they are still attached to the mother palm or they have been detached or transplanted. The disease also caused premature death of fronds in older palm. In offshoots the disease may be expressed in distinct types (1) the fungus may infect and kill the outside fronds, leaving the younger shoots and bud alive for sometimes before finally causing their death as well (2) the central leaf cluster and terminal bud may die the older fronds. The fungus usually enters the palm through wounds made during the pruning or through the cuts made during removing the offshoots from the mother palm. Faulty irrigation causing some roots to die back to the base of the palm may also be observed to contribute the infection. For control measures of this disease infected leaves and dead tissue should be removed. Pruning tools should be disinfected by dipping them in a formalin or bleach solution after each palm is treated. Offshoots /suckers of date palm at the time of transplanting must be treated with carbendazim 0.1% + chlorpyriphos 0.1% + IBA1000 ppm coupled with alternate day irrigation up to 30 days for better establishment and better survival of suckers.

## MARKETING AND EXPORT POTENTIAL

Harvest season of fresh dates in India starts from the month of June and continues up to the August. The production period of other main supplying countries including Iran, U.A.E, Saudi Arabia, Iraq, Israel, Tunisia etc. is from August to November. The dates are harvested and marketed at three stages of their development. The choice for harvesting at one or another stage depends on varietals characteristics, climatological conditions and market demand. The three stages of fruit ripening are as follows:

**Doka Stage (Khalal):** physiological mature, hard and crisp, moisture contents 50- 85 per cent, yellowish in color.

**Dang Stage (Rutab):** Partially browned, reduced moisture contents (30-35 %), softened.

**Pind Stage (Tamar):** Colour from amber to dark brown, moisture contents reduced below 25 to 10 per cent, texture from soft pliable to firm.

Like other horticulture crops, dates trade is with the private sector. However, Government largely facilitates the system by providing physical infrastructure especially wholesale markets and communication, market intelligence, market promotion and regulatory measures for smooth business operations. The world market for dates is expanding. The exporting countries are striving to further expand their market share as new markets open up with world trade liberalization. The World Dates export market is about 7.51 million tons per annum. The Saudi Arabia, UAE, Iran are the top of the list with main share in the world's top ten exporting countries. The world Date import is about 0.63 million tons per annum. The top ten dates importing countries of the world out of which India is the largest importer with market share of about 38 per cent while France and UK are second and third largest importers with shares of 4 per cent and 2.5 per cent, respectively. There is need to consolidate position in the existing markets especially USA, UK, Canada and Germany through strong market promotion and consistently supplying quality produce shipped by reefer containers, and also to identify new potential markets like Russia. Besides, Indonesia and Malaysia are important non-date producing countries. They are amongst top ten dates importing countries of the world and import dates for consumption in Ramadan and for other purposes in quite substantial quantities. In India there is a great scope of fresh date cultivation, before 2010 the main centre for production of fresh dates in India was Kutch district of Gujarat. Now the area under tissue culture date palm is increasing in the state of Rajasthan and Punjab. So there great potential of fresh dates marketing in domestic as well as export from India.

## FUTURE STRATEGIES

### Suggestion for area expansion

(i) The suitable plant material of varieties having the characteristics of early ripening, high yielding and high sugar content and resistant to rain and *Graphiola* should be imported, acclimatized, evaluated and promising types should be multiplied for commercial cultivation.

(ii) The quality planting materials raised by tissue culture by private companies should be purchased and provided to the farmers under different horticulture programmes.

(iii) The date palm being tolerant to high soil salinity, its plantation can be promoted under the problematic area where water table and salinity level is high.

(iv) In the waste land area, date palm can be grown through seedling, though the quality of fruit will be poor to its parent and sex ratio will be other problem but the plantation done under such conditions will reduce the degradation of soil, improve environmental conditions and it will also improve the economic conditions of the rural people.

**Propagation technology**: For getting superior quality plant materials, date palm may be propagated by offshoots or by micro propagation.

(i) **Offshoot propagation:** The multiplication through offshoots is a natural phenomenon of the plant but the availability through natural development shoots is very limited. Therefore, the offshoots may be initiated by using growth regulators, supplementary nutrition and providing sufficient irrigation, which require research support.

(ii) **Micro propagation:** The present demand of plant material can be materialized by standardization of protocol for micro propagation and also proper hardening of tissue culture plants before planting at the desired site.

**Improvement:** Selection of indigenous elite clones suitable for different regions and their multiplication, import of the elite varieties and their acclimatization and evaluation should be undertaken. Studies on physiology of flowering and fruit, breeding and physiological manipulation for delayed flowering for different uses including soft date preparation need to be initiated.

**Disease management:** The control of Graphiola leaf spot through integrated management including manipulation of cultural practices, sanitation, host resistance and use of fungicide is required.

**Insect- pests management:** Lesser date moth has been the newly record pest of the date palm, which damage the flowers and fruits of the date palm, therefore the studies on the behavior of this insect is important.

**Cropping system**: Date palm is tall growing trees which have sufficient space between the rows, which can very well utilized for intercropping with other fruit crops, legumes, fodder and vegetables. If this system is found suitable, it will increase the economic return of growers from the same piece of land.

**Cultivation under marshy waste land:** There is sufficient area lying waste under canal irrigation, which have high water table and high salinity level. Date palm being salt tolerant and require high amount of water, it is feasible to study the performance of date palm under such conditions.

**Micro irrigation:** Studies on the use of micro irrigation system in date palm should be under taken.

## LITERATURE CONSULTED


Aljuburi, H. J .1995. Fruits(Paris), 50 : 153-158.

Al-Ogaidi, H. K. and Kutlak, H. H. 1986. *Date Palm .J,* 4 : 191-203.

Anon, 2000. Biennial report of XI group workers meet. AICRP AZF, Rajasthan Agri. University, Bikaner.

Anon,2016. Annual report of XX group workers meet. AICRP AZF, RAU , Bikaner ( Rajasthan).

Arias-Jimenez,E.J.2002. Date palm cultivation Food and Agri. Organization of United Nations, p 107-227.

Bal,J.S.2014. Fruit Growing.3rd edition. Kalyani Publishers, New Delhi.

Becha, M. A .and Shaheen ,M. A. 1983. *Proc. Ist Sym. Date palm* in Saudi Arabia, pp: 4 174-180.

Chandra, A , Chandra, Anju and Gupta,I.C .1992. Date palm Research in Thar desert, Scientific Publishers, Jodhpur (Rajasthan).

Chandra, A., Swaminathan, R., Chaudhary , N.L ., Manohar, M S and Pareek, O.P.1994.A note on the performance of date palm cultivars in Thar Desert . *Indian J. Hort.*, 41: 28-33.

Cheng and Tsui, 2014. https://www.cheng-tsui.com>blog>th-. The importance of date palms/

Chundwat, B.S. 1990. Arid Horticulture. Oxford and IBH publishing Co., New Delhi.

Cheng and Tsui,2014. https://www.cheng-tsui.com>blog>th-. The importance of date palms/ Cheng &Tsui.

Hilgeman, R.H. 1954. The differentiation, development and anatomy of the axillary bud, inflorescence and off shoot in the detepalm. *Date Growers Inst. Rep.* 31:6-10.

Jawanda, J. S., Munshi, S.K. and PaL, R.N.1972. The nutritive value of date. *Punjab Hort.J.*,12: 54-57.

Jawanda,J.S. and Kalra,S.K. 1971. Date improvement work at Abohar. The Date seminar at Regional Research Station, Abohar 6th August 1971, pp 33–38.

Manohar, M.S. and Chandra, A.1995. Date palm culture in Rajasthan, Dir Res. RAU, Bikaner.

Pareek, O. P . 1984. Date palm growing potential of Indian Arid zone . *Indian Horticulture*, 29(2-5),8.

Pareek,O.P. 2000 .Date palm . In:Tropical and Subtropical Fruits:. 163-208

Pundir J. P.S.,Rathore, A R., Naqvi, A. R. and Porwal, R. 2004. Date palm Vol II Advances in Arid Hort.

Pundır, J,P.S. and Porwal, R. 1998. Performance of different date palm cultivars under hyper arid supplementary irrigated western plains of Rajasthan (India) .In: *Proc. Intl. Conf. date palms* , Al-Ain (UAE) from 8 -10 March: 329-336.

Randhawa, G. S.1980. A History of Agriculture in India. Vol 1-4, ICAR, New Delhi,

Rathore, G. S. and Naqvi, A. R. 2009. Production Technology of Date palm in Rajasthan, pp 1-18.

Singh,. R S ., Bhargava, R. and Dhandar, D. G. 2005. Performance of date palm introductions under hot arid environment. *Indian J. Plant Genetic Resources*, 18 (I): 65-66.

Thatai,S.K., Josan, J.S.and Monga,P.K.1994. Studies on weed management in date palm. *Indian J.Hort.*, 51(4):358-61.

# 12

# Fig

*A. P. Gaikwad and S. R. Lohate*

## INTRODUCTION

Fig (*Ficus carica* L.) is commonly known as Anjeer. It has a luscious taste as a fresh fruit. The fruit of fig have been prized over centuries for the medicinal and dietary properties. The figs are consumed fresh, dried, preserved, candied, jam and canned.

## NUTRITIVE AND CULTURAL SIGNIFICANCE

**Medicinal Uses:**The latex is widely applied on warts, skin ulcers and sores, and taken as a purgative and vermifuge, but with considerable risk. In Latin America, figs are much employed as folk remedies. A decoction of the fruits is gargled to relieve sore throat. Figs boiled in milk are repeatedly packed against swollen gums. Similarly, the fruits are much used as poultices on tumors and other abnormal growths. The leaf decoction is taken as a remedy for diabetes and calcifications in the kidneys and liver. Fresh and dried figs have long been appreciated for their laxative action. The food value of edible portion of fig (Morton, 1997) is given in detail in Table 1.

**Food Uses:**Some people peel the skin back from the stem end to expose the flesh for eating out of-hand. The more fastidious eater holds the fruit by the stem end, cuts the fruit into quarters from the apex, spreads the sections apart and lifts the flesh from the skin with a knife blade, discarding the stem and skin. Commercially, figs are peeled by immersion for one minute in boiling lye water or a boiling solution of sodium bicarbonate. In warm, humid climates, figs are generally eaten fresh and raw without peeling, and they are often served with cream and sugar. Peeled or unpeeled, the fruits may be merely stewed or cooked in various ways, as in pies, puddings, cakes, bread or other bakery products, or added to ice cream mix. Home owners preserve the whole fruits in sugar syrup or prepare them as jam, marmalade, or paste. Fig paste (with

**Table 1**: Food value per 100 g of edible portion

| Ingredient | Fresh | Dried | Ingredient | Fresh | Dried |
|---|---|---|---|---|---|
| Calories | 80 | 274 | Sodium | 2.0 mg | 34 mg |
| Moisture | 77.5-86.8g | 23.0g | Potassium | 194 mg | 640 mg |
| Protein | 1.2-1.3g | 4.3g | Carotene | 0.013-0.195 mg | — |
| Fat | 0.14-0.30g | 1.3g | Vitamin A | 20-270 I.U. | 80 I.U. |
| Carbohydrates | 17.1-20.3g | 69.1g | Thiamine | 0.034-0.06 mg | 0.10 mg |
| Fiber | 1.2-2.2 g | 5.6 g | Riboflavin | 0.053-0.079 mg | 0.10 mg |
| Ash | 0.48 0.85 g | 2.3 g | Niacin | 0.32-0.412 mg | 0.7 mg |
| Calcium | 35-78.2 mg | 126 mg | Ascorbic Acid | 12.2-17.6 mg | 0 mg |
| Phosphorus | 22-32.9 mg | 77 mg | Citric Acid | 0.10-0.44 mg | — |
| Iron | 0.6-4.09 mg | 3.0 mg | Malic, boric & oxalic acids | Small amounts | |

*Source*: https://www.hort.purdue.edu>moton

added wheat and corn flour, whey, syrup, oils and other ingredients) forms the filling for the well known bakery product, "Fig Newton". The fruits are sometimes candied whole commercially. In Europe; western Asia, northern Africa and California, commercial canning and drying of figs are industries of great importance.

Some drying is done in Poona, India, and there is currently interest in solar-drying in Guatemala. Usually, the fruits are allowed to fully ripen and partially dehydrate on the tree, then are exposed to sulphur fumes for about a half hour, placed out in the sun and turned daily to achieve uniform drying, and pressed flat during the 5 to 7 day process.

'Black Mission' and 'Kadota' figs are suitable for freezing whole in syrup, or sliced and layered with sugar. Dried cull figs have been roasted and ground as a coffee substitute. In Mediterranean countries, low-grade figs are converted into alcohol. An alcoholic extract of dried figs has been used as a flavoring for liquors and tobacco.

**Toxicity:**The latex of the unripe fruits and of any part of the tree may be severely irritating to the skin if not removed promptly. It is an occupational hazard not only to fig harvesters and packers but also to workers in food industries, and to those who employ the latex to treat skin diseases.

## Other Uses

**Seed oil:** Dried seeds contain 30% of a fixed oil containing the fatty acids: oleic 18.99%; linoleic 33.72%; linolenic 32.95%; palmitic 5.23%; stearic 2.1 8%; arachidic 1.05%. It is edible oil and can be used as a lubricant.

**Leaves:** Fig leaves are used for fodder in India. They are plucked after the fruit harvest. Analyses show: moisture 67.6%; protein 4.3%; fat 1.7%; crude fiber 4.7%; ash 5.3%; N-free extract 16.4%; pentosans 3.6% and carotene on a dry weight basis 0.002%. Also present are bergaptene, stigmasterol, sitosterol, and tyrosine.

In southern France, there is some use of fig leaves as a source of perfume material called "fig-leaf absolute". It a dark-green to brownish-green, semi-solid mass or thick liquid of herbaceous-woody-mossy odor, employed in creating woodland scents.

**Latex:** The latex contains caoutchouc (2.4%), resin, albumin, cerin, sugar and malic acid, rennin, proteolytic enzymes, diastase, esterase, lipase, catalase, and peroxidase. It is collected at its peak of activity in early morning, dried and powdered for use in coagulating milk to make cheese and junket. From it can be isolated the protein-digesting enzyme *ficin,* which is used for tenderizing meat, rendering fat, and clarifying beverages. In tropical America, the latex is often used for washing dishes, pots and pans. It was an ingredient in some of the early commercial detergents for household use but was abandoned after many reports of irritated or inflamed hands in housewives.

***Health Benefits of Figs:*** A few of the health benefits derived from figs include…

**Prevention of constipation:** There are 5 grams of fiber in every three-fig serving. That high concentration of fiber helps promote healthy, regular bowel function and prevents constipation. Fiber works to add bulk and mass to bowel movements, so it not only prevents constipation, but also eliminates diarrhea and unhealthy or irregular bowel movements.

**Weight loss:** The fiber in figs also helps to reduce weight and is often recommended for obese people. However, their high calorie count can also result in weight gain, especially when consumed with milk. A few figs are enough to get the recommended amount of nutrients.

**Lower cholesterol:** Figs contain pectin, which is a soluble fiber. When fiber moves through the digestive system, it basically mops up excess clumps of cholesterol and carries them to the excretory system to be eliminated from the body. As a soluble fiber, pectin from figs also stimulates healthy bowel movements. Figs can have a laxative effect, as they are one of the most fiber-dense foods available. High amounts of fiber in your diet can benefit your overall health by preventing certain types of abdominal cancer, as well as colon cancer.

**Prevention of coronary heart disease:** Dried figs contain phenol, Omega-3 and Omega-6. These fatty acids reduce the risk of coronary heart disease. Furthermore, the leaves of figs have a significant effect on the level of triglycerides in a person's system. Fig leaves have an inhibitory effect on triglycerides, and make the overall number of triglycerides drop. Triglycerides are another major factor behind various heart diseases.

**Prevention of colon cancer:** The presence of fiber helps to stimulate the elimination of free radicals and other cancer causing substances, particularly in the colon, since fiber increases the healthy movement of the bowels.

**Protection against post-menopausal breast cancer:** Fiber content in figs have been known to protect against breast cancer, and after menopause, the hormonal balance in women can often fluctuate. The body's systems are so interconnected that hormones affect the immune system, which in turn affect the ability of antioxidants to fight free radicals. Free radicals are prime factors behind the development of cancer, so figs take care of one extra line of defense by providing its wealth of fiber.

**Good for diabetic patients:** The American Diabetes Association recommends figs as a high fiber treat that helps promote functional control of diabetes. Fig leaves reduce the amount of insulin needed by diabetic patients who have to regularly take insulin injections. Figs are rich in potassium, which helps to regulate the amount of sugar which is absorbed into the body after meals. Large amounts of potassium can ensure that blood sugar spikes and falls are much less frequent, so figs can help diabetics live a much more normal life.

**Prevention of hypertension:** People usually take in sodium in the form of salt, but low potassium and high sodium level may lead to hypertension. Figs are high in potassium and low in sodium, so they are a perfect defense against the appearance and effects of hypertension, making figs a relaxing food as well, which can settle the nerves and bring some calmness to your day.

**Bronchitis:** The natural chemicals in fig leaves make it an ideal component for a tea base. Fig leaf tea has been popularly prescribed for various respiratory conditions like bronchitis, and it is also used as a way to prevent and lessen the symptoms of asthmatic patients.

**Venereal Diseases:** Figs have been traditionally used in the Indian subcontinent and a few other areas of the world as a calming salve for venereal diseases. Ingestion or topical application both work for relief from sexually transmitted diseases, although further research needs to be done.

**Sexual Dysfunction:** For centuries, figs have been recommended as a way to correct sexual dysfunction like sterility, endurance, or erectile dysfunction. It has been a major part of mythology and culture, and most of the time; it is referred as a powerful fertility or sexual supplement. Its actual success as an aphrodisiac is questionable, but the huge amount of valuable vitamins and minerals might result in the sudden boost in energy and stamina that people mistake for a sexual surge. Soak 2-3 figs in milk overnight and eat them in the morning to enhance your sexual abilities.

**Strengthens Bones:** Figs are rich in calcium, which is one of the most important components in strengthening bones, and reducing the risk of osteoporosis. It is also rich in phosphorus, which encourages bone formation and spurs re-growth if there is any damage or degradation to bones.

**Urinary calcium loss:** People that maintain a high-sodium diet may be affected by increased urinary calcium loss. The high potassium content in figs helps to avoid that condition and regulates the content of waste in your urine. It minimizes the calcium you lose, while increasing the amount of uric acid and other harmful toxins, which you want to get out of your body

**Prevention of macular degeneration:** Vision loss in older people is normally due to macular degeneration. Fruits and figs are particularly good at helping you avoid this very common symptom of aging.

**Relief for throat pain:** The high mucilage content in figs helps to heal and protect sore throats. The soothing nature of figs and their natural juices can relieve pain and stress on the vocal chords.

**Precaution:** It is possible to have too much of a good thing and eating too many figs can cause diarrhea. Furthermore, dried figs are high in sugar and can potentially cause tooth decay. Also, there are those who are allergic to figs, or certain chemical components within them, and the resulting allergic reactions can be mild to severe. As always, before making a major change in your behavioral patterns or lifestyle, speak with your doctor or usual medical professional.Finally, it is best not to consume too many figs in the week or two leading up to a surgery, because it can occasionally cause bleeding in the digestive tract in sensitive individuals.

## ORIGIN, HISTORY AND DISTRIBUTION

The common fig (*Ficus carica* L.) belongs to the Eusyce section of the family Moraceae, with over 1,400 species classified into about 40 genera (Watson and Dallwitz, 2004). The genus *Ficus*, comprised of about 700 species, is found mainly in the tropics and is currently classified into six subgenera, which are characterized by a particular reproductive system (Berg, 2003). The fig is indigenous to Persia, Asia Minor, and Syria and currently grown in most of the Mediterranean countries (Datwyler and Weiblen, 2004). Remnants of figs have been found in excavations of sites traced to at least 5,000 B.C. While the ancient history of the fig centers around the Mediterranean region, and it is most commonly cultivated in mild-temperate climates. It nevertheless has its place in tropical and subtropical horticulture. As time went on, the fig-growing territory stretched from Afghanistan to southern Germany and the Canary Islands. Pliny was aware of 29 types. Figs were introduced into England sometime between 1525 and 1548. It is not clear when the common fig entered China but by 1550 it was reliably reported to be in Chinese gardens. European types were taken to China, Japan, India, South Africa and Australia (Condit, 1947).

The first figs in the New World were planted in Mexico in 1560. Many special varieties were received from Europe and the eastern United States where the fig reached Virginia in 1669. Later, figs were introduced into California when the San Diego Mission was established in 1769. The Smyrna fig was brought to California in 1881-82 but it was not until 1900 that the wasp was introduced to serve as the pollinating agent and make commercial fig culture possible. From

Virginia, fig culture spread to the Carolinas, Georgia, Florida, Alabama, Mississippi, Louisiana and Texas. The tree was planted in Bermuda in early times and was common around Bahamian plantations in Colonial days. It became a familiar dooryard plant in the West Indies, and at medium and low altitudes in Central America and northern South America. There are fair-sized plantations on mountain sides of Honduras and at low elevations on the Pacific side of Costa Rica. From Florida to northern South America and in India only the common fig is grown. Chile and Argentina grow the types suited to cooler zones (Condit, 1947).

The tree is known almost universally simply as fig, common fig, or edible fig. The name is very similar in French (figue), German (feige), and Italian and Portuguese (figo). In Spanish it is higo or brevo. In Marathi it is called by a famous name as *Anjir*. Haitians coined the name figue in France, to distinguish it from the small, dried bananas called "figs" (Condit, 1947).

Figs were widespread in ancient Greece, and their cultivation was described by both Aristotle and Theophrastus. Aristotle noted that as in animal sexes, figs have individuals of two kinds, one (the cultivated fig) that bears fruit, and one (the wild caprifig) that assists the other to bear fruit. Further, Aristotle recorded that the fruits of the wild fig contain *psenes* (fig wasps); these begin life as larvae, and the adult *psen* splits it's "skin" (pupa) and flies out of the fig to find and enter a cultivated fig, saving it from dropping. Theophrastus observed that just as date palms have male and female flowers, and that farmers (from the East) help by scattering "dust" from the male on to the female, and as a male fish releases his milt over the female's eggs, so Greek farmers tie wild figs to cultivated trees. They do not say directly that figs reproduce sexually (Flaishman *et al.*, 2008).

The Food and Agriculture Organization (Anon, 2006) estimated that figs are harvested from 4,27,000 hectares worldwide, producing yearly over one million metric tones of figs around the world, with Turkey, Egypt, Iran, Greece, Algeria, Morocco, the United States, Syria and Spain producing 70 per cent of the crop and Turkey alone producing nearly 25 per cent of the total. The top three exporters of dried figs in the world are Turkey, Iran, and Greece. Turkey, the largest producer, supplies more than half of world export volume while Iran accounts for 12 and Greece for 5 per cent.

The economic importance of fig production is likely to continue into the future. In the world market, there is an increasing demand for fresh figs and a stable demand for dried figs. The most important trade aspects of this species are the short commercial life of the fresh fruits, and for the dried fruits, the market competition of Turkish production, where production costs are lower than for other countries (Europe).

In India, fig cultivation is mostly confined to western part of Maharashtra, Gujarat, Uttar Pradesh (Lucknow and Saharanpur), Karnataka (Bellary, Chitradurga and Srirangapatnam) and Tamil Nadu (Chadha, 2010). Among these states, Maharashtra is the leading state with an area of 1332 ha, production of 8219 metric tones and productivity of 6.17 tones/ha/year. In Maharashtra fig is grown in Pune, Ahmednagar and Aurangabad districts, wherein Pune is the leading producer of fig with an area of 1052 ha with production of about 7373 metric tons and productivity of 7.0 tons/ha/year (Anon, 2011). In Pune district, fig crop is mostly confined to Purandar and Haveli tahsils.

## TAXONOMICAL AND BOTANICAL DESCRIPTION

**Scientific classification**

| | |
|---|---|
| Kingdom | : Plantae |
| (Unranked) | : Angiosperms |
| (Unranked) | : Eudicots |
| (Unranked) | : Rosids |
| *Family* | : *Moraceae* |
| Tribe | : Ficeae |
| Genus | : *Ficus* |
| Sub-genus | : *Ficus* |
| Species | : *Ficus carica* L. |
| Common Names | : Fig (English), *Higo* (Spanish), *Figue* (French), *Feige* (German), *Fico* (Italian), *Anjir* (marathi). |
| Related Species | : Cluster fig (*Ficus racemosa*), Sycomore fig (*Ficus sycomorus*). |
| Distant Affinity | : Mulberry (*Morus* spp.); Bread fruit (*Artocarpus altilis* Fosb.); Jakfruit (*Artocarpus heterophyllus* Lam.) Che; Chinese Mulberry (*Cudrania tricuspidata*). |

## Botanical Description

**Growth habit:** The fig is a pleasing deciduous tree, up to 50 ft tall, but more typically to a height of 10 - 30 ft (3-9 m) with numerous branches that are muscular and twisting, spreading wider than they are tall. The trunk is rarely more than 7 inch (17.5 cm) in diameter. It contains copious milky latex that is irritating to human skin. Fig wood is weak and decays rapidly. The trunk often bears large nodal tumors, where branches have been shed or removed. The twigs are pithy rather than woody. Fig trees often grow as a multiple-branched shrub, especially where subjected to frequent frost damage.

**Root system:** The root system is typically shallow and spreading, sometimes covering 50 ft (15 m) of ground, but in permeable soil some of the roots may descend to 20 ft (6 m).

**Foliage:** The deciduous leaves are palmate, deeply divided into 3 to 7 main lobes, these more shallowly lobed and irregularly toothed on the margins. Fig leaves are bright green, single, alternate and large, measuring 10 to 12 inch (up to 1 ft) in length and width. They are rough, hairy on the upper surface and soft hairy on the underside. In summer their foliage lends a beautiful tropical feeling.

**Flowers:** The tiny flowers of the fig are out of sight, clustered inside the green "fruits", technically a synconium. In the case of the common fig, the flowers are all female and need no pollination. There are 3 other types:

a. The ''Caprifig", which has male and female flowers requiring visits by a tiny wasp, *Blastophaga grossorum*;

b. The "Smyrna" fig, needing cross pollination by Caprifigs in order to develop normally; and

c. The "San Pedro" fig which is intermediate, its first crop independent like the common fig, its second crop dependent on pollination.

Pollinating insects gain access to the flowers through an opening at the apex of the synconium.

**Fruits:** A "fruit" is technically a synconium, that is, a fleshy, hollow receptacle with a small opening at the apex partly closed by small scales. It may be obovoid, turbinate, or pear-shaped, 1 to 4 inch (2.5-10 cm) long. The matured "fruit" has a tough peel that varies in color from yellowish-green to coppery, bronze, or dark-purple, often cracking upon ripening, and exposing the pulp beneath. The interior is a white inner rind containing a seed mass bound with jelly-like flesh. The edible seeds may be large, medium, small or minute and range in number from 30 to 1,600 per fruit and generally hollow, unless pollinated. Pollinated seeds provide the characteristic nutty taste of dried figs. The common fig bears a first crop, called the breba crop in the spring or *Khatta bahar* in rainy season (in India) on last season's growth. The second crop is borne on the new growth and is known as the main crop or *Mitha bahar*.

## CLIMATIC AND SOIL ADAPTABILITY

### Climatic Adaptability

Fig is basically a subtropical crop that favours area having arid or semi arid environment with high summer temperature, plenty of sunshine and moderate winter. The plant can with stand high limit for temperature than for the lower.

Although plants can survive temperature as high as 45°C, the fruit quality deteriorates beyond 39°C. Mature trees can withstand temperature up to 4°C, but young ones need protection. However, deciduous nature of fig allows the plant to resist temperature as low as 10°C, when in dormancy. In mild climate, plants remain evergreen, lack well defined flowering and fruiting season, and sometimes produce long barren limbs.

Climate plays very important role in bearing of fruits and their size, shape colour of skin and pulp. A relatively cool climate stimulates production of larger and elongated fruits. Climatic conditions during fruit development considerably influence the fruit quality. Very high temperatures (>39°C) induce premature fruit ripening. High humidity results in fruit splitting, while hot breeze during ripening leads to sweet but small fruits.

In southern India, 'Marseilles' flourishes on hills above 5,000 ft (1,525 m). In tropical areas generally, figs thrive between 2,600 and 5,900 ft (800-1,800 m). The tree can tolerate 10° to 20° of frost in favourable sites. It should have a dry climate with light early rains if it is intended for the production of fresh fruit. Rains during fruit development and ripening are detrimental to the crop, causing the fruits to split. The semi arid tropical and subtropical regions of the world are ideal for fig-growing if means of irrigation are available. But very hot, dry spells will cause fruit-drop even if the trees are irrigated. In general, climate rather than soil is a limiting factor for its cultivation.

**Effect of climate on vegetative growth and development:** Fig growth and production are strongly dependent on climatic conditions. Generally, fig grows and produces high-quality fruit in Mediterranean and dryer warm temperate climates. The decrease of temperature in the cold winter conditions, and the growth temperature and rain all affect tree growth and crop production. The lower winter temperatures between November and mid February slow down fruit maturation at this period, while the rise of temperature at the end of March leads to resumed growth and early fruit maturation (Flaishman and Al Hadi, 2002). Fig tree has limited requirements for chilling units, and the length of the dormant period depends on the local climatic conditions (Erez and Shulman, 1982).

**Effect of climate on reproductive growth and development:** In fig growing areas of India, the minimum temperature does not go below 7-10°C, which is not affecting the shoot growth considerably like that of California where there is drop of temperatures in autumn that arrests shoot growth, and, as a result, a typical terminal bud develops. This process affects the late autumn crop production. In areas with night temperatures above 12°C, such as the India, Imperial Valley in California and in the coastal area in Israel, fig trees produce fruits in November and December and fruit maturation will continue until leaf fall (Flaishman and Al 1-ladi, 2002). Climate markedly affects the size, shape, and skin and pulp colour of figs (Condit, 1947). Cooler climates produce greener, as opposed to yellow skins, more vivid pulp colors, and larger, more elongated fruits. Crane (1986) has suggested that the larger individual size of first breba crop, which competes with shoot growth and second crop for available carbohydrates, is due to its development during a cooler period. In addition, Crane (1986) suggested that climate may also affect pollination requirement.

Other environmental conditions such as rain, hail, and wind can reduce fruit quality and production. Rain may cause fruit splits. Splitting is the result of sudden changed in the internal fruit pressure brought on by cool temperatures and/or high humidity as the fruit matures (Freguson *et al.*, 1990). Splitting in 'Calimyrna' and other varieties may also result from excessive pollination and the growth of too many developing heeds during fruit ripening.

**Season:**Fig trees usually bear two crops a year, the early season ("breba" or "*Khatta bahar*") fruits being inferior and frequently too acidic. Only the fruits of the second, or main ("*Mitha bahar*"), crop are of actual value. In India, Colombia and Venezuela, some fruits are borne throughout the year. But there are two principal crops, one in May and June and the other in December and January in Colombia and Venezuela. In India also two crops are taken as *Khatta bahar* in winter (September to January) and main i.e. *Mitha bahar* in summer season (February to May).

## Soil Adaptability

The fig can be grown on a wide range of soils ranging from light sandy, rich loam, heavy clay or limestone, providing there is sufficient depth and good drainage. However, medium to heavy, calcareous well drained, deep (about 1m) soil is ideally suited for fig cultivation. Deep soils encourage better root establishment and thus produces good plant canopy and yield. Sandy soil that is medium-dry and contains a good deal of lime is preferred when the crop is intended for drying. The pH should be between 6.0 and 6.5, but can grow well up to 8 pH. Highly acid soils are unsuitable. The crop can tolerate drought, salts (chlorides and sulphates) but is sensitive to sodium carbonate and boron salts.

## RECOMMENDED AND POPULAR CULTIVARS

There are many cultivated varieties in each class of figs. About 700 varieties of fig are known to be grown in the world. Varieties vary for different attributes like vegetative vigour, pollination requirement, yield, fruit size, shape, skin color, pulp quality and colour. Large sized figs belong to 'common fig 'group. The most popular varieties of fig are given in following Table 2.

**Table 2:** Horticultural classification of fig varieties

| Type | Popular varieties | Flower type | Mode of pollination | No.of crops | Listed varieties | Other features |
|---|---|---|---|---|---|---|
| Edible fig | Poona fig, Conardia, Mission, Kadota, Brown Turkey | Long Styled Pistillate Flowers | Fruit develops parthenocarpically | 1 | 470 | Seeds are hollow without inner kernels and the embryo. Some varieties produce a small breba or first crop in addition to main or second crop. |
| Smyrna (lopinjir) | Calimyrna(sari lop), Zidi, Taranimt | Long Styled Pistillate Flowers | Female wasps emerging from the spring caprifig enter Smyrna fig for oviposition and in the process effect pollination. | 1 | 116 | Originated from the caprifig. The fertile seeds contribute to the excellent fruit quality. |
| San pedro | King, Gentile, San Pedro, Dauphine, Lampeiria | Long Styled Pisitillate | First (breba) crop Fruit develops without pollination but not second (main) crop | 2 | 21 | Commercially not very important, some white large fruited types are grown in mediterranean countries for drying |
| Wild fig Caprifig (male or goat fig) | Roading 3, Samson, Stanford, Brawley | Flowers short styled, pistillateflowers and functional staminate flowers near the ostiole | Self fertile(persistent) syconia | 3 | 20 | A primitive type. Fruits have almost no edible value, but serves as and abode for fig wasp and Smyrna and San Pedro figs |

(*Source:* Chadha, 2010)

Poona fig is most popular cultivar grown in India. In addition, the varieties *viz*, Bangalore, Bellary, Coimbatore, Daulatabad, Dindigul, Ganjam, Hindupur, Lucknow and Saharanpur, have clearly acquired the name from the location in which they are cultivated. Most of them resemble in plant and fruit morphology to that of Poona fig. Possibly these are either clones or ecotypes and hardly they warrant varietal status. Black Ischia, Shahi, Maisram and Brown Turkey have not achieved prominence. Dinker, an improvement over Daultabad for yield and fruit quality, is gaining commercial significance.

Some well known fig hybrids from California have performed well in India in comparison to Poona fig under Bangalore conditions. They produce fruits parthenocarpically. Excel and Conardia figs that develop smaller canopies are suitable for high density planting. The fruits do not split like Poona and Conardia fig. Conardia, Excel and Deanna are good for drying, canning and table purposes, respectively.

**Caprifigs:** Caprifigs only produce male flowers and never bear fruit. Their only purpose is to pollinate female fig trees.

**Smyrna:** Smyrna figs bear all female flowers. They have to be pollinated by a Caprifig.

**San Pedro:** San Pedro figs bear two crops, one on leafless mature wood that requires no pollination and one on new wood that requires pollination by a male flower.

**Common figs:** Common figs are the type usually grown in home landscapes. They don't need another tree for pollination. Figs that require pollination have an opening that allows the pollinating wasps' entry to the internal flowers. Common figs don't need an opening, so they are less susceptible to rot caused by insects and rainwater entering the fruit.

One of the best ways to find a variety suitable to your area is to visit a local nursery. They will carry fig types suitable for your climate and can make recommendations based on local experience.

Here are some different types of figs in the common group that perform well in the fields.

**Poona fig:** Fruits are bell-shaped, of medium size, weighing about 42 g; thin-skinned; light-purple with red flesh, sweet, and good flavour. The pulp is pinkish or light red. Cultivar is suitable for both i.e., *Khatta bahar* and *Mitha bahar*. The leaves are very broad without knotches on leaf margin.

Poona fig

**Dinkar:** It is a selection from Daulatabad fig released by Dr. Vasantrao Naik Marathwada Agricultural University, Parbhani, Maharashtra. Fruits are bigger in size. Average fruit weight is 80-90 g with brix of 18-20$^0$ and average yield of 60-70 kg/plant.

Dinkar

**Deanna:** It produces large green figs that ripen to yellow or greenish yellow. The strawberry to amber flesh is considered to be of very good quality. The small eye helps to minimize spoiling. It was developed to replace Calimyrna and it also has good cold hardiness. Fig trees do best where they get at least 8 hours of direct sun per day. Once well established they are fairly drought tolerant but extended dry periods can cause leaf and fruit drop as well as early dormancy.

Deanna

**Excel:** One of the best all round white figs. It is very hardy and adapted to a wide range of climates. Amber flesh is very sweet and full flavored. Very Resistant to cracking, medium closed eye. Great for all fig purposes. Ideal in containers.

Excel

(See colour version on page 316 )

**Conadria:** It produces large green, thin-skinned figs that ripen to light greenish yellow and droop when ripe. The flesh is white and the pulp can range from red to amber with a rich, sweet, and distinctive flavor. The large figs are good for fresh eating, drying, and preservation. The trees grow to average size and are considered to be long lived and are highly productive producing large crops. It is self-pollinating and is resistant to spoiling and splitting.

Conadria

**Brown Turkey:** The fruits are broad-pyriform, usually without neck; medium to large and copper-coloured. The pulp is whitish shading to pink or light red; of good to very good quality and with few seeds. The main crop, beginning in mid-July is large, while the early breba, crop is small. This cultivar is well adapted to warm climates. It is grown largely on all the islands of Hawaii. It produces a crop of large, tasty figs over a long season. The fruit has attractive flesh and few seeds. The tree is prolific.

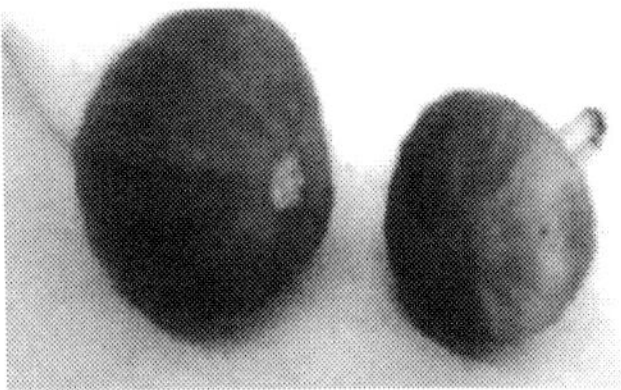

Brown Turkey

**Black Ischia:** It is an Italian variety. The tree is particularly ornamental; the leaves are glossy, only shallowly 3 lobed and a heavy bearer. Main crop has elongated pear shaped fruits with many noticeable ribs with short neck and short to medium stalk. Fruits are large, 2.$^1/_2$ inch (6.35 cm) long and 1 1/2 inch (3.8 cm) wide; dark purple-black except at the apex where it is lighter and greenish. There are many golden flecks; skin is wholly coated with thin, dark-blue bloom and eye open with red-violet scales. Pulp is violet-red of good quality.

Black Ischia

**Phule Rajewadi (JWF 6):** It is a selection from Poona fig. Released by Mahatma Phule Krishi Vidyapeeth, Rahuri, Maharashtra in 2015. It has bigger sized fruits and suitable for *Khatta* and *Mitha* bahar. Attractive purple colour with thick skin fruits. Pinkish red pulp colour suitable for processing. Moderately resistant to rust disease and major pests under field conditions. Average fruit weight is 65-70 g with brix of 18-20$^0$ and average yield of 80-85 kg/plant

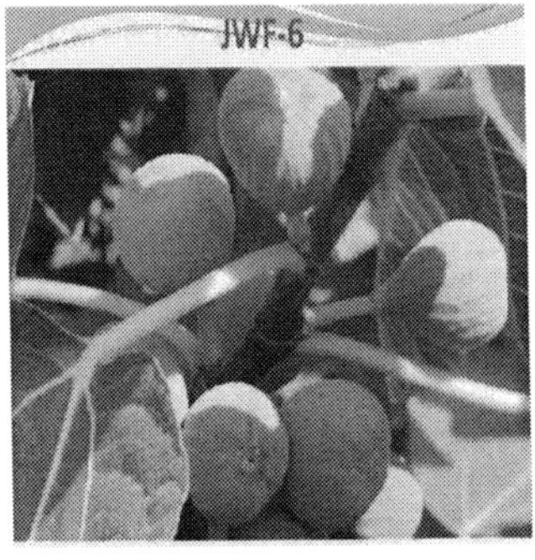


Phule Rajewadi

(See colour version on page 316 )

**Fegra fig** (*Ficus palmata*): This is small fruited fig of excellent taste that grows wild in the mid-hill region of the Western Himalayas. It is on a par with the cultivated figs in taste and flavour, however, size is rather small in this case. Fegra plants are of very common occurrence at places up to 1,550 metres above the sea-level. These trees are rarely found in the forests, but grow around the villages, in wastelands, fields, etc. The fruits are liked very much by the people. They are also offered for sale.

Fegra fig (*Source:* Parmar, 2013)

## PROPAGATION TECHNOLOGY

Although it is possible to propagate fig from seeds, layers, grafts and tissue culture plants and commercially cuttings are used for multiplication.

**Cuttings:** About 25 cm long cuttings having 3-6 nodes and 1/2 to 3/4 inch (1.25 - 2 cm) thick are usually made from mature wood of previous season. They are planted in moist sand either in seed pans or in nursery beds. Planting must be done within 24 hours. First, the upper slanting end of the cutting should be treated with a fungicide to protect it from disease, and the lower, flat, end with a root-promoting hormone. This can be taken up during pruning or just after the onset of monsoon. The cuttings are raised in shade with regular watering. After about 75 days, they are transplanted to polythene bags containing garden soil, sand and farmyard manure (1:1:1), and planted in the field about 4-6 months later. Trees of unsatisfactory varieties can be top worked by shield- or patch-budding, or cleft- or bark-grafting.

As per Chadha (2010) higher success can be achieved by

i. Using cuttings with short internodes, collected from basal portion of the shoots located in the lower part of the crown

ii. Storing of cuttings in moist sawdust or sphagnum moss for about 4 weeks at room temperature

iii. Treating with growth regulators like IBA

iv. Pre girdling at the base of canes (removing 2.5 cm width bark) a month prior to taking cuttings

v. Planting cuttings in a slightly slanting ($80^{o}$) position.

**Other Vegetative Methods:** Side grafting on *F. glomerata* and *F. palmata* may be adopted for circumventing nematode problem in the soil. Brown Turkey, as rootstock, imparts vegetative vigour in fig Excel, Conardia and Deanna. Shield or patch budding, cleft or bark grafting enables to top work a desirable genotype on established but inferior tree. Protocols are now available for micro propagation of fig shoot tips.

## PLANT AND FRUIT PHYSIOLOGY

**Vegetative morphology and development:** The fig is an unusual tree as it may produce multiple crops of fruits each year, and certain fig types need pollen from their pollinator caprifigs. The breba crop (*Khatta bahar*), which is not produced in all cultivar, is borne laterally on the growth of the previous season from buds produced in leaf axils. These buds develop in the following rainy season, and the fruit matures between September to January. The main crop of figs is produced laterally in the axils of leaves on shoots of the current season. Fruit maturation starts in February and last until May. At the end of the growth period, most of the leaves fall. Reproductive buds that do not produce fruit during the growing season remain dormant to give rise to the first rainy season *Khatta bahar* crop. In some cultivars and in appropriate environments, largely developed main crop figs remain on the tree during the winter and complete development in summer.

**Root system:** The fig tree has a fibrous root system that spreads up to three times the diameter of the canopy and is naturally very shallow and without a taproot. Once plants are established, they are relatively drought tolerant, probably due to their very extensive and widespread root system. The passionate ease of rooting figs has made its cultivation easy over thousands of years and is routinely used to establish new orchards from cuttings. Fig plants are fairly tolerant of poor soil and moderate salinity.

**Shoot and leaf systems:** Histological examination of the terminal bud in the spring shows that the apical meristem elongates to produce lateral outgrowth, which is the meristems of scales, leaves, inflorescence, and lateral vegetative buds (Crane 1986). Each terminal bud generally contains four or five primordial leaves flanked on either side by a scale. One vegetative and two inflorescences primordia are present toward the base of the bud. A primordium intends to become vegetative that has three or four scales that are laid down to cover the bud axis. The vegetative primordia continue during tree growth to initiate scales and leave. As the bud elongates, the cover scales abscise, and the apical meristem develops into a shoot that produces leaves and new inflorescences. Fig trees vary in their growth habits, ranging from open and drooping to upright and compact. Fig growth habit is characteristic of the cultivar. Individual trees in a favorable environment often reach large size, while fig trees in orchards are usually more compact. Fig has typical bright green, single, alternate and large leaves. Leaf characters are quite stable and serve as an important parameter in cultivar identification (Ferguson *et al.,*1990). They start to develop in the early spring and will continue to form of new leaves until temperature drops in autumn. Toward the end of the growing season, environmental conditions such as low temperature, photoperiod, wind, and water stress cause leaf fall.

**Latex cells:** Fig trees and fruit contain characteristic latex secreting cells, producing milky exudates that attribute to all cultivars. Latex is the cytoplasmic fluid of lactiferous tissues that contain the usual organelles of plant cells such as nucleus, mitochondria, vacuoles; ribosomes, and Golgi apparatus. Rubber (cis-1,4-polyisoprene) is produced in latex at the expense of high energy cost and is considered a secondary metabolite. Although it is not fully understood why plants produce rubber, it has been suggested that latex secretion is a defense against mechanical wounding and/or herbivores such as insects, vertebrates, microorganisms, and fungi. This view of latex as a defense system is based on the observations that latex contains a variety of defense-related proteins.

Recently Flaishman *et al.* (2008) isolated two major rubber particle proteins in *F. carica.* Kim *et al.* (2003) identified three rubber particles and latex genes from *F. carica* that includes trypsin inhibitor, chitinase, and peroxidase. Peroxidase is widely known to participate in a variety of plant defense mechanisms in which hydrogen peroxide is often supplied by an oxidative burst (Lamb and Dixon, 1997). The transcript level of peroxidase in *F. carica* was increased following treatment with various types of abiotic stresses or hormones, including wounding, drought, jasmonic acid, and abscisic acid. Also, the transcript level of the trypsin inhibitor was increased remarkably by wounding treatment and slightly by jasmonic acid treatment. In leaves, the expression of chitinase was remarkably induced by wounding or jasmonic acid treatment. The presence and expression of stress-related genes on the surface of rubber particles and latex in *F. carica* further support a possible role of rubber particle sand latex in defense mechanisms in this species. The identification and characterization of the three latex fig genes could be further used to investigate the physiological and biochemical traits of the fig tree cultivated in temperate zones.

**Reproductive development:** Primordia intended to develop into reproductive or vegetative organs that are identical at their initial stage of development. Generally, there are four scales laid down, which cover the bud axis. Following this stage, the vegetative meristem continues to initiate scales and leaves, unlike the reproductive meristem, which broadens and elongates and then begins the initiation of ostiolar scales. Hence, the first visual microscopic evidence of inflorescence differentiation is the elongation of the axis and the initiation of scales that eventually surround the ostiole of the syconium. Further development of the inflorescence primordial consists of continued broadening of the apex and cell division around the periphery, giving rise to many ostiolar scales forming a cup shaped structure lined with floral primordia. The formation of the syconium is complete when the apical portion of the cup shaped structure grows toward the center and forms an ostiole, which is partly closed with numerous scales. *F. carica* is gynodioecious, bearing either hermaphroditic or "female" figs on separate plants (Crane and Brown, 1950; Crane and Baker, 1953 and Crane,1986)

**Fruit growth and development:** The terminal bud unfolds and growth occurs and the fig fruits are borne in the axils of the leaves. Two inflorescences and one vegetative bud are present at the same lateral position in the leaf axils. In some cultivars (e.g., Mission and Brown Turkey), usually only one inflorescence develops into a syconium, while in few cultivars (e.g. Kadota and Calimyrna), often both inflorescences at a node may develop. Similar to other fruits, fig syconium development has three defined growth periods represented by a double signioid curve. The first period of growth - stage I - is characterized by a very rapid diameter increase and slower rate of fresh and dry weight, with almost no change in sugar accumulation. The second period stage II is a quiescence stage that is marked with almost no change in fruit diameter, dry and fresh weights, and sugar content. Stage III is characterized by accelerated rate of increase in diameter, in fresh and dry weights, in water as well as in sugar content. During this phase of growth, over 70 per cent of the total dry weight and 90 per cent of the total sugar content is accumulated in the fruit. Dramatic pigment changes occur during this period in many dark cultivars as chlorophyll content in the fruit skin decreases rapidly and the fruit skin turns from green to bluish black. In addition, fruit size increases and tissue softening occurs during the last stage of fig fruit development. Since the inflorescence buds begin developing as associated leaves emerge along the branch, fruit maturation is sequential, beginning with the basal fruit and progressing toward the branch apex, and harvest can last for a long period.

Different fig cultivars can set fruit with or without pollination. A consistent difference in nitrate levels has been detected in persistent versus non persistent fig cultivars. The average nitrate content of persistent figs is triple that of non persistent ones during stages I and II of main crop figs. By stage III nitrate is not found in non persistent fruit. Crane (1986) showed that indole acetic acid (IAA) is inhibited as nitrate levels rise and suggested that the reason nitrate levels differ so greatly in persistent versus non persistent figs has to do with the regulation of indole acetic acid oxidase. Therefore, persistent cultivars are expected to have higher auxin levels. Indeeed, auxin application was found to stimulate fruit set in non persistent Smyrna-type figs (Crane and Overbeek,

1965 and Crane, 1986). Various applied growth regulators, including auxins, gibberellins, and cytokinins can induce persistence in the Smyrna-type 'Calimyrna' cultivar (Crane, 1986). The maturation process of auxin-induced persistent fruits is somewhat longer than that of caprified ones, but their morphology is similar.

Common fig cultivars are facultatively persistent and may produce both persistent and pollinated main crop figs. Morphologically, the pollinated fig syconium creates true fruits, while the non pollinated fig syconium presents an enlarged inflorescence with multiple long- styled pistillate flowers. The number of crops produced by fig trees directly influences its carbohydrate balance. Smyrna type figs, producing a single main crop, have maximum starch concentration at early spring, midsummer, and late fall. The sugar content of fruits decreases in spring and simultaneously starch concentration occur when shoot and breba fruit initiate in March. Shortage in available carbohydrates and competition between the new foliage and the breba syconium can cause syconium drop and elimination of the breba crop (Crane, 1986).

**Fruit maturation:** The fig is a highly perishable climacteric fruit subject to rapid physiological breakdown. The postharvest life of the fruit is considered to range from 7 to 10 days even when stored at low temperatures. Profound cell wall modification processes occur within the tissues during maturation. Basic studies on processes that occur during ripening are essential for studying systems in which the biological and physiological processes linked to maturation are involved in postharvest deterioration. Application of ethylene to fig fruits during late stage II of their development stimulates growth and ripening. In mature and ripe fig fruit, the receptacle tissue and the pulpy tissue of the drupelets within it are clearly distinct. Therefore, analysis of both tissues as a single mixture may obscure some cell wall changes that are crucial in understanding the cell wall modification processes during ripening.

Characteristic changes in cell wall polysaccharides taking place within the distinct and separate tissues of the receptacle and the pulpy drupelets during sequential ripening in fig fruit. The pectic extracts had high uronic acid contents in addition to high amounts of neutral sugars. At the fig-ripening onset, the amounts of both uronic acid and total sugars were more pronounced in the drupelets than in the receptacle. The data suggest that even though quantitative and qualitative changes in cell wall polysaccharides occur during ripening in both tissues, qualitative variations between tissues occur only in the pectic polymers, not in the hemicellulosic polymers.

## PLANING AND PLANTING

**Planting;**The best time for planting is the onset of the rainy season. Cuttings are raised in nursery beds and are set out in the field after 12 or 15 months. The layout for planting can be either square or hexagonal system. The square system is more common and desirable. They may be spaced from 6 to 25 ft (1.8-7.5 m) apart depending on the cultivar and the fertility of the soil. A spacing of 13 x 13 ft (4x4 m) allows 260 trees/acre (625 trees/ha). The recommended spacing for Poona fig is 5 x 5 m (400 plants /ha). It is 2.5 x 2.5 m (1,600 plants/ha) for Excel and Conardia. The Pits of 60 x 60 x 60 cm size are dug and exposed to sun for about 15 days, and then filled with a mixture of compost, top soil and sand (1:1:1) and 2 kg of neem or castor cake/ pit.

Planting can preferably be taken up on an overcast day. When grafts are used the graft joint should remain above the ground level. Once the tree is planted the soil around the plant should be tamped firmly. Water is applied immediately after planting. Fruiting will commence in less than a year from planting out. Young plants need shading until they are well established. If there are heavy rains, drainage ditches should be dug to prevent water-logging. Fig trees remain productive up to 12 or 15 years of age and thereafter the crop declines though the trees may live to a very advanced age.

**Aftercare:**After the plants are set in the field, regular watering is essential until they are well established. A basin of 60 cm diameter should be prepared around the plant and is widened as the canopy size increases. Basin cleaning is taken up regularly to keep it weed free. The side shoots and suckers should be removed as and when they emerge.

Maintenance of weed free orchard is very important. During early years of orchard, raising green manure or intercrop is recommended. Green manure sun hemp suppresses the weed growth and augments the supply of organic matter in the soil. Intercropping vegetables and legumes is beneficial.

## SOIL CULTURAL PRACTICES TECHNOLOGY

### Water Need

Young fig trees should be watered regularly until fully established. Though, fig plants can sustain heat and drought, the commercial fig production is possible only if plants are timely irrigated. Adequately irrigated plants produce better shoot growth and higher yields of superior quality fruits. Loose and sandy soils require larger quantities of water than heavy soil. Either drip or flood irrigation can be practiced. The drip irrigation minimizes water requirement and allows fertilizer application through irrigation water. Desert gardeners may have to water more frequently. Mulch the soil around the trees to conserve moisture. If a tree is not getting enough water, the leaves will turn yellow and drop. Also, drought-stressed trees will not produce fruit and are more susceptible to nematode damage.

The frequency of irrigation depends on the soil type and weather. Flood irrigation may be given at every 10 days in summer months. Excessive irrigation during fruit development should be avoided, because it leads the terminal buds to initiate growth at the expense of fruit development. Also excessive irrigation or heavy rains during ripening result in fruit cracking and production of insipid fruits. The fruit development is affected, resulting in small and hard fruits due to the absence of adequate and regular irrigation. The irrigation may be reduced after harvesting of fruits is completed and thereafter, regular schedule be resumed after pruning. If drip irrigation is adopted, 15-20 liters of water/ day/ plant may be supplied. The thumb rule is to replenish 50 per cent of pan evaporation losses.

## Nutritional Need

Nutrition plays very important role in the quantity and quality of fruit yield. Regular fertilizing of figs is usually necessary when they are grown on sands or in light soils. A general manure and fertilizer requirement for fig is given in the table 3, which varies according to the variety and soil type. For young plants fertilizers can be applied with the onset of monsoon and, just after pruning for those which have commenced yielding. The annual requirement can be best divided into two applications, half after pruning and remaining two months later when the syconia are developing. Nitrogen is essential for rapid growth of foliage and development of syconia, fruit color and maturation and K for yield and quality. Better fruit quality can be achieved if N and K are applied in the form of ammonium sulphate and sulphate of potash, respectively.

**Table 3:** Recommended dose of manures and fertilizers for fig

| Age of plant (year) | Organic manure (kg/plant) | | Inorganic manure (g/plant) | | |
|---|---|---|---|---|---|
| | Farmyard manure | Oil cake* | N | P | K |
| 1-2 | 15.0 | 0.5 | 75 | 50 | 50 |
| 3-5 | 25.0 | 1.0-1.5 | 150 | 100 | 100 |
| Above 5 | 40.0 | 2.0 | 300 | 200 | 200 |

(*Source:* Chadha, 2010), *Neem, Pongamia or castor

Excess nitrogen encourages plant growth at the expense of fruit production, and the fruit that is produced often ripens improperly, if at all. As a general rule, fertilize fig trees if the branches grew less than a foot the previous year. Apply total nitrogen, divided into three or four splits beginning at the time of bahar treatment with subsequent applications at monthly interval.

Several soils are deficient in minor nutritional elements that need to be supplemented with required micronutrients. However, it is always advised to get the soil tested and consult the soil specialist for specific nutrient requirement. General guidelines for correcting micronutrient deficiencies are given in Table 4. Application of compost, which is done mostly in the beginning of monsoon, also supplies micronutrients to some extent.

**Table 4:** Micronutrients to be applied for correcting deficiencies

| Micronutrients | Soil application with | Quantity | Foliar application* |
|---|---|---|---|
| Zinc | $ZnSO_4$ | 30kg/ha | 3-4 sprays of 0.25% $ZnSO_4$ (unneutralized) at 10 day interval |
| Iron | - | - | 3-4 sprays of 0.5% $FeSO_4$ at 10 days interval |
| Boron | Borax | 12kg/ha | - |
| Magnesium | $MgSO_4$ | 50 kg/ha | 2-3 sprays of 0.5% $MgSO_4$ at 10 day interval |

(*Source:* Chadha, 2010), *Foliar spray is applied when the plants are flushing.

## PLANT CULTURAL PRACTICES TECHNOLOGY

### Training

Trees should be trained according to use of fruit, such as a low crown for fresh-market figs. Fig trees are trained initially to single stem to encourage a wide, symmetrical crown with a mechanically strong framework having evenly distributed laterals. The tree is allowed to grow for about a meter and then it is topped, which induces side branches all round the main stem. The interior of the bush should be maintained free of suckers, dry and sick branches.

### Pruning

Fig trees are productive with or without heavy pruning. Preferably, it is essential mostly during the initial years. Pruning in fig is practiced annually to stimulate production of new growth, and bearing fruits. The time and type of pruning vary with location, variety and number of crops harvested annually. The best time to secure a mature crop is hot, dry summer. Therefore, pruning may be done 4-5 months in advance. Generally, a single marketable crop is harvested yearly in India. Either heavy or light pruning can be adopted in fig. When heavy pruning is practiced, trees are headed back severely every year, leaving about 2 buds on each one year old shoot. If light pruning is adopted, shoots which have yielded fruits are lightly headed back after harvesting.

Since the crop is borne on terminals of previous year's wood, once the tree form is established, avoid heavy winter pruning, which causes loss of the following year's crop. It is better to prune immediately after the main crop is harvested, or with late-ripening cultivars, summer prune half the branches and prune the remainder the following summer. If radical pruning is done, whitewash the entire tree or copper fungicides should be used to protect the cut ends.

### Notching

Notching is practiced sometimes in Poona fig for activating dormant buds before the start of vigorous growth. Usually 1-2 buds are selected for notching in the middle portion of about 8 month old canes. Notching involves removing of small slice of bark immediately above the dormant bud, giving 2 slanting cuts as deep as the bark. Notch should be about 2.5 cm long and the breadth depends on thickness of the shoot. The cut checks the free flow of sap and stimulates the bud just below it to throw out a fruiting shoot. The technique is useful for induction of fruiting laterals on vigorous upright braches and to increase the total bearing area of the plant.

## SPECIAL PROBLEMS

**Physiological disorder:** Fig is susceptible to sun burn, fruit splitting and fruit drop.

**Sun burn :** Sun burn is noticed mostly in young plants and those subjected to excessive pruning. The trunk and shoots that are exposed to direct sun are prone to sun burn. The affected parts crack and the bark peels off, providing easy access for fungi and other infection. Developing a good canopy by proper pruning and coating the exposed limbs with lime protect the plants from sunburn.

**Fruit splitting:**Fruit splitting is attributed to sudden change in atmospheric humidity during ripening. This makes the fruit unfit for consumption as the pulp is exposed to insect and microbial infection.

**Fruit drop:** Fruit drop may result from excessive drought and heat, cold nights or light frost. Lack of pollination also causes fruit drop in figs. For management of fruit drop avoid water stress and use recommended dose of fertilizers. In winter season, when the temperatures are very low give flood irrigation.

## HARVESTING AND PRODUCTION OF FRUITS

Harvesting of figs depends on their use. About 90 per cent of the figs produced in the world are dried. But figs produced in India are mostly sold as fresh. Figs must be allowed to ripen fully on the tree before they are picked. They will not ripen if picked when immature. Fresh figs should be harvested when they are soft and slightly wilted at the neck and droop and little or no milky latex flow at the cut end of the stalk. Harvest the fruit gently to avoid bruising. Sudden increase in fruit size and opening of ostiole are other maturity indices.

Harvesting process is mechanized in some parts of the world. But in our country, figs are handpicked from the trees by cutting or twisting the neck at the stem end. Workers must wear gloves and protective clothing because of the latex. The fruits are collected and spread in shallow trays. Then they are transported to packing shed. Harvested fruits are spread out in the shade for a day so that the latex will dry a little. Since fresh figs are very delicate, extra care is required in handling. Fresh figs do not keep well and can be stored in the refrigerator for only 2 - 3 days.

Environmental factors such as temperature, photoperiod, and humidity affect the development and yield of the fig tree. Growing figs in unsuitable conditions may cause crop loss and various types of fruit damage. Bearing in fig commences a year after planting, the life span of the tree being 35 year. The harvesting season varies with region and the yield depends on variety and cultivation practices. The second crop is mostly of poor quality fruits. In India, a fig tree bears 180 to 360 fruits per year. Venezuelan growers expect 132 to 176 lbs (6-8 kg) per tree.

## POST-HARVEST FRUIT TECHNOLOGY

### Packing of Fruits

Post- harvest aspects like packing and marketing of fig are very important in order to obtain good price. Fig is a climacteric fruit and to some extent ripening continues once the fruit is harvested. After picking, figs are carefully sorted. The diseased and damaged ones are culled. Fruits are graded for size as more than 50 g, 40-50g and 30-40g. They are packed in a corrugated box carton of 3 ply having 12 holes for ventilation. Fruits are arranged in the carton in two layers, each of 28 (4 rows of 7 figs in a line). Fig leaves are used for cushioning. Owing to perishable nature of fruits, growers prefer to sell their produce to some extent in local or nearby markets. Figs can be held for a short period (7-10 days), at 0°C and 85-90 per cent relative humidity.

## Commercial potential

Because of losses in transport and short shelf life, figs are a high-value fruits of limited demand. The best outlet is direct sale at roadside or farmers markets, but do not permit handling of the fruit. Figs for shipping are collected daily just before they reach the fully ripe stage, but yield to a soft pressure, usually indicated by small cracks in the skin. They should be immediately refrigerated. For commerce, choose a cultivar that parts readily from the branch and does not tear the neck.

Large-scale fig producers in California spray ethephon to speed up ripening and then wind-machines are drawn past the trees or helicopter over flights are made to hasten fruit drop, thus shortening the harvest period by as much as 10 days in order to avoid impending rain and insect attack. Proper timing of the growth regulator is crucial to fruit quality.

## Processing of Fruits

Figs are one of the first fruits to be preserved by drying. Apart from drying and canning, figs are processed into paste and jelly. Some fig varieties are delicious when dried. When figs are grown for drying, they are allowed to ripen and to dry partially on the tree and fall naturally to the ground. Hence, during this period, area beneath the canopy should be maintained clean and dry. Once in 2-3 days, the figs are gathered for further processing. They take 4 - 5 days to dry in the sun and 10 -12 hours in a dehydrator. Dried figs can be stored for six to eight months.

## Keeping Quality

Fresh figs are very perishable. At 40° to 43°F (4.44°-6.11°C) and 75 per cent relative humidity, figs remain in good condition for 8 days but have a shelf life of only 1 to 2 days when removed from storage. At 50°F (10°C) and relative humidity of 85 per cent, figs can be kept no longer than 21 days. They remain in good condition for 30 days when stored at 32° to 35° F (0°-1.67° C). If frozen whole, they can be maintained for several months.

## INSECT-PESTS AND MANAGEMENT

The fig tree is infected by the pests like stem borer, nematodes, lepidopterous pest like fig borer (*Azochis gripusalis*), coleopterous insects of the genera *Epitrix* and *Colaspis* that perforate and severely damage the leaves and shoots. Other pests include scale insects (*Asterolecanium* sp.), dried fruit beetle or sour bug (*Carpophilus* spp.), euryphid mites and the lesser enemy, *Saissetia haemispherica*.

1. **Stem-borer:**In India, a stem-borer (*Batocera rufomaculata*) feeds on the branches. The pest makes holes in the stem and feeds on the internal parts. The presence of excreta at the opening of hole is the sign of insect damage. The severe infection may leads to killing of the tree.

   **Management:** The pest can be very well controlled by adopting following control measures as suggested by Alka and Fenemore (2006), Bhusal (2006) and Baiq *et al.* (2013).

   **Biological Control:** Exclude alternative host trees, such as silk cotton, mango tree and remove the infested branches from the garden to prevent the spread of the pest. Use neem based and *Lantana camera* leaf extracts @ 5 per cent biopesticides inside the hole.

   **Physical Control:** Insert long wire into the borer hole to kill the larvae, pupae and put in the cotton soaked in kerosene or petroleum in the borer hole and plasters them from outside with mud.

   **Chemical Control:** Keeping Carbofuran 3G @ 5 g per hole or Aluminium Phosphide tablets in the holes and plug with mud. Or apply Imidacloprid 17.8% SL @ 1.0 ml/L of water or Thiomethoxam 25% WG @ 1.0g/L of water in the hole for 5 times starting from $2^{nd}$ week of July at 15days interval and then plug it with copper oxychloride paste or mud.

2. **Nematodes:** Fig trees are prone to attack by nematodes (especially *Meloidogyne* spp.). Nematodes, particularly in sandy soils, attack roots, forming galls and stunting the trees. Heavy infection leads to wilting of the plants. For management of nematodes it is suggested to use resistant rootstock. Trees grafted on '*Zidi*' cultivar were vigorous under different field conditions and usefully tolerant to soil sickness (Akihiro *et al.*, 2002). In addition, the pest can be managed very well by inter cropping with marigolds, use of heavy mulch and proper application of nematicide like carbofuron 3G @ 10 kg/ha when the crop is not in fruit bearing stage.

3. **Fig borer (*Azochis gripusalis*):** The larvae feed on the new growth, tunnel down through the trees to the roots and kill the tree. Control is difficult when the larvae are in the tree. Squirt insecticide in the tunnels with a syringe. Enclose the lower portion of the tree in shade netting to prevent the female from laying the eggs in the bark. Make sure that the netting does not touch the tree. Branches can be removed that suffer borer damage.

4. **Scales (*Asterolecanium* sp.):** It attacks the bark of trees weakened by excessive humidity or prolonged drought, and the lesser enemy. The pest can be managed by collection of infested plant parts and their destruction,

application of well rotten sheep or poultry manure @ 10 t/ ha in two splits, and control of ants are some of the cultural measures suggested for management of scales. In addition, the biological control measures like field release of Vadalia and Australian ladybugs and sprays with insecticides like dimethoate 30 EC or chlorpyrifos 20 EC or malathion 50 EC are suggested.

5. **Fruit beetle:** *Mitadulid* spp. and *Carpophilus* spp., the dried fruit beetle, or sour bug enters the fruit through the eye and leads to souring and smut caused by *Aspergillus niger*. This fungus may attack ripening fruits. They frequently breed in fallen citrus fruits. Hence, keep a clean orchard by destroying fallen fruits and do not grow near citrus trees.

6. **Eriophyid mites:** They cause little damage but are carriers of mosaic virus from infected to clean trees. For control of mites spray insecticides like dimethoate 30 EC or chlorpyriphos 20 EC.

## DISEASES AND MANAGEMENT

One of the main constraints for low productivity of fig is the prevalence of number of destructive diseases at different stages of its growth and development. Fig, like any other fruit crops suffers from many diseases caused by fungi, bacteria and viruses. The most important diseases observed on fig are rust (*Cerotelium fici*), anthracnose (*Colletotrichum gloeosporioides*), Armilleria rot (*Armillaria mellica*), bacterial canker (*Pseudomonas fici*), leaf spots (*Cylindrocladium scoparium* and *Cercospora fici*), aerial web blight (*Rhizoctonia solani*), sooty mold (*Capnodium* spp.), pink limb blight (*Corticum salmonicolor),* nectriella stem gall (*Kutilakesa pironii*), fruit rot (*Phytopthora palmivora* and *Ceratocystis fimbriata*), Rosellinia root rot (*Rosellinia necatrix*), smut (*Aspergillus* spp.), cankar (*Phomopsis cinerascens*), fig endosepsis (*Fusarium moniliforme* var. *fici*), crown gall, leaf blotch, stem canker, fig mosaic (fig mosaic virus), etc. (Ferguson *et al.,* 1990 and Palmateer *et al.,* 1999).

**1. Rust [*Cerotelium fici* (Cast.) Arth.]:** Among the several diseases infecting fig, the rust is the most important fungal disease and was reported in India for the first time in 1914 (Butler, 1914). The rust symptom appears on foliage, young growing shoots, petioles and fruits. However, the leaves are found to be severely infected by disease. In the beginning, symptoms appear as pin-point small, yellow to yellowish green spots on the adaxial surface of leaves. Subsequently, small, circular, silvery to whitish or golden yellow blisters/ pustules are formed on the abaxial surface of leaves. Simultaneously, the spots enlarge in to roughly circular to irregular brownish spots which are quite conspicuous

Rust symptoms on fig leaf

Rust symptoms on fig fruit

(See colour version on page 317)

on the adaxial surface. Meanwhile, the pustules burst open by rupturing the epidermal wall with thin delicate silvery white peridium to release golden yellow to brown spore masses on the lower leaf surface. Likewise, several pustules are produced on the dorsal side with corresponding brownish necrotic spots on adaxial surface. With an advancement of disease, spots/pustules turn dark brown in colour and coalesce together. Under severe infection stage, entire leaf blade is covered with rust lesions that results in to yellowing, drying and fall off leaves prematurely.

Under severe conditions, the rust also appears on petioles, tender shoots and green fruits. On petioles and shoots, several light brown elongated, narrow pustules are noticed. The pustules did not erupt as that on leaves. Further, with advancement of disease, mostly circular, dark brown to black lesions/pustules are observed on well developed green fruits, which leads to blackening and deterioration of fruit quality.

At present, the disease has become a major limiting factor in fig cultivation. The rust is an endemic disease and severe in all parts where the crop is grown. The fig growers are much worried about the worst situation caused by this disease. About 50 per cent losses have been reported by earlier workers (Gaikwad and Nimbalkar, 2004 and Anon, 2004) indicating the seriousness of rust disease in fig.

**2. Anthracnose (*Colletotrichum gloeosporioides*) and leaf spots (*Cylindrocladium scoparium* and *Cercospora fici*):** Anthracnose and leaf spots cause circular, small to big (2 to 10 mm) spots on the leaves, which are conspicuous on the upper surface. The spots are ashy to light brown in colour with or without distinct dark margins. Many times, shot hole appearance is also noticed. Under severe condition, the leaves turn yellow, dry and fall down.

## Management of rust, anthracnose and leaf spots

- Collection and destruction of affected plant parts.
- Avoid overcrowding in the orchard.
- Carry out pruning of plants, which removes the diseased plant parts and helps for boosting the plant growth and thereby yield.
- Remove excess water from the orchard.
- Spray the crop with carbendazim (0.1%) + chlorothalonil (0.2%) when disease is observed. Five to six sprays at 10 days interval, starting first spray at disease appearance controls the disease very effectively (Gaikwad and Nimbalkar, 2004).

**3. Mosaic:** It is caused by fig mosaic virus (FMV). The symptoms resemble potassium deficiency. FMV results in a variety of symptoms, such as appearance of a leaf mosaic pattern, defoliation, decreased fruit yield, vein banding, ring spots, distortion and chlorosis of leaves, and yellow spotting on fruits. Leaves are marbled with yellow spots, and the veins are light coloured. Symptoms are often not apparent until the tree is older or when it becomes heat or water-stressed.

Symptoms of mosaic on fig leaves (Source: https://en.wikipedia.org/wiki/Fig_mosaic_virus) (See colour version on page 317)

**Management:** Do not purchase infected trees, isolate those which show symptoms before planting, destroy diseased plants from the field and spray insecticides for control of vectors (mites).

**4. Fruit rots:** The fungi like spp. of *Rhizopus, Fusarium, Aspergillus* and *Penicillium* are responsible for various soft or dry rots in fig fruits. Smut caused by *Rhyzopus* spp. attacks ripened fruits on the tree, causing charcoal black coating inside the fruit, and is worst on cultivars with large, open eyes. Most ripe fruit losses are caused by *Fusarium* and *Aspergillus* rot, which are introduced by insects, even pollinating wasps. The fruit appears to burst, or a ropy, mucus-like exudates drain from the eye, rendering the fruit uneatable.

**Management:** The best control is to destroy all crop for one year, apply diazinon granules beneath trees to eliminate insect vectors, and destroy adjacent wild trees. *Penicillium* fungus attacks dried fruits in storage but can be controlled by keeping them dry, or sulfuring before storage.

**5. Fig canker:** Fig canker is caused by a bacterium which enters the trunk at damaged zones, causing necrosis and girdling and loss of branches. It usually starts at sunburned areas, so it is important to keep exposed branches white washed. Also cut the infected branches with some healthy portion to ward off the pathogen.

## FUTURE STRATEGIES

The research on the following aspects needs to be undertaken for further improvement in fig production

i. Studies on organic fig fruit production.

ii. Water management and fertigation.

iii. Standardization of pruning and *bahar* techniques.

iv. High density planting.

v. Pre and post harvest management.

vi. Use of different rootstocks for abiotic and biotic stress.

vii. Studies on mechanical harvesting of fig fruits.

viii. Improvement in the self life of fruits.

ix. Use of improved attractive packaging and export marketing.

x. Processing for value addition.

## LITERATURE CONSULTED

Akihiro, Hosomi,Masayuki, Dan, Akihiro Kato, 2002. Screening of fig varieties for rootstocks resistant to soil sickness. *Journal of the Japanese Society for Horticultural Science,* 71 (2): 171-176.

Alka, Prakash and Fenemore,P.G. 2006. Applied Entomology, New Age International Publisher,s 2nd edition, 2006.

Anonymous. 2004. Chemical control of fig rust. *Joint Agresco Report.*, 2004. RFRS, Ganeshkhind, Pune.

Anonymous. 2006. Agricultural data. FAOSTAT, FAO stat agriculture, Aug. 2006.

Anonymous. 2011. Area and production of fruits in Maharashtra. Directorate of Horticulture, Maharashtra State, Pune.

Baiq Nurul Hidayah, Muji Rahayu, Mujiono, Brian Thistleton, Sohail Qureshi, Ian Baker. 2013. Effectiveness of pesticides in controlling major pest and disease of mangoes in West Nusa Tenggara Province – Indonesia, 3rd International Conference on Chemical, Biological and Environment Sciences (ICCEBS'2013), January 8-9, 2013, Kuala Lumpur (Malaysia), pp: 53-55.

Berg, C. C. 2003. *Flora malesiano* precursor for the treatment of Moraceae 1: The main subdivision of *Ficus:* The subgenera. *Blumea.*, 48:167-78.

Bhusal, S. J. 2006. Insect pest complex of mango fruit in Koshi East Terai Region. *Annual Report RARS*, 2005-2006, Tarahara, pp: 62-65.

Butler, E. J. 1914. Notes on some rusts in India. *Annual Mycology*, 12: 76-82.

Chadha, K. L. 2010. Handbook of Horticulture. Indian Council of Agricultural Research, New Delhi, pp: 176.

Condit, I. J. 1947. *The Fig*. Waltham, Mass., Chronica Botanica Co., 1947.

Crane, J. C. 1986. Fig in S. P. Monselise (Ed.). Handbook of fruit set and development. CRC Press, Boca Raton, FL.

Crane, J. C. and Baker,R.E. 1953. Growth comparisons of the fruits and fruitlets of figs and strawberries. *Proc. Am. Soc. Hort. Sci.,* 62:142-153.

Crane, J. C. and Brown,J.B. 1950. Growth of the fig fruit, *Ficus carica* var. Mission. *Proc. Am. Soc. Hort. Sci.,* 56:93-97.

Crane, J. C. and J. Van Overbeek. 1965. Kinin-Induced Parthenocarpy in the Fig, *Ficus carica* L. *Science*, 147:1468-1469.

Datwyler, S. L. and Weiblen,G.D. 2004. On the origin of the fig: phylogenetic relationships of Moraceae from *ndliF* sequences. *Am. J. Bot.,* 91:767-777.

Erez, A. and Shulman,Y. 1982. Growing breba figs in a meadow orchard: A possibility for commercial fig growing. (In Hebrew). Hassadeh, 63:969-971.

Ferguson, L., J. Marchailidas, Themis, and Shor,H.. 1990. The California fig industry. *Horticulture Rev.*, 12: 448-451.

Flaishman, M. A. and Al Hadi,F.A. 2002. Fig growth in Israel.(In Hchrew). *Alon Hanote,* 56: 56-57.

Flaishman, M. A., Rodov,V. and Stover,E. 2008. The fig: botany, horticulture, and breeding.

Gaikwad, A. P. and Nimbalkar,C.A. 2004. Management of fig rust with fungicides. *J. Mycol. Plant Pathol.*, 34 (2): 418-419.

Kim, J. S.,Kim,Y.O. Ryu, H.J.Kwak,Y.S. Lee,J.Y. and Kang,H. 2003. Isolation of stress-related genes of rubber particles and latex in fig tree (*Ficus carica)* and their expressions by abiotic stress or plant hormone treatments. *Plant Cell Physiol.*, 44:412-414.

Lamb, C. and Dixon,R.A 1997. The oxidative burst in plant disease resistance. Annu, Rev. Plant Physiol. *Plant Mal. Biol.,* 48:251-275.

Morton, J. 1987. Fig P.47-50. In: Fruits of warm climates, Miami, F.L. https://hort. purdue. edu>mooton

Palmateer, A. J.,Tara,L., Tarnowski,B. and Robert,P.D 1999. Florida Plant Disease Management Guide: Fig (*Ficus carica*). Publication of Plant Pathology Department, Florida Cooperative Extension Service, Institute of Food and Agricultural Sciences, University of Florida. PDMG-V3-14:1-4.

Parmar, Chiranjit. 2013. Fruitipedia. Encyclopedia of the Edible fruits of the world. *Fruit World Online Magazine.* (http://fruitipedia.com/fegra.htm).

Watson, L. and Dallwitz,M.J. 2004. The families of flowering plants: Descriptions, illustrations, identification, and information retrieval. [http://biodiversity.uno.edu/ delta (accessed June 2004)].

# About the Editor

**Dr. J.S.Bal,** born on 8 October 1951,belongs to Ajnala in district Amritsar did B.Sc (Agri.) in 1973 from Khalsa College, Amritsar and M.Sc (Hort.) in 1975 from Punjab Agricultural University, Ludhiana. He joined PAU, Ludhiana as Res.Assoc. (Hort.) in September 1977 and Asstt. Prof. Horticulture in April 1980. He completed Ph.D (Hort.) in 1991 from GNDU, Amritsar. Dr. Bal elevated to the position of Professor of Horticulture in 2001 and served as Head, Department of Horticulture, PAU in 2009.

Dr. Bal has written 5 books in horticulture viz. Fruit Growing, Raising Fruit Nursery, Mulching in Ber Orchards, Ber Varieties and Post-harvest Handling of Fruits and Vegetables which are very popular among students, growers, nurserymen, scientists and extension workers. He has also compiled 4 under-graduate and 3 post-graduate Practical Manuals in horticulture for students and contributed 6 Book Chapters. He has taught 528 credit hours to UG/PG students at Punjab Agricultural University which is a record itself in teaching career. Dr. Bal has guided 5 UG Advisory Groups, 18 M.Sc / Ph.D students as Major Advisor and 78 M.Sc / Ph.D students as Minor Advisor. He has evaluated and conducted viva-voce of 24 M.Sc and 23 Ph.D students of different universities as external expert. He has selected 10 Asstt.Prof., 9 Assoc.Prof.,24 Prof. in various universities and 25 HDO in other states as an expert member. He acted as Chairman Academic Affairs Committee (Teaching) in the Department of Horticulture from 2004 to 2009.

Dr. Bal was team leader of *ber* crop at PAU for 19 years. He has published192 Research Papers in National and International Journals and compiled & edited 4 Research Reports. His seventeen recommendations are included in Package of Practices for the benefits of fruit growers. He has earned distinction for evaluating a draft report on genetic resources of *ber* by International Bureau of Plant Genetic Resources, Rome (Italy). He has attended 29 National and International Seminar/ Symposia/Conference in horticulture. Dr. Bal has attended 2nd International Jujube Symposium at Xinzheng, Henan (China) from 3-7 Sept. 2011 and presented Keynote Report on *ber* in Inaugural Session and Chaired Session on "Post-harvest Physiology, Nutrition and Processing". He has attended 3rd International Jujube Symposium at Brisbane(Australia) from 17-22 August 2014 and delivered Keynote Address and Chaired Oral Session. Acted as a member, International Scientific Committee, Jujube 2016 Workshop, Bucharest, Romania, October 9-12,2016. Recently nominated member, International

Scientific Committee, XXX. International Horticultural Congress. S-8 Jujube (4th International Symposium) to be held on 12-16 August 2018, Istanbol- Turkey. He was instrumental in rejuvenation of 440 years old Historic *ber* plants at Golden Temple, Amritsar.

He has edited monthly "The Horticultural Bulletin" in English and Vernacular for 7 years and monthly "Horticulture Newsletter" from January 2008 to October 2011 for the benefits of fruit growers. He has published 109 Extension Articles, 9 Bulletins and 4 Newsletters in horticulture. He was Incharge of PAU orchard for 8½ years. He also participated in the course "Parliament Process and Procedures" at Lok Sabha Secretariat, New Delhi. Dr. Bal has also actively participated in various curricular activities of PAU such as NCC Officer, Programme Officer NSS, President Arts and Literature / Young Writers Association for 10 years in College of Agriculture and Hostel Warden.

Dr. Bal was awarded Shri Hans Raj Pahwa Award 1998 by PAU; honoured by Punjab Minister at his home town Ajnala in September 2001; Distinguished Alumni Award 2002 by Khalsa College Amritsar; Shri G.L.Chadha Memorial Gold Medal 2004 by 'The Horticultural Society of India'; The Best Citizen of India Award 2006 by IPH New Delhi and Best Paper Award by Punjab Academy of Sciences in February 2007 and by PAU in October 2010 in National Seminar on 'Impact of Climate Change on Fruit Crops'.

Dr. Bal has total 337 publications to his credit and completed 35 years service in horticulture (34 in PAU and 1 in LPU).